218
Advances in Polymer Science

Editorial Board:
A. Abe · A.-C. Albertsson · R. Duncan · K. Dušek · W. H. de Jeu
H.-H. Kausch · S. Kobayashi · K.-S. Lee · L. Leibler · T. E. Long
I. Manners · M. Möller · O. Nuyken · E. M. Terentjev
B. Voit · G. Wegner · U. Wiesner

Advances in Polymer Science
Recently Published and Forthcoming Volumes

Self-Assembled Nanomaterials II
Nanotubes
Volume Editor: Shimizu, T.
Vol. 220, 2008

Self-Assembled Nanomaterials I
Nanofibers
Volume Editor: Shimizu, T.
Vol. 219, 2008

Interfacial Processes and Molecular Aggregation of Surfactants
Volume Editor: Narayanan, R.
Vol. 218, 2008

New Frontiers in Polymer Synthesis
Volume Editor: Kobayashi, S.
Vol. 217, 2008

Polymers for Fuel Cells II
Volume Editor: Scherer, G. G.
Vol. 216, 2008

Polymers for Fuel Cells I
Volume Editor: Scherer, G. G.
Vol. 215, 2008

Photoresponsive Polymers II
Volume Editors: Marder, S. R., Lee, K.-S.
Vol. 214, 2008

Photoresponsive Polymers I
Volume Editors: Marder, S. R., Lee, K.-S.
Vol. 213, 2008

Polyfluorenes
Volume Editors: Scherf, U., Neher, D.
Vol. 212, 2008

Chromatography for Sustainable Polymeric Materials
Renewable, Degradable and Recyclable
Volume Editors: Albertsson, A.-C., Hakkarainen, M.
Vol. 211, 2008

Wax Crystal Control · Nanocomposites Stimuli-Responsive Polymers
Vol. 210, 2008

Functional Materials and Biomaterials
Vol. 209, 2007

Phase-Separated Interpenetrating Polymer Networks
Authors: Lipatov, Y. S., Alekseeva, T.
Vol. 208, 2007

Hydrogen Bonded Polymers
Volume Editor: Binder, W.
Vol. 207, 2007

Oligomers · Polymer Composites Molecular Imprinting
Vol. 206, 2007

Polysaccharides II
Volume Editor: Klemm, D.
Vol. 205, 2006

Neodymium Based Ziegler Catalysts – Fundamental Chemistry
Volume Editor: Nuyken, O.
Vol. 204, 2006

Polymers for Regenerative Medicine
Volume Editor: Werner, C.
Vol. 203, 2006

Peptide Hybrid Polymers
Volume Editors: Klok, H.-A., Schlaad, H.
Vol. 202, 2006

Interfacial Processes and Molecular Aggregation of Surfactants

Volume Editor: Ranga Narayanan

With contributions by
J. M. Andérez · R. E. Antón · A. Bose · C. Bracho · S. Brown
A. V. Chengara · H. Hoffmann · K. Holmberg · V. John
S. I. Karakashev · D. Lundberg · C. A. Miller · J. D. Miller
P. K. Misra · B. Moudgil · A. V. Nguyen · A. D. Nikolov
G. Ramanath · J.-L. Salager · J. Sarkar · P. Sharma
P. Somasundaran · M. Stjerndahl · M. Varshney · F. Vejar
D. T. Wasan · A. Zapf

The series *Advances in Polymer Science* presents critical reviews of the present and future trends in polymer and biopolymer science including chemistry, physical chemistry, physics and material science. It is adressed to all scientists at universities and in industry who wish to keep abreast of advances in the topics covered.

As a rule, contributions are specially commissioned. The editors and publishers will, however, always be pleased to receive suggestions and supplementary information. Papers are accepted for *Advances in Polymer Science* in English.

In references *Advances in Polymer Science* is abbreviated *Adv Polym Sci* and is cited as a journal.

Springer WWW home page: springer.com
Visit the APS content at springerlink.com

ISBN 978-3-540-69809-8   e-ISBN 978-3-540-69810-4
DOI 10.1007/978-3-540-69810-4

Advances in Polymer Science ISSN 0065-3195

Library of Congress Control Number: 2008933502

© 2008 Springer-Verlag Berlin Heidelberg

This work is subject to copyright. All rights are reserved, whether the whole or part of the material is concerned, specifically the rights of translation, reprinting, reuse of illustrations, recitation, broadcasting, reproduction on microfilm or in any other way, and storage in data banks. Duplication of this publication or parts thereof is permitted only under the provisions of the German Copyright Law of September 9, 1965, in its current version, and permission for use must always be obtained from Springer. Violations are liable to prosecution under the German Copyright Law.

The use of general descriptive names, registered names, trademarks, etc. in this publication does not imply, even in the absence of a specific statement, that such names are exempt from the relevant protective laws and regulations and therefore free for general use.

Cover design: WMXDesign GmbH, Heidelberg
Typesetting and Production: le-tex publishing services oHG, Leipzig

Printed on acid-free paper

9 8 7 6 5 4 3 2 1 0

springer.com

Volume Editor

Ranga Narayanan
University of Florida
Dept. Chemical Engineering
Director Center for Surface Science
and Engineering
P.O.Box 116005
Gainesville FL 32611, USA
ranga@ufl.edu

Editorial Board

Prof. Akihiro Abe
Department of Industrial Chemistry
Tokyo Institute of Polytechnics
1583 Iiyama, Atsugi-shi 243-02, Japan
aabe@chem.t-kougei.ac.jp

Prof. A.-C. Albertsson
Department of Polymer Technology
The Royal Institute of Technology
10044 Stockholm, Sweden
aila@polymer.kth.se

Prof. Ruth Duncan
Welsh School of Pharmacy
Cardiff University
Redwood Building
King Edward VII Avenue
Cardiff CF 10 3XF, UK
DuncanR@cf.ac.uk

Prof. Karel Dušek
Institute of Macromolecular Chemistry,
Czech
Academy of Sciences of the Czech Republic
Heyrovský Sq. 2
16206 Prague 6, Czech Republic
dusek@imc.cas.cz

Prof. Dr. Wim H. de Jeu
Polymer Science and Engineering
University of Massachusetts
120 Governors Drive
Amherst MA 01003, USA
dejeu@mail.pse.umass.edu

Prof. Hans-Henning Kausch
Ecole Polytechnique Fédérale de Lausanne
Science de Base
Station 6
1015 Lausanne, Switzerland
kausch.cully@bluewin.ch

Prof. Shiro Kobayashi
R & D Center for Bio-based Materials
Kyoto Institute of Technology
Matsugasaki, Sakyo-ku
Kyoto 606-8585, Japan
kobayash@kit.ac.jp

Prof. Kwang-Sup Lee
Department of Advanced Materials
Hannam University
561-6 Jeonmin-Dong
Yuseong-Gu 305-811
Daejeon, South Korea
kslee@hnu.kr

Prof. L. Leibler

Matière Molle et Chimie
Ecole Supérieure de Physique
et Chimie Industrielles (ESPCI)
10 rue Vauquelin
75231 Paris Cedex 05, France
ludwik.leibler@espci.fr

Prof. Timothy E. Long

Department of Chemistry
and Research Institute
Virginia Tech
2110 Hahn Hall (0344)
Blacksburg, VA 24061, USA
telong@vt.edu

Prof. Ian Manners

School of Chemistry
University of Bristol
Cantock's Close
BS8 1TS Bristol, UK
ian.manners@bristol.ac.uk

Prof. Martin Möller

Deutsches Wollforschungsinstitut
an der RWTH Aachen e.V.
Pauwelsstraße 8
52056 Aachen, Germany
moeller@dwi.rwth-aachen.de

Prof. Oskar Nuyken

Lehrstuhl für Makromolekulare Stoffe
TU München
Lichtenbergstr. 4
85747 Garching, Germany
oskar.nuyken@ch.tum.de

Prof. E. M. Terentjev

Cavendish Laboratory
Madingley Road
Cambridge CB 3 OHE, UK
emt1000@cam.ac.uk

Prof. Brigitte Voit

Institut für Polymerforschung Dresden
Hohe Straße 6
01069 Dresden, Germany
voit@ipfdd.de

Prof. Gerhard Wegner

Max-Planck-Institut
für Polymerforschung
Ackermannweg 10
55128 Mainz, Germany
wegner@mpip-mainz.mpg.de

Prof. Ulrich Wiesner

Materials Science & Engineering
Cornell University
329 Bard Hall
Ithaca, NY 14853, USA
ubw1@cornell.edu

Advances in Polymer Science
Also Available Electronically

For all customers who have a standing order to Advances in Polymer Science, we offer the electronic version via SpringerLink free of charge. Please contact your librarian who can receive a password or free access to the full articles by registering at:

springerlink.com

If you do not have a subscription, you can still view the tables of contents of the volumes and the abstract of each article by going to the SpringerLink Homepage, clicking on "Browse by Online Libraries", then "Chemical Sciences", and finally choose Advances in Polymer Science.

You will find information about the

- Editorial Board
- Aims and Scope
- Instructions for Authors
- Sample Contribution

at springer.com using the search function.

Color figures are published in full color within the electronic version on SpringerLink.

Professor Dinesh O. Shah

Preface

Springer is pleased to bring out this edited volume on interfacial and molecular aggregation phenomena. The volume is timely given the enormous interest of surface phenomena in areas ranging from materials science to applications in biology.

This book is dedicated to the life achievements of Professor Dinesh O. Shah, a professor of chemical engineering and anesthesiology at the University of Florida. Dinesh Shah is one of the most distinguished scientists in his field having made renowned contributions in the physical chemistry of surfactants, monomolecular films, enhanced oil recovery, combustion of coal dispersion in oil and aqueous media, surfactant–polymer interaction, boundary lubrication and surface phenomena in magnetic media, membranes, lungs, vision and anesthesia. After receiving a Doctorate under the guidance of the famous professor J.H. Schulman at Columbia he joined the University of Florida in 1970, rising from a post doctoral associate to become the first Charles A. Stokes Professor of Chemical Engineering. He has received international recognition for his outstanding work in many fields associated with surfactant physics and molecular aggregation. Dinesh is the author of over 300 papers and several books in this field and has delivered hundreds of invited lectures and seminars throughout the world. A dedicated teacher and educator he has received the university's highest awards including the Florida Blue Key distinguished faculty award and the Teacher-Scholar award.

It is a treat to listen to Dinesh. His common sense approach to science and willingness to look at new ideas constantly has made his laboratory the congregating place for world-class scientists. Dinesh's depth and breadth of outstanding scholarly and high-quality contributions are truly remarkable. We honor ourselves as we dedicate this volume of nine essays, written by experts, to this unusual individual who has shown us the way to academic enlightenment.

We look forward to many more years of contributions from Dinesh O. Shah, scholar, visionary and true academic.

Gainesville, FL, August 2008   Ranga Narayanan

Contents

Part I

Dissolution Rates of Surfactants
C. A. Miller 3

Equilibrium Adsorption of Surfactants at the Gas–Liquid Interface
S. I. Karakashev · A. V. Nguyen · J. D. Miller 25

Surfactants Containing Hydrolyzable Bonds
D. Lundberg · M. Stjerndahl · K. Holmberg 57

Practical Surfactant Mixing Rules Based on the Attainment
of Microemulsion–Oil–Water Three-Phase Behavior Systems
R. E. Antón · J. M. Andérez · C. Bracho · F. Vejar · J.-L. Salager 83

Part II

New Paradigms for Spreading of Colloidal Fluids on Solid Surfaces
A. V. Chengara · A. D. Nikolov · D. T. Wasan 117

Fluorescence Probing of the Surfactant Assemblies
in Solutions and at Solid–Liquid Interfaces
P. K. Misra · P. Somasundaran 143

Surfactant-Mediated Fabrication of Optical Nanoprobes
P. Sharma · S. Brown · M. Varshney · B. Moudgil 189

Directed Synthesis of Micro-Sized Nanoplatelets of Gold
from a Chemically Active Mixed Surfactant Mesophase
J. Sarkar · G. Ramanath · V. John · A. Bose 235

Colloid Chemistry: The Fascinating World of Microscopic Order
A. Zapf · H. Hoffmann 251

Subject Index 261

Part I

Dissolution Rates of Surfactants

Clarence A. Miller

Department of Chemical and Biomolecular Engineering, Rice University,
Houston, TX 77251-1892, USA
camill@rice.edu

1	Introduction	4
2	Pure Nonionic Surfactants	6
3	Nonionic Surfactant Mixtures	9
4	Mixtures of Nonionic and Anionic Surfactants	13
5	Pure Surfactants with Low Solubility: Myelinic Figures	16
6	Summary	23
	References	23

Abstract Information on dissolution rates of surfactants is reviewed and discussed. One or more viscous liquid crystalline phases form during dissolution in water of pure liquid nonionic surfactants of the linear alcohol ethoxylate type, as expected from known equilibrium phase diagrams. Dissolution rates are controlled by diffusion at temperatures below a critical solution point T_C, where the aqueous micellar solution begins to separate into dilute and concentrated phases. Analysis of linear penetration and drop dissolution experiments yields values of effective diffusion coefficients for each phase. Once the temperature exceeds T_C, solubility of surfactant in the aqueous phase falls rapidly to very low values and complete dissolution does not occur. Particularly striking is formation by a swelling process at temperatures somewhat above T_C of filaments of the lamellar liquid crystalline phase known as myelinic figures. Phospholipids and other rather lipophilic surfactants also form myelinic figures when contacted with water. A model of their growth is described and applied to the pure anionic surfactant Aerosol OT. Using data from linear penetration experiments, one can calculate effective diffusivities in the liquid crystalline phases (other than the myelins) and the composition of the myelins. Preferential dissolution of more hydrophilic species occurs during dissolution of drops of surfactant mixtures, causing dissolution rates to decrease as the drops become less soluble. Moreover, drop compositions can reach states where new phases form, leading to intriguing behavior such as spontaneous emulsification, anisotropic swelling, and formation of jets.

Keywords Myelinic figures · Surfactant dissolution · Surfactant penetration experiments

1
Introduction

Surfactants are used in a variety of applications, frequently in the form of dilute aqueous solutions. However, it is not cost effective to transport, store, and display in retail outlets surfactant products such as household detergents in this form. Accordingly, it is important to have products that dissolve quickly and to understand what aspects of surfactant composition and structure promote rapid dissolution. The dissolution process is more complex for surfactants than for most other materials because it typically involves formation of one or more concentrated and highly viscous liquid crystalline phases, which are not present initially and which could potentially hinder dissolution. In this article the rates and mechanisms of surfactant dissolution are reviewed and discussed.

Dissolution of surfactants cannot be understood without some knowledge of relevant equilibrium phase diagrams. We use the binary diagram for the pure nonionic surfactant $C_{12}(EO)_5$ and water to illustrate relevant aspects of phase behavior. This surfactant is made by adding five ethylene oxide (EO) groups to 1-dodecanol. At low temperatures and surfactant concentrations an aqueous micellar solution L_1 exists, as shown in Fig. 1 [27]. With increasing surfactant concentration one sees three viscous lyotropic liquid crystalline phases: the normal hexagonal phase H_1, which consists of long, parallel rodlike micelles in a regular hexagonal array; the bicontinuous viscous isotropic phase V_1; and the lamellar phase L_α, made up of a regular array of surfactant bilayers alternating with wa-

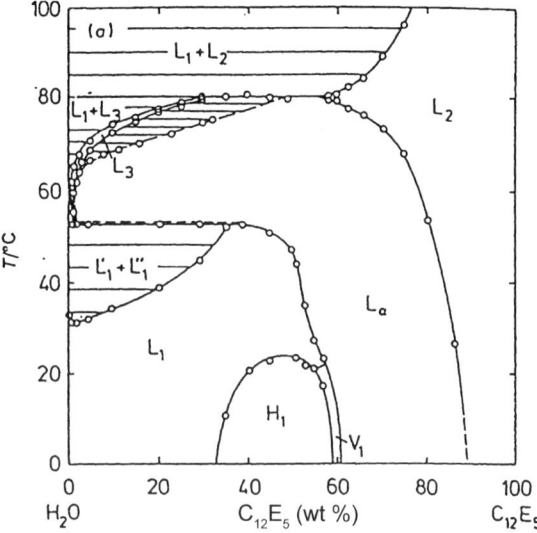

Fig. 1 $C_{12}(EO)_5/H_2O$ phase diagram (Strey et al. 1990)

ter layers. At very high surfactant concentrations is the L_2 phase, in which the pure surfactant, an isotropic liquid, has incorporated small amounts of water as reverse micelles. At low surfactant concentrations in the L_1 phase the micelle shape is that which minimizes the free energy of isolated micelles. At higher concentrations interaction among micelles becomes important and dictates formation of the liquid crystalline phases where aggregate shapes are long rods and bilayers, the latter being saddle-shaped in the V_1 phase.

The interaction of EO groups with water decreases with increasing temperature, which reduces hydration and modifies the configuration of the EO chains of the surfactant molecules. As a result, the area a occupied by an EO chain at the surface of a micelle decreases with increasing temperature, causing the packing parameter (v/la) to increase. Here v is the volume of the surfactant's hydrocarbon chain and l its length when extended. When $(v/la) = 1$, a surfactant film has no tendency to curve, favoring formation of surfactant bilayers. When $(v/la) < 1$, "normal" micelles are favored with interiors composed of hydrocarbon chains. In contrast, when $(v/la) > 1$, reverse micelles with aqueous interiors are favored.

At low surfactant concentrations Fig. 1 shows that as temperature increases, the L_1 phase of normal micelles separates into two isotropic phases above a critical solution temperature T_C. This separation occurs because the reduced hydration and changes in EO chain configuration decrease repulsion between micelles and because the larger value of (v/la) leads to rodlike micelles where phase separation occurs more readily than for spherical micelles. Indeed, the more concentrated phase consists of a network of interconnected rodlike micelles over a wide composition range in many nonionic surfactant systems [31]. At still higher temperatures the lamellar phase forms, then the L_3 phase, an isotropic liquid often called the "sponge" phase because its microstructure resembles that of a sponge, albeit on a very small scale, with surfactant bilayers forming the sponge backbone and water filling the pores. It can be shown that (v/la) slightly exceeds 1 for this phase. At still higher temperatures and higher values of (v/la) the L_2 phase with reverse micelles exists. A more extensive discussion of phase behavior in surfactant/water systems may be found in the book by Laughlin [19].

In the remainder of this article, discussion of surfactant dissolution mechanisms and rates proceeds from the simplest case of pure nonionic surfactants to nonionic surfactant mixtures, mixtures of nonionics with anionics, and finally to development of myelinic figures during dissolution, with emphasis on studies in one anionic surfactant/water system. Not considered here are studies of rates of transformation between individual phases or aggregate structures in surfactant systems, e.g., between micelles and vesicles. Reviews of these phenomena, which include some of the information summarized below, have been given elsewhere [7, 15, 29].

2
Pure Nonionic Surfactants

Chen et al. [9] studied the dissolution of pure liquid surfactants $C_{12}(EO)_n$ for $n = 3$–6. In a "linear penetration" experiment the surfactant was carefully layered (to minimize mixing) on deionized water in a vertically oriented rectangular glass capillary cell having a thickness of 400 μm (Fig. 2a). The cell was maintained at constant temperature in a thermal stage. The dissolution process was followed by videomicroscopy and recorded on videotape, so that the positions of various interfaces involving the initial phases and any intermediate phases that formed were tracked as a function of time. In the case of pure $C_{12}(EO)_5$ at 28 °C the lamellar liquid crystalline phase formed between the aqueous (L_1) and surfactant (L_2) liquids and increased in thickness over time, as shown schematically in Fig. 2b. The equilibrium phase behavior of Fig. 1 is consistent with this behavior. It is noteworthy that at this temperature the miscibility gap, i.e., the composition difference between phases in equilibrium, is minimal. Similar behavior was seen for $C_{12}(EO)_6$ at temperatures below about 35 °C, except that in this case three intermediate liquid crystalline phases (H_1, V_1, L_α) developed as expected from the corresponding phase diagram [21]. According to Fig. 1, similar behavior should occur for $C_{12}(EO)_5$ below 20 °C. That is, by increasing a and thereby reducing (v/la), the additional EO group in $C_{12}(EO)_6$ makes the surfactant more hydrophilic and raises the temperatures at which various phases and phase transitions occur from those of Fig. 1. For both surfactants the distances of all the interfaces from the surface of initial contact between surfactant and water were found to be proportional to the square root of time, indicating that the dissolution process was controlled by diffusion.

The observations were interpreted using a well-known similarity solution based on binary diffusion. In each phase j surfactant volume fraction Φ_j is given by

$$\Phi_j = a_{1j} + a_{2j} \text{erf}(x/4D_{ej}t), \tag{1}$$

where x is the distance from the initial surface of contact, t is time, and D_{ej} is an effective binary diffusivity in phase j. The constants a_{ij} are found from the applicable initial and boundary conditions, i.e., compositions of the surfactant and aqueous phases at $t = 0$ and $x = \pm\infty$, local equilibrium at the interfaces as specified by the known phase diagram, and a mass balance of surfactant at each interface.

Because there are j phases but only $(j - 1)$ interfaces, this experiment alone cannot provide sufficient information to calculate all the D_{ej} given the phase diagram and measurements of interfacial positions as a function of time. However, when its results are combined with measured times for complete dissolution in water of individual drops of surfactant of known initial radius injected through glass micropipettes into similar cells placed on a conven-

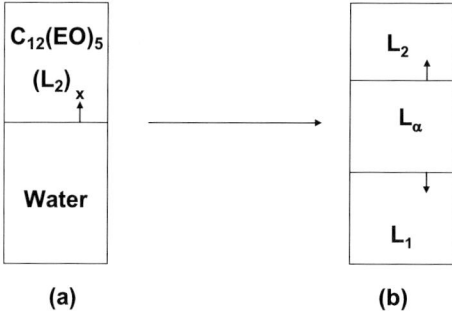

Fig. 2 Schematic diagram of vertically oriented linear penetration experiment for $C_{12}(EO)_5/H_2O$ at 28 °C. **a** Initial contact; **b** later showing intermediate lamellar phase

Table 1 Effective diffusion coefficients in the various phases for pure nonionic surfactants

Surfactant	Bulk solution	Temp. [°C]	Diffusion coeff. [10^{-10} m^2/s]				
			L_1	H_1	V_1	L_α	L_2
$C_{12}(EO)_5$	Water	28	0.73	–	–	2.0	1.0
$C_{12}(EO)_6$	Water	30	0.86	1.08	2.32	2.19	1.0

tional (horizontal) microscope stage, the diffusivities can be calculated. The analysis for the drop included formation and subsequent dissolution of the liquid crystalline intermediate phases assuming diffusion control, which was confirmed by the observation that dissolution time t_D was proportional to the square of initial drop radius R_0 [9]. Diffusivities were found to be of the order 10^{-10} m^2/s for all phases, as shown in Table 1. Results of a thin-layer penetration experiment as described in Sect. 5 could be used instead of those of the drop dissolution experiment in the calculations.

Prinsen et al. [23] and Warren et al. [31] used dissipative particle dynamics to simulate dissolution of a pure surfactant in a solvent. Tuning surfactant–surfactant, surfactant–solvent, and solvent–solvent interactions to yield an equilibrium phase diagram similar to Fig. 1 at low temperatures except for the absence of the V_1 phase, they found that the kinetics of formation of the liquid crystalline phases at the interfaces was rapid and that the rate of dissolution was controlled by diffusion, in agreement with the above experimental results.

It bears repeating that the D_{ej} values are effective diffusivities and that, in fact, diffusivity is a function of surfactant concentration, as shown by interferometry for the L_1 phase of $C_{12}(EO)_5$ [9]. For the anisotropic phases diffusivity is also a function of orientation, and D_{ej} depends on the number and orientation of domains of the phase as formed during dissolution. Thus, the value shown in Table 1 for D_{eH_1} of $C_{12}(EO)_6$ is intermediate between the diffusivities parallel and perpendicular to the rodlike micelles measured in fully oriented samples of the hexagonal phase H_1 for this surfactant [26].

Since their effective diffusivities are of the same magnitude as those of micellar solutions, the liquid crystalline phases, though viscous, do not significantly hinder surfactant dissolution for these rather hydrophilic surfactants. Indeed, a drop of $C_{12}(EO)_6$ having $R_0 = 78$ μm dissolved completely in only 16 s at 30 °C. Rapid dissolution is favored because free energy decreases as the surfactant is transferred from the liquid surfactant phase L_2 to liquid crystal(s) to aqueous micellar solution and the aggregate shape approaches that of a dilute L_1 phase, where its free energy is minimized at this temperature.

Provided that the temperature remains below T_C, where the micellar solution L_1 separates into L_1' and L_1'' phases (see Fig. 1), the rate of dissolution of pure nonionic surfactants increases with increasing temperature. For example, dissolution time t_D for a drop of $C_{12}(EO)_6$ with $R_0 = 73$ μm was 11 s at 35 °C. As indicated above, t_D is proportional to R_0^2, so that t_D would be about 13 s at this temperature if R_0 were 78 μm. As indicated in the preceding paragraph, a drop with $R_0 = 78$ μm dissolved more slowly, taking 16 s, when the temperature was reduced to 30 °C.

For nonionic surfactants above T_C dissolution does not yield a single phase at equilibrium except at extreme dilutions, as may be seen from Fig. 1. For instance, at 35 and 40 °C both lamellar and surfactant-rich liquid appeared as intermediate phases when drops of $C_{12}(EO)_5$ were injected into water, the former phase dissolving by a shrinking core mechanism to leave a drop of the latter at equilibrium, as expected from Fig. 1. Pure $C_{12}(EO)_3$ and $C_{12}(EO)_4$ are considerably less hydrophilic with phase transitions shifted to lower temperatures. Indeed, T_C pure for $C_{12}(EO)_4$ is near 10 °C and pure $C_{12}(EO)_3$ exhibits no T_C, i.e., it would occur below 0 °C if freezing of water did not intervene [21]. At 30 and 35 °C the lamellar phase appeared for both these surfactants but in the intriguing form of multiple filaments known as myelinic figures growing into the aqueous phase when drops of these surfactants were injected into water (see Fig. 3), in contrast to the above situations where interfaces between the lamellar and aqueous phases were relatively smooth. The phase behavior corresponding to these experiments was similar to that near 60 °C for $C_{12}(EO)_5$ (Fig. 1).

Solubility of surfactant in the aqueous phase becomes quite low at temperatures only slightly above T_C, as may be seen from Fig. 1, causing surfactant transfer to this phase to be very slow. Moreover, there are substantial miscibility gaps between the dilute and more concentrated phases, e.g., the two-phase region denoted $L_1' + L_1''$ in Fig. 1 for the $C_{12}(EO)_5$ system above about 30 °C. As discussed further below, the myelinic figures are a nonequilibrium morphology which is observed when there is a substantial miscibility gap between the lamellar phase and a dilute surfactant phase. It results not from diffusion but from swelling of the lamellar phase as it takes up water. Swelling is nonuniform owing to the anisotropic microstructure of the phase.

When a drop of $C_{12}(EO)_3$ was injected into water at 40 °C, myelinic figures formed initially but later dissolved to form drops of the L_3 (sponge) phase, as

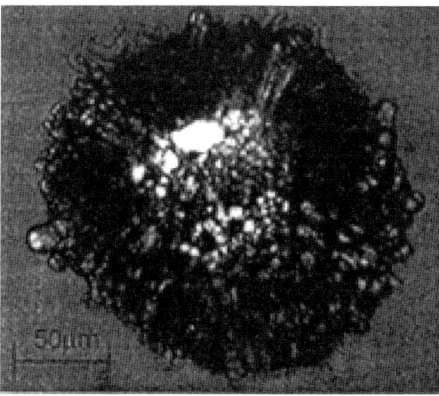

Fig. 3 Myelinic figures for a drop of $C_{12}(EO)_4$ injected into water at 30 °C

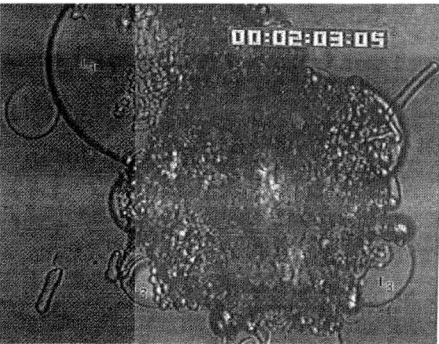

Fig. 4 Sponge (L_3) phase growing as myelinic figures dissolve for a drop of $C_{12}(EO)_3$ injected into water at 40 °C

shown in Fig. 4 [9]. At equilibrium a relatively large drop of L_3 was seen dispersed in a dilute aqueous phase, as expected from the phase diagram, which is similar to that near 70 °C in Fig. 1 for $C_{12}(EO)_5$. It is noteworthy that if diffusion had occurred with local equilibrium at all interfaces, the L_3 phase would have started to form immediately. It did not appear until later because local equilibrium did not exist at the surfaces of the growing myelinic figures. Presumably the surfactant concentration of the myelinic figures decreased with time, eventually reaching that at which transformation to L_3 occurred.

3
Nonionic Surfactant Mixtures

Commercial nonionic surfactants are mixtures of multiple species with different degrees of ethoxylation and typically with some distribution in hy-

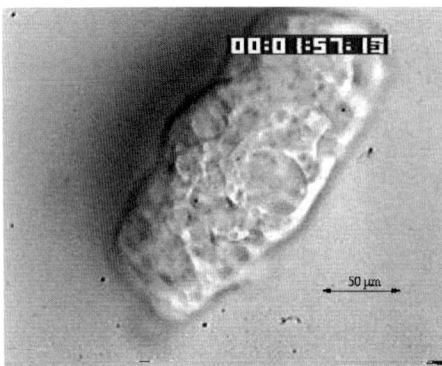

Fig. 5 Emulsification at approximately 2 min after injection for an elongated drop of Tergitol 15-S-7 at 30 °C

drocarbon chain lengths as well. At temperatures well below that where L_1 first separates into two liquid phases, such surfactants dissolve rapidly like the pure nonionics of the preceding section. For simplicity we shall refer to this temperature as the cloud point. In contrast to the situation for pure nonionic surfactants, dissolution time increases substantially as temperature approaches the cloud point.

Consider, for example, Neodol 25-7, a linear alcohol ethoxylate made by Shell having hydrocarbon chain lengths between 12 and 15, an average ethylene oxide (EO) number of 7.3, and a cloud point of 50 °C. At 35 °C a drop of this surfactant with an initial radius of 61 μm took 10 s to dissolve. At 45 °C the corresponding time was 27 s [10]. The initial behavior was basically the same as at 35 °C, i.e., an intermediate lamellar phase formed and dissolution was rapid. Subsequent behavior was different, however. Droplets of another phase appeared within or at the surface of the remaining drop, which then typically broke up into two or three small but rather elongated drops. These dissolved slowly while retaining their elongated shape.

A significant increase in t_D and similar elongated drops and emulsification were seen as the cloud point of 37 °C was approached for Dow's commercial secondary alcohol ethoxylate Tergitol 15-S-7 (Fig. 5). Its hydrophobe consists of various double-chain species with the sum of the chain lengths ranging between 11 and 15, and its average EO number is 7.3. In some cases a conical projection developed on the elongated drop, and a jet was emitted, which broke up into small droplets (Fig. 6).

Further videomicroscopy investigations were conducted for model systems consisting of binary mixtures of pure nonionic surfactants [10]. The results indicated that the common source of the intriguing phenomena described above is that the more hydrophilic species dissolve faster, causing the remaining drop to become enriched in the less hydrophilic species, which are less soluble and dissolve more slowly. Indeed, the drop can achieve compositions

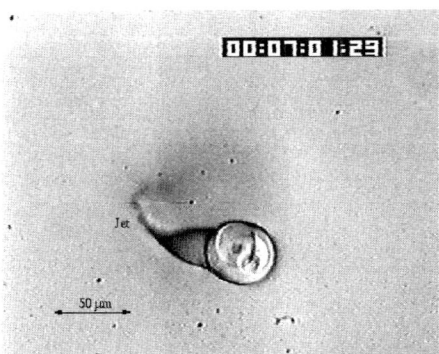

Fig. 6 Jet at approximately 7 min after injection for an elongated drop of Tergitol 15-S-7 at 30 °C

where phase separation (droplet formation) occurs at the temperature of the experiment. As shown more clearly later for another system (Fig. 12), the drops are likely surfactant-rich liquid L_1'' forming as a nonwetting phase between the L_α and L_1' phases. Moreover, the L_α phase shifts from being concentrated and highly viscous to being dilute and more fluid, the analog to similar behavior for pure $C_{12}(EO)_5$ near 50 °C in Fig. 1. The reason for this shift appears to be a sharp decrease in bilayer rigidity, which permits thermal fluctuations of bilayers to have greater amplitude, i.e., the cost in elastic energy of deforming the lamellae becomes small compared to the gain in system entropy. The result is a substantial "undulation" repulsive force between bilayers, which increases equilibrium spacing between them [17] and leads to uptake of water and associated swelling to reduce free energy. Figure 7 shows an example of a dilute lamellar phase swelling anisotropically during dissolution of an 85/15 mixture of pure $C_{12}(EO)_8$/1-decanol at 35 °C, just 2 °C below its cloud point. Note that the drop's surface is relatively smooth compared to swelling of a more concentrated lamellar phase as myelinic figures in Fig. 4. The small drops within the lamellar phase are not round, consistent with its anisotropic microstructure.

The mechanism of formation of jets such as that in Fig. 6 is not clear but apparently is associated with swelling of the L_α or L_3 phase (the latter can also exist at very low surfactant concentrations, as shown in Fig. 1). The phenomenon resembles the "tip streaming" observed in drops of liquids subjected to shear or extensional flows with surfactants present [12, 13]. In these cases shear stresses from the flow in the external phase cause the drop to elongate and form a jet with a conical shape similar to that seen in Fig. 6. No such external flow is present here, but perhaps flow inside the drop accompanying the swelling process produces a similar effect.

One way to visualize the source of the behavior described in the preceding paragraphs in terms of Fig. 1 is to recognize that temperatures of the various phase boundaries for the drop decrease as it becomes less hydrophilic during dissolution, eventually falling below the experimental temperature. The shift

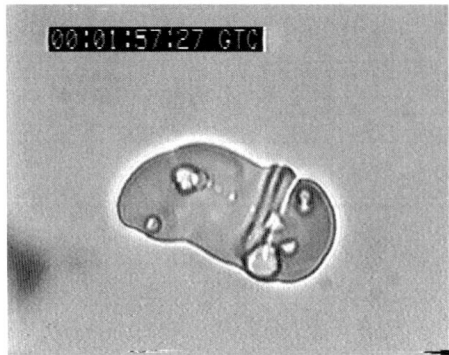

Fig. 7 Swelling of dilute lamellar phase for 85/15 $C_{12}(EO)_8$/decanol mixture at 35 °C

is similar to that discussed in the preceding section in changing from pure $C_{12}(EO)_5$ to the less hydrophilic surfactants $C_{12}(EO)_3$ and $C_{12}(EO)_4$.

In addition to dissolution rates and mechanisms of the neat surfactants themselves, there is interest in whether liquid nonionic surfactants incorporated into powder detergent formulations can influence the performance of some of the solid constituents. For instance, viscous liquid crystalline phases formed during nonionic surfactant dissolution might encapsulate solid particles such as zeolites, thereby preventing them from removing calcium and magnesium ions from solution by ion exchange. Bai et al. [2] used optical microscopy to observe the behavior when flooded by water of individual granules some 500 μm in diameter containing approximately 70% micrometer-sized particles of (nonswelling) Zeolite 4A and 30% liquid nonionic surfactant. Using the "hanging drop" cell shown in Fig. 8, they observed

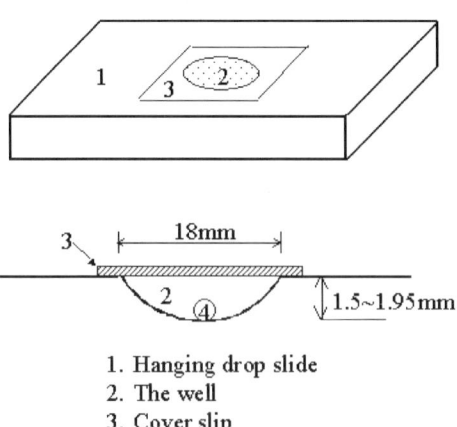

1. Hanging drop slide
2. The well
3. Cover slip
4. Single granule

Fig. 8 "Hanging drop" slide with well used to study granule disintegration

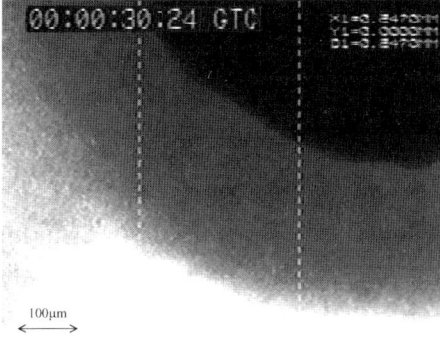

Fig. 9 Disintegration of granule of Zeolite 4A/pure $C_{12}(EO)_6$ in water at 30 °C

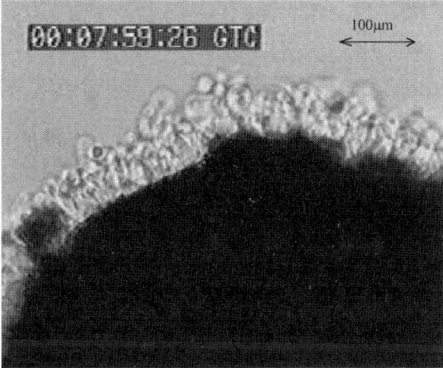

Fig. 10 Granule of Zeolite 4A/pure $C_{12}(EO)_3$ in water at 30 °C. Myelinic figures form, and no disintegration occurs

rapid disintegration of granules for conditions where complete dissolution of neat surfactant drops was observed by the above techniques (Fig. 9) but no disintegration otherwise (e.g., Fig. 10). Little difference in granule disintegration was seen between pure nonionic surfactants below T_C and nonionic surfactant mixtures slightly below their cloud points, even though the overall dissolution times of drops of the latter were greater, as discussed above. A possible reason for this behavior is that much of the mixed surfactant dissolves rapidly, allowing most of the Zeolite 4A particles to be released from the granule, with only the last amount dissolving slowly.

4
Mixtures of Nonionic and Anionic Surfactants

Many anionic surfactants are too hydrophilic to be effective in removing nonpolar oils from synthetic fabrics, which requires conditions where hydrophilic and lipophilic properties are nearly balanced [20]. Nonionic surfactants are typically less hydrophilic and hence better suited for this purpose. It is usually desirable to use a mixture of nonionic and anionic surfactants because anionics are often less expensive and provide a charge to oily and particulate soil removed from the fabric, thereby hindering redeposition. Accordingly, dissolution rates of anionic/nonionic surfactant mixtures are of interest.

Chen [8] studied mixtures of the pure surfactants $C_{12}(EO)_4$ and sodium dodecyl sulfate (SDS) at 30 °C. At this temperature the former is a liquid which does not dissolve in water (see Fig. 3), and the latter is a solid. The SDS was doubly recrystallized from ethanol to remove n-dodecanol and other impurities. The solubility of SDS in pure $C_{12}(EO)_4$ at 30 °C was found to be approximately 9 wt. %. When small drops of an 8 wt. % mixture were injected into water at 30 °C, complete dissolution was observed, the time required being a linear function of the square root of initial drop radius. For instance, a drop having an initial radius of 70 μm required approximately 100 s to dissolve, significantly more than the 16 s cited above for a slightly larger drop of pure $C_{12}(EO)_6$. Behavior was similar to that of nonionic mixtures below their cloud points discussed previously in that most of the drop dissolved rapidly, but the final small volume dissolved rather slowly with some observable emulsification.

Bai [2] performed similar drop dissolution experiments with sodium oleate (NaOl) and $C_{12}(EO)_4$. For drops initially containing 7 and 10 wt. % NaOl (particle size < 38 μm) the behavior was similar to that described above for drops having 8 wt. % SDS. However for drops with 15 and 17 wt. % NaOl dissolution was faster—comparable to that of the pure nonionics—and neither a surfactant-rich liquid immiscible with water nor emulsification was seen. Instead a concentrated liquid crystalline phase transformed directly into a micellar solution, as seen for the pure nonionics and nonionic mixtures well below their cloud points.

The explanation for this behavior is similar to that given in the preceding section for nonionic surfactant mixtures. Adding a hydrophilic anionic surfactant raises the temperature at the cloud point and other phase transitions above those for pure $C_{12}(EO)_4$. If the amount of anionic added exceeds only slightly that needed for complete solubility, the final stages of the dissolution process are slow because preferential dissolution of the anionic causes the remaining drop to rise above its cloud point and nucleate small droplets of surfactant-rich liquid. But if the amount added is sufficiently large, drop composition remains below the cloud point in spite of preferential dissolution, with the result that dissolution is fast as with pure nonionic surfactants below their cloud points.

Surfactant solutions used in cleaning processes are often alkaline. It is thus of some interest to study the dissolution of a mixture of a nonionic surfactant with an organic acid, which is converted to the corresponding soap in the alkaline solution. As soaps of multivalent ions such as calcium have low solubility in water, this approach is most effective when hardness is low, either naturally or owing to the action of additives to the surfactant solution. Of course, the same limitations on hardness apply if a soap such as NaOl is added to a nonionic surfactant as discussed above, and an alkyl sulfate such as SDS or, even better, an alkylethoxy sulfate would be preferable.

Bai [2] used videomicroscopy with the technique described previously to observe behavior when drops consisting of mixtures of $C_{12}(EO)_4$ and oleic acid were injected into aqueous solutions buffered at various alkalinities and maintained at 30 °C. The phases present at equilibrium at the end of the experiments are shown as a function of initial oleic acid content and solution pH in Fig. 11. An increase in either variable increased the ratio of oleate soap to $C_{12}(EO)_4$, thereby making the surfactant mixture more hydrophilic. Complete dissolution to an aqueous micellar solution L_1 was seen when values of both variables were sufficiently high. An increase in either variable increased the rate of dissolution in this L_1 region. As with the $C_{12}(EO)_4$/NaOl mixtures discussed above, the fastest dissolution was observed when, after a brief initial transient, the surface of the drop became a liquid crystal, which dissolved directly into the aqueous phase with no additional surfactant-rich liquid or emulsification. Such behavior was observed for a drop of an 80/20 $C_{12}(EO)_4$/oleic acid mixture injected into a buffer solution at pH 12. However, for the same mixture at pH 10 dissolution was incomplete. Figure 12a shows surfactant-rich liquid forming at the interface between the lamellar phase and the external aqueous phase in the early stages of dissolution. The discrete drops show that the surfactant-rich phase is nonwetting. As indicated above, such behavior is likely responsible for much of the emulsification observed

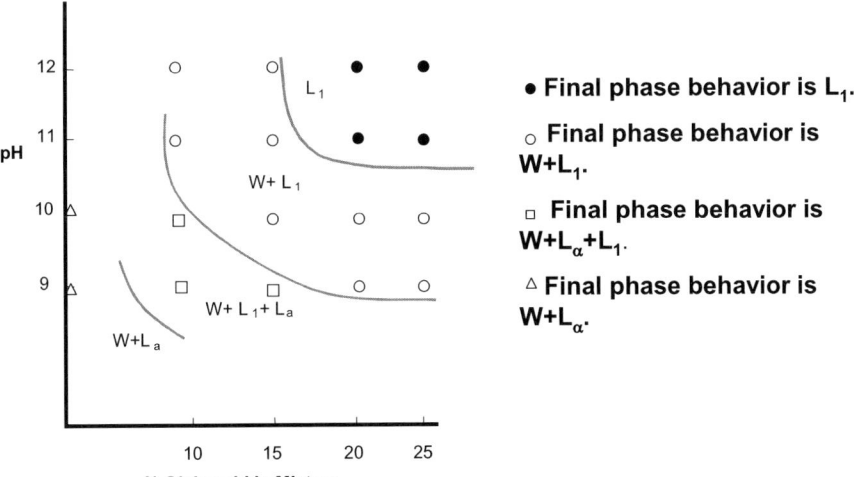

Fig. 11 Final state for $C_{12}(EO)_4$/oleic acid drops injected into buffer solutions at 30 °C

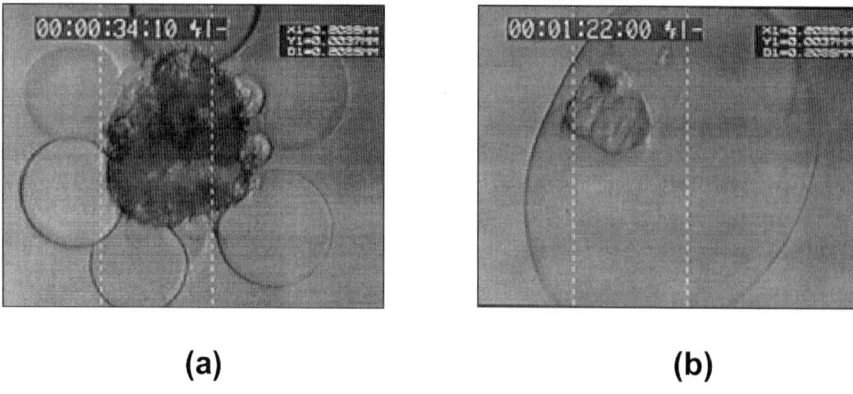

(a)   (b)

Fig. 12 Dissolution of an 80/20 $C_{12}(EO)_4$/oleic acid drop in buffer solution at pH 10. **a** 34 s, **b** 1 min 22 s after injection

for nonionic surfactant mixtures. Later the drops coalesced and dissolution of the lamellar phase proceeded by a shrinking core mechanism, as shown in Fig. 12b.

5
Pure Surfactants with Low Solubility: Myelinic Figures

As indicated above, miscibility gaps are small and intermediate lamellar liquid crystalline phases dissolve rapidly into the aqueous phase if the surfactant or surfactant mixture is rather hydrophilic with a high spontaneous curvature (low (v/la)), for instance at temperatures below T_C for pure nonionic surfactants. In this case dissolution, which converts lamellae of zero curvature to aggregates with significant curvature as surfactant concentration decreases, occurs spontaneously because it reduces system free energy.

In contrast, there is little or no driving force for curvature to increase when the surfactant or surfactant mixture comprising a lamellar phase is lipophilic, and dissolution proceeds slowly. Individual surfactant molecules can still dissolve in the aqueous phase and diffuse away from the liquid crystal, but molecular solubility under these conditions is low, and there is a wide miscibility gap. Although system free energy cannot be reduced by increasing the curvature of individual lamellae, it can usually be lowered by increasing the thickness of the water layers between lamellae because repulsive forces due to electrostatic or undulation effects are thereby reduced. Hence, if defects or other morphological features of the lamellar phase are such that water can enter, the phase swells. Being an anisotropic phase, however, it does not swell uniformly. Often the swelling takes the form of "myelinic figures", such as those shown in Fig. 3 for pure $C_{12}(EO)_4$ at 30 °C.

Lamellar phases of phospholipids often exhibit myelinic figures when contacted with water. Electron micrographs [24, 26] showed that each tubular myelinic figure in the egg-yolk phosphatidylcholine/water system consisted of a water core surrounded by many concentric bilayers. More recently Raman spectroscopy techniques have confirmed the concentric bilayer arrangement [1, 18]. Myelinic figures are not equilibrium structures, however, and eventually break up to form vesicles or other lamellar structures. Indeed, adding water to a vessel whose inner walls are coated with a thin layer of a lamellar phase of low water content is a well-known way of forming vesicles.

One feature of myelinic figure growth when surfactant and water are contacted along a plane interface in a thin cell of uniform thickness is that many parallel, closely packed myelinic figures grow toward the aqueous phase at nearly the same rate, as shown in Fig. 13. After an initial transient period, the mean length L is found experimentally to be proportional to the square root of time for nonionic surfactants [6], phospholipids [23], and the anionic surfactant Aerosol OT (AOT) discussed further below [4, 16]. Although such behavior is often indicative of growth controlled by diffusion, swelling is dominant in growth of the myelinic figures themselves.

Confirmation of the swelling mechanism was most clearly provided by Buchanan et al. [6], who studied myelinic figure growth in the $C_{12}(EO)_3$/water system at 20 °C. They placed small (1 µm) latex particles in the aqueous phase before contacting the pure surfactant and water. Using optical microscopy, they observed particles moving toward the base of the myelinic figures through the channels separating adjacent myelins, thereby confirming water flow along these channels (see Fig. 14A). That is, growth by diffusion as in Fig. 14B was ruled out. Moreover, Buchanan et al. [5] conducted other ex-

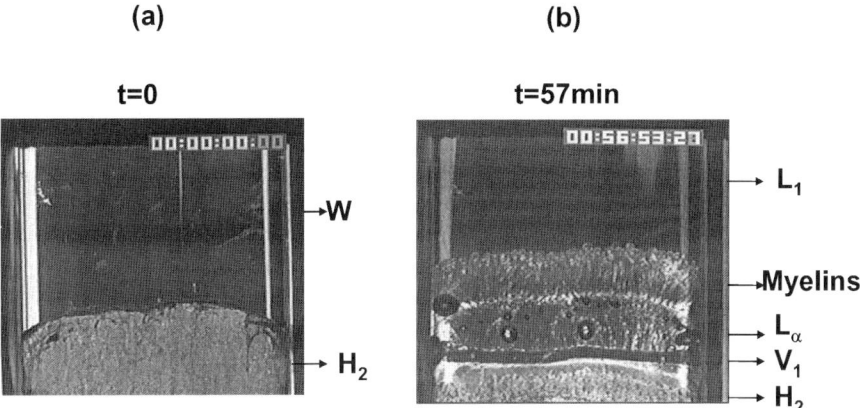

Fig. 13 Video frames of AOT/water semi-infinite experiment at 30 °C. **a** AOT before water addition; **b** 57 min after water addition. Width of cell is 4 mm

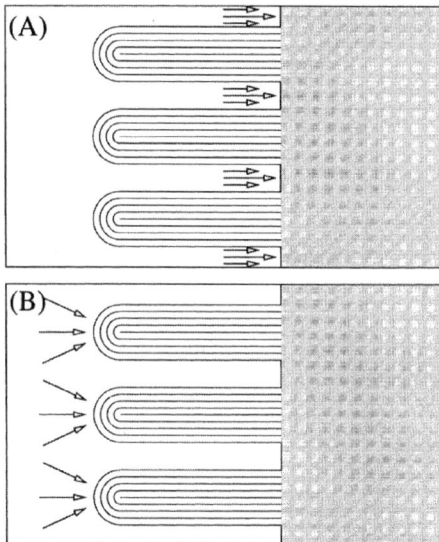

Fig. 14 A Schematic diagram of motion of latex particles toward the base of myelinic figures, confirming that their growth is due to swelling, not to diffusion as in (**B**)

periments with a lamellar phase which had been subjected to high shear so that it took the form of multiple "onions", i.e., spherical particles with multiple concentric bilayers. In this case no myelinic figures were observed as long as the onions persisted, presumably because they were free of defects where water could enter and produce swelling. Further information on the mechanism of growth was provided by Zou and Nagel [34], who were able to grow individual myelinic figures by creating a single defect in a well-ordered layer of the lamellar phase of a phospholipid.

The equilibrium phase diagram for the pure anionic surfactant AOT (sodium bis(2-ethylhexyl) sulfosuccinate)/water system at 25 °C is shown in Fig. 15 [14]. It exhibits a miscibility gap between the L_1 and L_α phases and indicates that intermediate lamellar L_α and viscous isotropic V_1 phases should form when pure AOT, which is a reverse hexagonal phase H_2 (similar to H_1 but with long rodlike reverse micelles), is brought into contact with water without stirring. Such behavior was, in fact, observed by Bai and Miller [4], as shown in Fig. 13, except that the portion of the lamellar phase above the surface of initial contact consisted of many vertically oriented, closely packed myelinic figures as indicated above. The base of the myelinic figures remained at the elevation of the surface of initial contact throughout the experiment. It was found that not only the length of the myelinic figures but also the displacements of the L_α/V_1 and V_1/H_2 boundaries from the initial surface of contact were proportional to the square root of time. As indicated in Fig. 13, the timescales of interest here are of the order of tens of minutes. At much

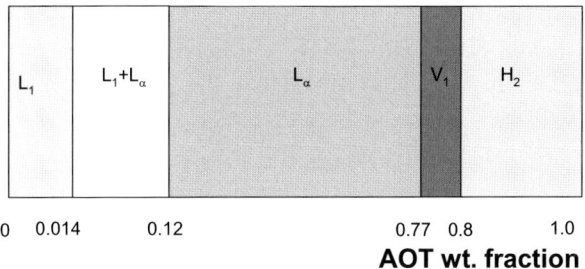

L_1: micellar solution phase
L_α: lamellar liquid crystalline phase
V_1: viscous isotropic phase
H_2: reverse hexagonal phase

Fig. 15 Phase diagram for AOT/water at 25 °C

shorter times while the myelins are first developing, Taribagil et al. [29] have reported a different time dependence in another system.

This behavior was analyzed using a similarity solution [4], which yielded Eq. 1 for the concentration distribution in each phase below the surface of initial contact ($x = 0$). Boundary conditions imposed included local equilibrium and conservation of mass of surfactant at each interface and composition approaching pure AOT at distances far below the initial surface of contact (large x). The novelty of the analysis is the form of the surfactant mass balance imposed at $x = 0$:

$$D_{elam}(\partial \Phi_{lam}/\partial x)(0) + j_{aq}\Phi_{L_1} = [\,d(\varepsilon_m t^{1/2}\Phi_{lam}(0)\phi)/\,dt\,]\,. \tag{2}$$

The first term on the left side of this equation represents the flux of surfactant to $x = 0$ from the portion of the lamellar phase below the myelinic figures. D_{elam} is the effective diffusivity in the lamellar phase. The second term represents surfactant dissolved in the aqueous phase, which flows between the myelins to their base at $x = 0$. j_{aq} is the total volumetric flux and Φ_{L_1} the surfactant content. That is, neat AOT could be contacted with a dilute AOT solution instead of with pure water. On the right side of the equation ($\varepsilon_m t^{1/2}$) is the length of the myelins with ε_m a constant, $\Phi_{lam}(0)$ is the surfactant volume fraction in the lamellar phase at the base of the myelins and in the myelins themselves, and ϕ is the fraction of the cross-sectional area occupied by the myelins. Surfactant volume fraction is assumed uniform along the myelins, i.e., diffusion along the myelins is neglected, since they are found to be of nearly uniform diameter and are forced out of the lamellar phase by the swelling process. For uniform close-packed cylinders it can be shown that ϕ is approximately 91%, in good agreement with the value found for myelinic figures of pure $C_{12}(EO)_3$ [6].

As the base of the myelins is fixed at the initial surface of contact, according to the experimental observations, the volumetric flow rate of aqueous solution entering the lamellar phase must equal the rate of increase in volume of the myelinic figures:

$$j_{aq} = [d(\varepsilon_m t^{1/2}\phi)/dt] . \tag{3}$$

Substitution of Eqs. 1 and 3 into Eq. 2 yields

$$(D_{elam}/\pi)^{1/2} a_{2lam} = [(\varepsilon_m \phi(a_{1lam} - \Phi_{L_1})/2)] . \tag{4}$$

One more boundary condition is required to specify the problem. Suppose that the rate of myelin growth is directly proportional to the osmotic pressure difference Δp between the bulk aqueous solution and the lamellar phase at the base of the myelins ($x = 0$) and inversely proportional to myelin length. For example, contact between the myelins could provide resistance to the swelling process proportional to myelin length. Hence

$$[d(\varepsilon_m t^{1/2})/dt] = [(k\Delta p)/\varepsilon_m t^{1/2}] , \tag{5}$$

where k is a proportionality constant. This equation is readily simplified to

$$\varepsilon_m = (2k\Delta p)^{1/2} . \tag{6}$$

For the experiments of Fig. 13, ε_m can be measured, so that it is not necessary to have independent information on $(k\Delta p)$. Moreover, with the known interfacial compositions given by Fig. 15 and the measured velocities of the two interfaces below the base of the myelins, it is possible to solve for the effective diffusivities D_{eV_1} and D_{eH_2} and $\Phi_{lam}(0)$ in terms of D_{elam}. Results of such calculations are shown in Table 2.

As in Sect. 2, another experiment is required to evaluate all the diffusivities, i.e., to obtain the correct value for D_{elam} for use with Table 2. Bai and Miller [4] repeated the contacting experiment of Fig. 13 except that only a thin layer of AOT was present initially. As a result, the similarity solution, which assumes a semi-infinite AOT phase, is not valid after a short transient. Instead the governing equations must be solved numerically with the boundary

Table 2 Effective diffusion coefficients and $\Phi_{lam}(0)$ (myelin composition) for assumed values of D_{elam} in the lamellar phase for AOT/water semi-infinite experiment

D_{elam} (m^2/s)	D_{eV_1} (m^2/s)	D_{eH_2} (m^2/s)	$\Phi_{lam}(0)$
1.2×10^{-10}	1.59×10^{-10}	1.94×10^{-11}	0.483
1.31×10^{-10}	1.70×10^{-10}	3.56×10^{-11}	0.497
1.44×10^{-10}	1.82×10^{-10}	5.6×10^{-11}	0.513
2.0×10^{-10}	2.20×10^{-10}	1.47×10^{-10}	0.562
3.0×10^{-10}	2.63×10^{-10}	2.87×10^{-10}	0.617

condition of zero flux at the bottom of the cell. The experiment was continued until the H_2 phase completely disappeared, and the value of D_{elam} for which the numerical solution best fitted the time-dependent behavior of the L_α/V_1 and V_1/H_2 interfaces was found (there was no observable change from the previous experiment in the rate of growth of the myelins). Figures 16a and 16b show that a good fit was obtained for $D_{elam} = 1.3 \times 10^{-10}$ m^2/s and other parameters as in the second row of Table 2. As indicated in the table, the surfactant mass fraction $\Phi_{lam}(0)$ at the base of the myelins is about 0.50.

As may be seen in Fig. 13, nearly all of the myelinic figures have diameters in the range 50–100 µm, i.e, there is a preferred diameter at which the rate of increase in myelin length is fastest. Dave et al. [11] studied the effects of cell

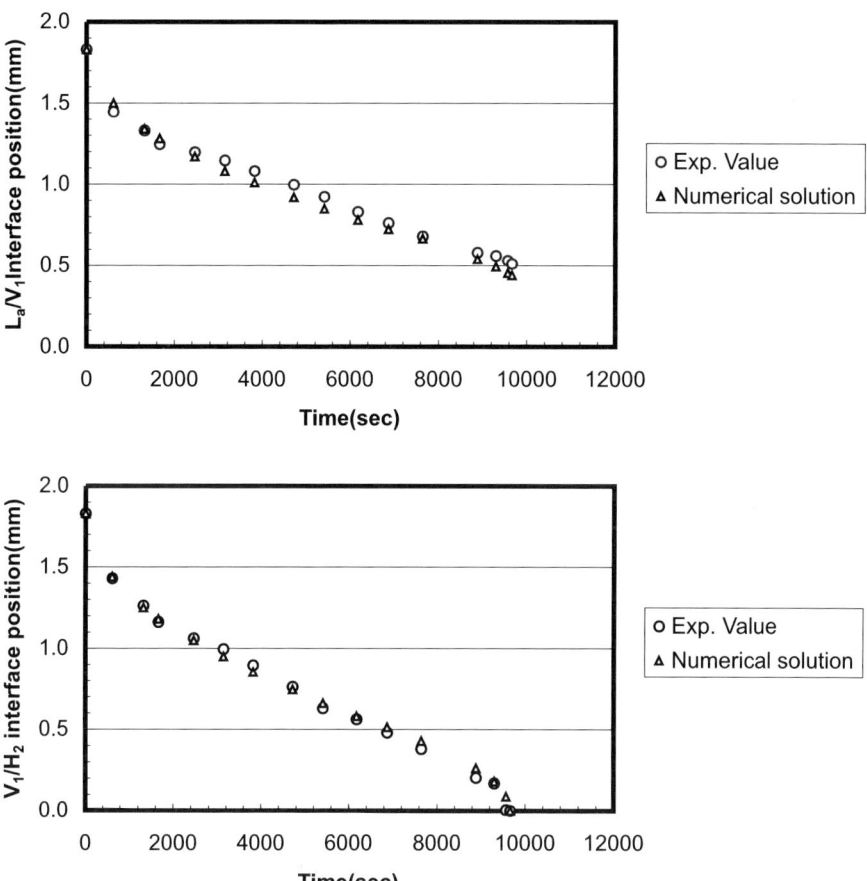

Fig. 16 Plots of interfacial position as a function of time for AOT/water thin-layer experiment and comparison with simulation for L_α/V_1 interface (*upper panel*), V_1/H_2 interface (*lower panel*)

thickness, i.e., the direction perpendicular to the plane of Fig. 13, on diameters of closely packed myelinic figures of phosphatidylcholine. They found that for thicknesses up to about 150 µm, only one layer of myelinic figures could be seen. Thus, myelins of very small diameter were not favored. One may speculate that the high resistance to flow through the small gaps between such myelins would cause them to grow more slowly than larger myelins, so that the latter would dominate the observed behavior. As a result, myelin diameter is dictated by cell thickness when the latter is small, as was long ago suggested by Lawrence. Yet the observation of Dave et al. [11] that more than one layer of myelins is seen at large cell thicknesses suggests that there is also a diameter above which growth slows, possibly because transport of water at the base of a myelin from its periphery to its entire cross section becomes limiting. Further studies are required to determine the validity of these ideas.

It is noteworthy that when a lamellar phase containing 34 wt. % AOT was contacted with water in the vertical cell penetration experiment, the L_α/L_1 interface remained flat but moved upward, displacement from its initial position being proportional to the square root of time. No myelinic figures were observed [2, 4]. Analysis based on diffusion control, i.e., a flat interface with local equilibrium, gave plausible values for D_{eL_1} when D_{eL_α} was assumed to be near that found in the pure AOT contacting experiment. As the maximum surfactant concentration in the L_1 phase is, according to Fig. 15, 1.4 wt. %—more than an order of magnitude larger than that for phospholipids and $C_{12}(EO)_3$—it is plausible that the lamellar phase would dissolve directly into the micellar solution under these conditions instead of forming myelinic figures much higher in water content and hence having less structural integrity than those seen during dissolution of pure AOT. The initial domain orientation in the lamellar phase containing 34 wt. % AOT could also be a factor in whether myelins could develop.

There may be limitations in applying the above model to other systems. For instance, the initial surfactant often exists as a lamellar phase as for phospholipids, so that there are no interfaces between various liquid crystalline phases whose velocities can be measured and used to determine effective diffusivities as in the AOT analysis above. As a result, the base of the myelinic figures must approach the base of the vertical cell as the volume of the lamellar phase shrinks, and the assumption made above that the composition of the myelins is independent of time may not be valid.

When layers of certain block copolymers of ethylene oxide and butylene oxide are contacted with water, there is an initial period when the position of the interface is proportional to t_m, where $m < 0.5$ [32]. That is, initial swelling is not controlled by diffusion but instead by hydration and rearrangement of the long molecules to form the various phases. In the case of $(EO)_{16}(BO)_{22}$ small-angle X-ray scattering did detect evidence of both reverse hexagonal and lamellar phases during this initial period, but it was not clear whether all the swollen block copolymer layer consisted of these phases or how the phases were arranged. Later in the process growth did become diffusion controlled [32].

Lamellar phases of phospholipids and other lipophilic surfactants can be dissolved, i.e., converted into micellar solutions, by contacting them with aqueous phases of hydrophilic surfactants, a phenomenon that is sometimes desirable and sometimes undesirable in studies of biological membranes. Often myelinic figures form during the dissolution process. Simoes et al. [33] reported rates of dissolution of various mixtures of phosphatidylcholine and the nonionic surfactant Tween 80.

6
Summary

The ability of surfactant/water systems to form multiple phases having various shapes of surfactant aggregates virtually assures that at least one liquid crystalline phase will be present transiently during dissolution of the neat surfactant in water. For rather hydrophilic surfactants and surfactant mixtures the kinetics of formation and disappearance of these phases is rapid, and the rate of the dissolution process is controlled by diffusion. For less hydrophilic surfactants more complex behavior may be seen, the formation of myelinic figures of the lamellar phase by a swelling process being a notable example. Owing to preferential dissolution of the more hydrophilic species in surfactant mixtures, the undissolved material becomes ever less hydrophilic, which can lead to phase transformations and intriguing behavior such as spontaneous formation of droplets, anisotropic swelling of dilute lamellar phases, and emission of jets.

References

1. Arunagirinathan MA, Roy M, Dua AK, Manohar C, Bellare JR (2004) Langmuir 20:4816
2. Bai J (2003) Ph.D. Thesis, Rice University, Houston, TX
3. Bai J, Miller CA, Wilson JE (2003) J Surfactants Deterg 6:7
4. Bai J, Miller CA (2004) Colloids Surf A 244:113
5. Buchanan M, Arrault J, Cates ME (1998) Langmuir 14:7371
6. Buchanan M, Egelhaaf SU, Cates ME (2000) Langmuir 16:3718
7. Buchanan M (2004) ACS Symp Ser 869:226
8. Chen BH (1998) Ph.D. Thesis, Rice University, Houston, TX
9. Chen BH, Miller CA, Walsh JM, Warren PB, Ruddock JN, Garrett PR, Argoul F, Leger C (2000) Langmuir 16:5276
10. Chen BH, Miller CA, Garrett PR (2001) Colloids Surf A 183–185:191
11. Dave H, Surve M, Manohar C, Bellare J (2003) J Colloid Interface Sci 264:76
12. De Bruijn RA (1993) Chem Eng Sci 48:277
13. Eggleton CD, Tsai TM, Stebe KJ (2001) Phys Rev Lett 87:048302

14. Franses EE, Hart TJ (1983) J Colloid Interface Sci 94:1
15. Gradzielski M (2003) Curr Opin Colloid Interface Sci 8:337
16. Haran M, Chowdhury A, Manohar C, Bellare J (2002) Colloids Surf A 205:21
17. Helfrich W (1978) Z Naturforsch 33a:305
18. Kennedy AP, Sutcliffe J, Cheng JX (2005) Langmuir 21:6478
19. Laughlin RG (1994) The aqueous phase behavior of surfactants. Academic, London
20. Miller CA, Raney KH (1993) Colloids Surf A 74:169
21. Mitchell DJ, Tiddy GJT, Waring L, Bostock T, McDonald MP (1983) J Chem Soc Faraday Trans I 79:975
22. Prinsen P, Warren PB, Michels MAJ (2002) Phys Rev Lett 89:148302
23. Sakurai I (1985) Biochim Biophys Acta 815:149
24. Sakurai I, Suzuki T, Sakurai S (1985) Mol Cryst Liq Cryst 180B:305
25. Sakurai I, Suzuki T, Sakurai S (1989) Biochim Biophys Acta 985:101
26. Sallen L, Oswald P, Sotta P (1997) J Phys II 7:107
27. Strey R, Schomäcker R, Roux D, Nallet F, Olsson U (1990) J Chem Soc Faraday Trans 86:2253
28. Taribagil R, Arunagirinathan MA, Manohar C, Bellare JR (2005) J Colloid Interface Sci 289:242
29. Warren PB, Buchanan M (2001) Curr Opin Colloid Interface Sci 6:287
30. Warren PB, Prinsen P, Michels MAJ (2003) Phil Trans R Soc Lond Ser A 361:665
31. Zilman A, Safran SA, Sottmann T, Strey R (2004) Langmuir 20:2199
32. Battaglia G, Ryan AJ (2006) J Phys Chem B 110:10272
33. Simoes SI, Tapadas JM, Marques CM, Cruz MEM, Martina MBF, Cevc G (2005) Eur J Pharm Sci 26:307
34. Zou LN, Nagel SR (2006) Phys Rev Lett 96:138301

Equilibrium Adsorption of Surfactants at the Gas–Liquid Interface

Stoyan I. Karakashev[1] · Anh V. Nguyen[1] · Jan D. Miller[2] (✉)

[1] Division of Chemical Engineering, School of Engineering,
The University of Queensland, Brisbane, 4072 Queensland, Australia

[2] Department of Metallurgical Engineering, University of Utah, 135 South 1460 East,
Salt Lake City, UT 84112, USA
Jan.Miller@utah.edu

1	Introduction	26
2	Adsorption of Nonionic Surfactants	27
2.1	General Approach of Butler and Lucassen–Reynders	27
2.2	Effect of Surfactant Orientation at the Interface	32
2.3	Effect of Surfactant Aggregation at the Interface	33
3	Adsorption of Ionic Surfactants in Presence of Inorganic Electrolytes	34
4	Adsorption of Nonionic and Ionic (1:1) Surfactant Mixtures	37
5	Comparison Between Theory and Experiment	38
5.1	Comparison Between Theory and Experiment	38
5.2	Ionic Surfactants	43
5.3	Mixtures of Nonionic and Ionic Surfactants	46
6	New Perspectives and Further Developments of the Adsorption Theories	47
7	Summary and Conclusion	52
	References	53

Abstract Theories on equilibrium adsorption of surfactants at the gas-liquid interface have been reviewed and validated. For the adsorption of nonionic surfactants, the thermodynamic approach of Butler has been used, in conjunction with the Lucassen-Reynders dividing surface, to describe the adsorption layer state and adsorption isotherm as a function of partial molar area. Applying the Butler–Lucassen–Reynders modeling approach provides the generalized adsorption isotherm and equation of state, which is capable of describing the effect of the surfactant orientational states and aggregation at the interface. For Langmuirian and Frumkinian surfactant adsorption, the Butler–Lucassen–Reynders modeling approach produces the same predictions for surface tension as described by the well-known Langmuir and Frumkin adsorption isotherms. The adsorption of ionic surfactants and ionic–nonionic surfactant mixtures has been described following the traditional approach with the Gibbs dividing surface and Gibbs adsorption isotherm, and the Gouy-Chapman electrical double layer electrostatics. The developed theories have been validated through comparison with the experimental data on surface tension. Regression analysis by minimizing the reduced chi-square has been used to best fit the models to the experimental data to obtain the model free parameters. For the surfactant homologous series of octaethyleneglycol-n-alkyl ethers $C_nH_{2n+1}O(CH_2CH_2)_8H$, the negative sign of the intermolecular interaction parameter obtained in the regression analysis of surface tension has not been resolved by the model for the surfactant orientational state at the interface. For the surfactant series, the surface aggregation model gives physically consistent fitting and parameters. The models for adsorption of ionic surfactants have been validated using the surface tension of a series of sodium n-hexadecylsulfates with the sulfate group located at the different positions in the hydrocarbon chain, a homologue series of sodium alkyl sulfates, and a series of alkali dodecylsulfates. Improved adsorption models for ionic surfactants have been developed through fundamental modeling of the adsorption processes and the molecular interactions in the adsorption layers. The improved predictions reduce the required number of free parameters and agree with the surface tension and surface potential data better than the conventional models.

Keywords Equilibrium adsorption of surfactants · Gas–liquid interface · Interfacial thermodynamics · Ionic surfactants · Nonionic surfactants · Surface tension · Surface potential

1
Introduction

Surfactants are amphiphilic organic compounds capable of adsorbing at interfaces and have attracted intense studies in the past decades due to the important role of surfactants in interfacial phenomena such as wetting, lubrication, adhesion, colloid stabilization, and molecular and biological recognition [1]. An improved understanding of surfactant adsorption at the gas–liquid and solid–liquid interfaces is of crucial importance in many industrial applications including particle separation by froth flotation, tribology, sensors, molecular electronic devices, and nanofabrications.

Adsorption of surfactants at the interface changes the interfacial properties including surface tension, which is important in many diverse areas of application, and in biological and biochemical processes. It affects foamability and foam stability, wettability, coating flows, foam drainage, and many other science and engineering processes. An accurate description of the thermodynamics of adsorption layers at liquid interfaces is the vital prerequisite for a quantitative understanding of the equilibrium or any non-equilibrium processes taking place at the fluid–liquid interface [1]. Both the Langmuir adsorption isotherm and the Szyszkowski equation for surface tension have been successfully used to explain the dependence of surface tension on surfactant concentration in the bulk solution. However, in a number of cases the Szyszkowski–Langmuir adsorption theory shows significant deviations, especially if there is strong attraction among the adsorbed molecules with long hydrocarbon chains. In this case, the Szyszkowski–Langmuir theory has been improved by accounting for the interaction between the adsorbed molecules, leading to the formulation of the Frumkin adsorption isotherm [2, 3]. Theoretically, the intermolecular interaction parameter in the Frumkin adsorption theory is positive due to the attraction among hydrophobic hydrocarbon chains. In practice, the parameter has often been found negative and has prompted endless debates [4]. Obviously, the organization of surfactant molecules adsorbed at the gas–liquid interface can be complicated. The deviations from the celebrated Szyszkowski–Langmuir adsorption theory can be due to a number of factors, which include the lateral interaction among the heads and the tails of the adsorbed molecules [5], the different orientations of the adsorbed molecules at the interface resulting in different projected areas at the interface [6, 7], the binding interactions between the polar heads of the adsorbed molecules and the counterions in the solution [8–11], and the aggregation of the adsorbed molecules at the interface [12, 13].

The deviations from the Szyszkowski–Langmuir adsorption theory have led to the proposal of a number of models for the equilibrium adsorption of surfactants at the gas–liquid interface. The aim of this paper is to critically analyze the theories and assess their applicability to the adsorption of both ionic and nonionic surfactants at the gas–liquid interface. The thermodynamic approach of Butler [14] and the Lucassen–Reynders dividing surface [15] will be used to describe the adsorption layer state and adsorption isotherm as a function of partial molecular area for adsorbed nonionic surfactants. The traditional approach with the Gibbs dividing surface and Gibbs adsorption isotherm, and the Gouy–Chapman electrical double layer electrostatics will be used to describe the adsorption of ionic surfactants and ionic–nonionic surfactant mixtures. The fundamental modeling of the adsorption processes and the molecular interactions in the adsorption layers will be developed to predict the parameters of the proposed models and improve the adsorption models for ionic surfactants. Finally, experimental data for surface tension will be used to validate the proposed adsorption models.

2
Adsorption of Nonionic Surfactants

2.1
General Approach of Butler and Lucassen–Reynders

The standard approach for describing surfactant adsorption at the gas–liquid interface is based on the Gibbs methodology [16]. The Gibbs dividing surface was introduced and is mathematically defined by the interface line that divides the surface excess of the solvent into two equal parts with opposite signs, and the total surface excess of the solvent is, therefore, equal to zero. The interfacial properties, including surface tension, can be simply described with respect to the Gibbs dividing surface. However, the Gibbs approach is difficult to use in investigating many interfacial structures such as the surfactant orientation at the interface. A more general thermodynamic approach of Butler [14] and Lucassen–Reynders [15] has been successfully applied to a number of systems with different interfacial structures and is described below to show generalized model equations for adsorption isotherms.

We commence with the adsorption of nonionic surfactants, which does not require the consideration of the effect of the electrical double layer on adsorption. The equilibrium distribution of the surfactant molecules and the solvent between the bulk solution (b) and at the surface (s) is determined by the respective chemical potentials. The chemical potential μ_i^s of each component i in the surface layer can be expressed in terms of partial molar fraction, x_i^s, partial molar area ω_i, and surface tension γ by the Butler equation as [14]

$$\mu_i^s = \mu_i^{0s} + RT \ln(f_i^s x_i^s) - \omega_i \gamma, \tag{1}$$

where the first term on the right hand side describes the standard chemical potential, f_i^s is the surface activity coefficient (for water, $i = 0$ and surfactant components, $i \geq 1$), and RT is thermal energy per mole. The physics underlying the Butler equation is that the adsorption requires work against the surface tension.

Equation 1 has been used to derive various adsorption isotherms and equations of state for the adsorption layer. For example, since the chemical potentials of the components in the bulk and on the surface are balanced at equilibrium, Eq. 1 yields

$$\mu_i^{0s} - \omega_i \gamma + RT \ln(f_i^s x_i^s) = \mu_i^{0b} + RT \ln(f_i^b x_i^b), \tag{2}$$

where the right hand side with the superscript b describes the chemical potentials in the bulk. Since the standard chemical potential is the chemical potential of the pure substance, for the solvent (water), $x_0^s = 1$, $f_0^s = 1$, $x_0^b = 1$, and $f_0^b = 1$ (for $i = 0$), Eq. 2 gives

$$\mu_0^{0s} - \omega_0 \gamma_0 = \mu_0^{0b}, \tag{3}$$

where γ_0 and ω_0 are the surface tension and molar area of the solvent. For the solvent, substituting Eq. 3 into Eq. 2 gives

$$\gamma = \gamma_0 + \frac{RT}{\omega_0} \ln \left(\frac{f_0^s x_0^s}{f_0^b x_0^b} \right). \tag{4}$$

For the surfactant components, similar equations relating the partial molar fraction and the activity coefficient with the surface tension can be established. Comparing with very dilute surfactant solutions, the surface tension of which is not significantly different from the pure solvent [17], one obtains

$$\gamma = \gamma_0 + \frac{RT}{\omega_i} \ln\left(\frac{f_i^s x_i^s f_{i0}^b}{f_i^b x_i^b f_{i0}^s}\right), \tag{5}$$

where f_{i0}^s and f_{i0}^b are the activity coefficients of the ith component in the dilute solution. Equations 4 and 5 are central to the adsorption thermodynamics of nonionic surfactants. They can be used to derive the adsorption isotherms and equations of state of adsorption layers. For example, assuming ideal behavior of surfactant molecules in the bulk solution, the equation of state for a non-ideal adsorption layer described by Eqs. 4 and 5 gives

$$\gamma = \gamma_0 + \frac{RT}{\omega_0} \ln(x_0^s f_0^s) \tag{6}$$

$$\ln\left(\frac{f_i^s x_i^s}{K_i x_i^b}\right) = \frac{\omega_i}{\omega_0} \ln(x_0^s f_0^s), \tag{7}$$

where K_i is the equilibrium adsorption constant of the ith component.

The equation of state can be furnished if the partial molar fraction x_i^s can be described in terms of the real surface excess Γ_i. The molar fraction is defined as the molar concentration of the component in the surface layer divided by the total molar concentration of all components in the layer. Since the molar concentrations can also be expressed in terms of the surface area, the molar fraction is equivalent to the partial coverage θ_i of the component. Now we have $x_i^s = \theta_i = \Gamma_i \omega_i$. In terms of the surface coverage θ_i and the bulk concentration C_i of the ith component, Eqs. 6 and 7 can be transformed to give

$$\gamma_0 - \gamma = -\frac{RT}{\omega_0}\left[\ln\left(1 - \sum_{j\geq 1}\theta_j\right) + \ln f_0^s\right] \tag{8}$$

$$K_i C_i = \frac{\theta_i f_i^s}{\left(1 - \sum_{j\geq 1}\theta_j\right)^{n_i} (f_0^s)^{n_i}}, \tag{9}$$

where $n_i = \omega_i/\omega_0$. The coefficient of surface activity f_i^s in the above equations is determined by the enthalpic and entropic non-idealities of the components in the surface layer [18–21]. The description of the non-idealities for many surfactant systems is complicated. For example, in the case of systems containing two surfactants, Eqs. 8 and 9 give [21]

$$\gamma_0 - \gamma = -\frac{RT}{\omega_0}\left[\ln\left(1 - \sum_{j=1}^{2}\theta_j\right) + \sum_{j=1}^{2}\left\{\theta_j\left(1 - \frac{1}{n_j}\right) + \beta_j\theta_j^2\right\} + \beta_{12}\theta_1\theta_2\right] \tag{10}$$

$$b_iC_i = \frac{\theta_i \exp\left[-n_i - 2\beta_i\theta_i - 2\beta_{12}\theta_j + (1-n_i)(\beta_1\theta_1^2 + \beta_2\theta_2^2 + \beta_{12}\theta_1\theta_2)\right]}{(1 - \theta_1 - \theta_2)^{n_i}}, \tag{11}$$

where β_i are the molecular interaction parameters and b_i is another adsorption constant (modified from K_i), which is independent of surface coverage and partial molecular area. Equation 11 is applied for i and $j = 1$ and 2, and $i \neq j$. Equations 10 and 11 present the generalized equations for two-component systems or mono-component systems in which the surfactant has two different adsorption states. These two equations can reduce to a number of the available adsorption models [15, 22–34].

For a one-surfactant system without molecular interaction, Eqs. 10 and 11 simplify into $\gamma_0 - \gamma = (RT/\omega_0)\ln(1 + bC)$. This simplified equation would be the same as that of the well-known Szyszkowski–Langmuir adsorption equation [3] if the partial molecular area ω_0 of the solvent could be equal to the reciprocal of the surface excess of the surfactant at saturation. However, this is not the case since the realistic surface area of the solvent for ω_0 is significantly small (approximately 10 Å^2 for water molecules). This contradiction was solved by Lucassen-Reynders [15] who positioned the dividing surface in the plane $\omega_0 = \omega$ such that the total surface excess of all components including the solvent is equal to the surface excess of surfactants at saturation, i.e., $\Sigma_{i=0}\Gamma_i = \Gamma_\infty$. For saturated monolayers, the Lucassen-Reynders dividing surface does not significantly change the surface excess values from those obtained for the Gibbs dividing surface ($\Gamma_0 = 0$). For very low surface excess of surfactants, the Lucassen-Reynders dividing surface is shifted towards the bulk solution by some fractions of the water molecule diameter. For surfactants of large size, the Lucassen-Reynders dividing surface practically coincides with Gibbs dividing surface.

The average $\overline{\omega}$ of the partial molar area for all components or all possible states at the interface is often used in conjunction with the Lucassen-Reynders dividing surface, which can be equivalently described as

$$\omega_0 = \overline{\omega} = \frac{\sum_{i\geq 1}\Gamma_i\omega_i}{\sum_{i\geq 1}\Gamma_i}. \tag{12}$$

The choice of the Lucassen-Reynders dividing surface has a number of advantages, including the fact that the contribution of non-ideality of entropy for mixing in the adsorption layer disappears in the equations of state and adsorption [21].

In the following, the Lucassen-Reynders dividing surface is used to obtain a number of significant adsorption models.

Firstly, assuming that the enthalpy of mixing in the surface layer is zero, i.e., $\beta_1 = \beta_2 = \beta_{12} = 0$, Eqs. 10 and 11 can be simplified into [21]

$$\gamma_0 - \gamma = -\frac{RT}{\omega}\left[\ln(1 - \theta_1 - \theta_2) + \theta_1\left(1 - \frac{1}{n_1}\right) + \theta_2\left(1 - \frac{1}{n_2}\right)\right] \quad (13)$$

$$b_i C_i = \frac{\theta_i \exp(-n_i)}{(1 - \theta_1 - \theta_2)^n} \quad \text{for } i = 1 \text{ and } 2. \quad (14)$$

These two equations represent the generalized Szyszkowski-Langmuir adsorption model.

Secondly, if only ideality of the mixing entropy is considered, $n_1 = n_2 = 1$ and the Frumkin generalized equation of state and adsorption isotherm are obtained, giving [15, 26]

$$\gamma_0 - \gamma = -\frac{RT}{\omega}\left[\ln(1 - \theta_1 - \theta_2) + \beta_1 \theta_1^2 + \beta_2 \theta_2^2 + \beta_{12} \theta_1 \theta_2\right] \quad (15)$$

$$b_i C_i = \frac{\theta_i}{1 - \theta_1 - \theta_2} \exp(-2\beta_i \theta_i - 2\beta_{12} \theta_j) \quad \text{for } i \text{ and } j = 1 \text{ and } 2, \text{ and } i \neq j. \quad (16)$$

In Eq. 16, b_i is another adsorption constant (independent of surface coverage) and is equal to the product of b_i in Eq. 11 and the base of natural logarithm (= 2.718). For systems containing only one surfactant, $\beta_2 = \beta_{12} = 0$, and Eqs. 15 and 16 reduce to the well-known Frumkin equation of state and adsorption isotherm described as

$$\gamma_0 - \gamma = -\frac{RT}{\omega}\left[\ln(1 - \theta) + \beta \theta^2\right] \quad (17)$$

$$bC = \frac{\theta}{1 - \theta} \exp(-2\beta\theta). \quad (18)$$

The equilibrium adsorption constant b in Eq. 18 is described as [35]

$$b = N_A v_m \exp\left(-\frac{\Delta \mu^0}{RT}\right), \quad (19)$$

where N_A is the Avogadro number, v_m is the molecular volume of the nonionic surfactant, and $\Delta \mu^0$ is the molecular free energy of adsorption.

Finally, assuming the ideality of both the enthalpic and entropic mixing gives $\beta = 0$ and Eqs. 17 and 18 simplify to the well-known Szyszkowski-Langmuir equation given by

$$\gamma_0 - \gamma = \frac{RT}{\omega} \ln(1 + bC). \quad (20)$$

The general approach of Butler and Lucassen-Reynders can now be employed for investigating additional effects of the adsorbed molecules. Two important effects are examined below [17].

2.2
Effect of Surfactant Orientation at the Interface

Surfactant molecules are often asymmetric, able to adsorb in two or more states. At low concentration (low surface pressure), adsorption with large partial molar surface area is preferential. On the other hand, at high surface pressure, adsorption takes place with different orientations having minimal partial molar areas. The effect of the surfactant reorientation on the surface can be described using Eqs. 10–11. For simplicity, the system is assumed to contain only one surfactant that has two possible orientations at the surface. The system is one-component in the bulk (excluding the solvent) and is two-component at the surface, leading to two possible partial molar areas, ω_1 and ω_2. Assuming that $b_1 = b_2 = b$, Eqs. 13 and 14 can be applied to the systems with one surfactant having two states with two different orientations at the interfaces, giving

$$\gamma_0 - \gamma = -\frac{RT}{\overline{\omega}} \ln(1 - \Gamma_t \overline{\omega}) \tag{21}$$

$$bC = \frac{\Gamma_i \overline{\omega}}{(1 - \Gamma_t \overline{\omega})^{\omega_i/\overline{\omega}}} \exp\left(-\frac{\omega_i}{\overline{\omega}}\right) \quad \text{for } i = 1 \text{ and } 2, \tag{22}$$

where $\Gamma_t = \Gamma_1 + \Gamma_2$ is the total surface excess of the adsorbed surfactant with two orientational states at the interface. Equation 21 was obtained from Eq. 13 using the fact that the contribution of non-ideality of entropy of mixing in the adsorption layer is zero for the Lucassen-Reynders dividing surface, i.e., $1 - \omega_0 \Sigma_{i \geq 0} \Gamma_i = 0$.

Equations 21 and 22 present the useful extension of the Szyszkowski–Langmuir model to the adsorption with two orientational states at the interface. If the molecular interactions are considered, a similar simplified model with $\beta_1 = \beta_2 = \beta$ and $b_1 = b_2 = b$ can be obtained from Eqs. 10 and 11, giving

$$\gamma_0 - \gamma = -\frac{RT}{\overline{\omega}} \left[\ln(1 - \Gamma_t \overline{\omega}) + \beta(\Gamma_t \overline{\omega})^2\right] \tag{23}$$

$$bC = \frac{\Gamma_i \overline{\omega}}{(1 - \Gamma_t \overline{\omega})^{\omega_i/\overline{\omega}}} \exp\left[-\frac{\omega_i}{\overline{\omega}} - 2\beta \Gamma_t \overline{\omega} + \beta \left\{1 - \frac{\omega_i}{\overline{\omega}}\right\} (\Gamma_t \overline{\omega})^2\right]. \tag{24}$$

These two equations present the extension of the Frumkin model to the adsorption of one-surfactant system with two orientational states at the interface. The model equations now contain four free parameters, including ω_1, ω_2, β, and b. The equations are highly nonlinear, and regression used in the analysis of surface tension data involves special combinations of Eqs. 23 and 24, which produces a special model function used in the least-square minimization with measured surface tension data. Since the model function also contains surface tension [6], the regression analysis does not follow the standard procedure of minimizing the chi-square, χ^2, as traditionally defined [36]. Although the non-standard regression analysis has produced useful results, the statistical certainties of the model predictions [36] are difficult to establish. As shown in a later section of this paper, Eq. 22 can be transformed into a useful equation suitable for the standard regression analysis[1].

2.3
Effect of Surfactant Aggregation at the Interface

At surfactant concentrations higher than the critical micelle concentration (CMC) micelles are formed in the bulk solution, reducing the number of the molecules in the bulk solution available for adsorption. Similar processes can occur at the air–water interface, expressed by the formation of interfacial aggregates [12, 13], which reduces the number of kinetic units in the adsorption layer, because the monomers and the aggregates decrease the surface tension equally. Subsequently, the formation of the aggregates increases the surface tension. If n monomers form one aggregate, the aggregation can be described as $nM \rightleftharpoons M_n$. The equilibrium between the monomers and the aggregates is described as $\mu_n^s = n\mu_1^s$, where μ_1^s and μ_n^s are the chemical potentials of the monomers and the aggregates. The chemical potential of the molecules adsorbed at the surface described by the Butler equation (Eq. 1) can be applied to the surfactant adsorption with aggregation, giving $x_n^s = K_n(x_1^s)^n \exp\{-\Pi\Delta\omega/RT\}$, where $\Pi = \gamma_0 - \gamma$ is the surface pressure, K_n is the equilibrium constant of aggregation, and $\Delta\omega = n\omega_1 - \omega_n$. For simplicity, an ideal mixing in the adsorption layer is assumed in deriving the equation for the aggregation state. For small aggregates (two or three monomers), the equation for the aggregation state yields: $\Gamma_n\overline{\omega} = K_n(\Gamma_1\overline{\omega})^n \exp\{-\Pi\Delta\omega/RT\}$. The aggregation constant of the formation of small aggregates (with $n\omega_1 = \omega_n$) can be determined in terms of the critical aggregation adsorption excess Γ_c (which is equal to the monomer adsorption at $\Gamma_1 = \Gamma_n$), giving

$$K_n = (\Gamma_c\overline{\omega})^{1-n} \tag{25}$$

$$\Gamma_n = (\Gamma_1)^n/(\Gamma_c)^{n-1}. \tag{26}$$

Substituting Eqs. 25 and 26 into Eqs. 8 and 9 gives the following equations for the adsorption state and adsorption isotherm:

$$\gamma_0 - \gamma = -\frac{RT}{\overline{\omega}} \ln\left\{1 - \Gamma_1\overline{\omega}\left[1 + (\Gamma_1/\Gamma_c)^{n-1}\right]\right\} \tag{27}$$

$$bC = \frac{\Gamma_1\overline{\omega}}{\left\{1 - \Gamma_1\overline{\omega}\left[1 + (\Gamma_1/\Gamma_c)^{n-1}\right]\right\}^{\omega_1/\overline{\omega}}}. \tag{28}$$

---

[1] The orientation and Frumkin adsorption models have recently been combined to describe adsorption layer for surfactant-protein mixtures in [37] and [38].

The average partial molar area is determined by Eq. 12 as

$$\overline{\omega} = \omega_1 \frac{1 + n(\Gamma_1/\Gamma_c)^{n-1}}{1 + (\Gamma_1/\Gamma_c)^{n-1}} \,. \tag{29}$$

Equations 27 and 28 present the extension of the Szyszkowski–Langmuir model to the adsorption of one-surfactant systems with aggregation at the interface. For the formation of dimmers on the surface, $n = 2$ and Eqs. 27 and 28 can be expanded to obtain the Frumkin equation of adsorption state. In general, the surface aggregation model described by Eqs. 27 and 28 contains four free parameters, including ω_1, n, b and Γ_c, which can be obtained by regression analysis of the data for surface tension versus surfactant concentration in the solution.

3
Adsorption of Ionic Surfactants in Presence of Inorganic Electrolytes

Ionic surfactants are electrolytes dissociated in water, forming an electrical double layer consisting of counterions and co-ions at the interface. The Gouy–Chapman theory is used to model the double layer. In conjunction with the Gibbs adsorption equation and the equations of state, the theory allows the surfactant adsorption and the related interfacial properties to be determined [9, 10] (The Gibbs adsorption model is certainly simpler than the Butler–Lucassen-Reynders model for this case.).

Assuming that there are N components adsorbed at the interface, the Gibbs adsorption equation at the constant temperature gives

$$d\gamma = -RT \sum_{i=1}^{N} \tilde{\Gamma}_i \, d\ln(a_{i\infty}) \,. \tag{30}$$

In Eq. 30, $a_{i\infty}$ and $\tilde{\Gamma}_i$ are the activity in solution and the surface excess of the ith component, respectively. The activity is related to the concentration in solution $C_{i\infty}$ and the activity coefficient f by: $a_{i\infty} = f C_{i\infty}$. The activity coefficient is a function of the solution ionic strength I [39]. The surface excess $\tilde{\Gamma}_i$ includes the adsorption Γ_i in the Stern layer and the contribution, $\int_0^\infty [C_i(x) - C_{i\infty}] \, dx$, from the diffuse part of the electrical double layer. The Boltzmann distribution gives $C_i(x) = C_{i\infty} \exp\{-z_i\phi(x)\}$, where z_i is the ion valence and $\phi(x)$ is the dimensionless potential (measured from the Stern layer) obtained by dividing the actual potential, $\psi(x)$, by the thermal potential, $k_B T/e (= 25.7$ mV at 25 °C). Similarly, the ionic activity in solution and at the Stern layer is inter-related as $a_{i\infty} = a_i^s \exp(z\phi_s)$, where ϕ_s is the scaled surface potential. Given that the sum of $\tilde{\Gamma}_i z_i$ is equal to zero due to the electrical neutrality of the system, Eq. 30 yields

$$d\gamma \simeq RT \sum_{i=1}^{N} \Gamma_i d\ln a_i^s - RT \sum_{i=1}^{N} \left(\int_0^\infty a_{i\infty}[\exp(-z_i\phi) - 1] \, dx \right) d\ln a_i^s. \quad (31)$$

The potential in Eq. 31 is readily obtained by solving the Poisson–Boltzmann equation

$$\varepsilon\varepsilon_0 \frac{d^2\psi}{dx^2} = -1000F \sum_{i=1}^{N} z_i C_{i\infty} \exp\left(-\frac{z_i e\psi}{k_B T}\right), \quad (32)$$

where ε_0 and ε is the permittivity of vacuum and the dielectric constant of the solution, respectively, F is the Faraday constant. The concentration in Eq. 32 is given in mol/L. With the boundary conditions far from the surface: $\psi(\infty) = 0$ and $d\psi(\infty)/dx = 0$, integration of Eq. 32 gives

$$\left(\frac{d\phi}{dx}\right)^2 = \frac{2000F^2}{\varepsilon\varepsilon_0 RT} \sum_{i=1}^{N} C_{i\infty}[\exp(-z_i\phi) - 1]. \quad (33)$$

Equation 33 can be used to determine the integral in the second term on the right hand side of Eq. 31. Using the approximation $a_{i\infty} \approx C_{i\infty}$ for not-too-high ionic strengths and carrying out the functional variation of the left hand side of Eq. 31 with respect to small changes in the solution composition [9, 10] yield the following equation of state of the adsorption layer:

$$\gamma_0 - \gamma = RT \sum_{i=1}^{N} \int_0^{a_i^s} \Gamma_i d\ln a_i^s$$

$$+ \sqrt{\frac{\varepsilon\varepsilon_0 (RT)^3}{500F^2}} \int_0^{\phi_s} \left\{ \sum_{i=1}^{N} a_{i\infty}[\exp(-z\phi) - 1] \right\}^{1/2} d\phi. \quad (34)$$

The right hand side of Eq. 33 can be integrated using the available adsorption isotherms and the known valences of the ions.

The condition of electrical neutrality for the whole system also leads to the following useful equation, referred to as the generalized equation of Graham,

$$\sum_{i=1}^{N} z_i \Gamma_i = \sqrt{\frac{\varepsilon\varepsilon_0 RT}{500F^2}} \operatorname{sgn}(\phi_s) \left\{ \sum_{i=1}^{N} C_{i\infty}[\exp(-z\phi_s) - 1] \right\}^{1/2}, \quad (35)$$

where sgn describes the sign function. Equation 35 provides the relationship between the density of charges from the adsorption layer and the surface potential. The left-hand side of the equation describes adsorptions in the Stern layer. For symmetrical 1 : 1 ionic surfactants and added 1 : 1 inorganic salts (e.g., sodium dodecyl sulfate, NaDS, and NaCl), Eq. 35 gives

$$\Gamma_1 - \Gamma_2 = \sqrt{\frac{\varepsilon\varepsilon_0 RT a_\infty}{125 F^2}} \sinh |\phi_s/2|, \tag{36}$$

where the subscript 1 describes the surface-active ion (DS$^-$) and the subscript 2 describes the counterion (Na$^+$), and a_∞ is the total concentration of salts in the bulk solution. For the same symmetrical system, the second integral in Eq. 34 yields

$$\int_0^{\phi_s} \left\{ \sum_{i=1}^{N} a_{i\infty} [\exp(-z_i\phi) - 1] \right\}^{1/2} d\phi = 2\sqrt{2a_\infty} \left[\cosh\left(\frac{\phi_s}{2}\right) - 1 \right]. \tag{37}$$

Equation 37 gives the analytical form of the free surface energy of the diffuse part of the double layer and has been derived by a number of authors [8, 9, 40] for 1 : 1 ionic surfactants. For systems with mixed valences, the integral in Eq. 34 is usually not available in close analytical expressions and numerical integration is often required.

Regarding the adsorption isotherm, the Frumkin isotherm is usually used for surfactant ions and the Stern isotherm for the counterion adsorption in the Stern layer. For these isotherms, the following equations can be derived.

- The Frumkin adsorption isotherm for the surfactant ions gives

$$K a_1^s = \frac{\theta_1}{1 - \theta_1} \exp(-2\beta\theta_1). \tag{38}$$

In Eq. 38, the partial surface coverage, θ_1, of the surfactant is defined as $\theta_1 = \Gamma_1/\Gamma_\infty$, where Γ_∞ is the surface excess of surfactant at saturation. K is the adsorption constant, which is a function of the surfactant and counterion adsorptions. The dependence is usually linear, yielding $K = K_1 + K_2 a_2^s$, where K_1 and K_2 are the equilibrium adsorption constants of the surfactant ions and their counterions.

- The Stern adsorption isotherm for the counterions gives [41]:

$$\theta_2 = \theta_1 \frac{K_2 a_2^s}{K_1 + K_2 a_2^s}, \tag{39}$$

where $\theta_2 = \Gamma_2/\Gamma_\infty$ is the surface coverage of the counterions.

- Equation of state for the adsorption layer is described by Eq. 34, which gives

$$\gamma = \gamma_0 + RT\Gamma_\infty \left[\ln(1 - \theta_1) + \beta(\theta_1)^2 \right] - \frac{8RT\sqrt{a_\infty}}{\kappa} \left[\cosh\left(\frac{\phi_s}{2}\right) - 1 \right]. \tag{40}$$

Equations 36–40 describe the entire adsorption behavior of 1 : 1 ionic surfactants in the absence of added salt or in the presence of salt with the

4
Adsorption of Nonionic and Ionic (1:1) Surfactant Mixtures

The description of a mixed adsorption layer of ionic and nonionic surfactants requires the appropriate adsorption isotherms. For example, the Frumkin isotherm gives

$$Ka_1^s = \frac{\theta_1}{1 - \theta_1 - \theta_{non}} \exp[-2\beta(\theta_1 + \theta_{non})], \qquad (41)$$

where the parameters for the nonionic surfactant are described by the subscript *non*. One has

$$K_{non} C_{non} = \frac{\theta_{non}}{1 - \theta_1 - \theta_{non}} \exp[-2\beta(\theta_1 + \theta_{non})], \qquad (42)$$

where K_{non} and C_{non} are the equilibrium adsorption constant and the bulk concentration of the nonionic surfactant. The equation of state as described by Eq. 34 gives

$$\gamma = \gamma_0 + RT\Gamma_\infty \left[\ln(1 - \theta_1 - \theta_{non}) + \beta(\theta_1 + \theta_{non})^2\right] \\ - \frac{8RT\sqrt{a_\infty}}{\kappa} \left[\cosh\left(\frac{\phi_s}{2} - 1\right)\right]. \qquad (43)$$

The surface excess Γ_∞ at saturation depends on the composition of the mixed adsorption layer. It can be described as [5]:

$$\Gamma_\infty = \frac{\Gamma_{1\infty} K_1 a_1 + \Gamma_{non\infty} K_{non} C_{non}}{K_1 a_1 + K_{non} C_{non}}. \qquad (44)$$

The model for the adsorption of the surfactant mixtures includes six parameters: $\Gamma_{1\infty}$, $\Gamma_{non\infty}$, β, K_1, K_2 and K_{non}, which can be obtained by regression analysis.

In this combined approach, water does not have any contribution to the entropy of mixing. In addition, this model considers only one interaction coefficient β, which presents an average value for all the interactions in the adsorption and Stern layers. β is determined by the molecular interactions

---

[2] A new analysis of the adsorption layer of ionic surfactants with new adsorption isotherms and equations of states was made in [42]. The effect of mono and bivalent anions on the adsorption of cethyltrimethyl ammonium salts was recently examined in [43].

(the second virial coefficient in the virial equation of state). It can only be assigned a positive value (for attractive interactions).[3]

5
Comparison Between Theory and Experiment

In this section, the described models are compared with experimental data. The comparison shows agreement between theory and experiment for some systems. The difference between theory and experiment for other systems is discussed to highlight the need of further development.

5.1
Comparison Between Theory and Experiment

The theories developed for predicting adsorption of nonionic surfactants have been validated using the available experimental data [45] for surface tension and surface properties of a homologous series of octaethyleneglycol-n-alkyl ethers $C_nH_{2n+1}O(CH_2CH_2)_8H$ (abbreviated to C_nE_8; $n = 9$ to 15). The purity and homogeneity of the octaethylene oxide chain was obtained by chromatography. Surface tension was measured by the Wilhelmy plate technique at room temperature $T = 298.15$ K. An example of the measured and predicted surface tension as a function of the solution concentration of $C_{11}E_8$ is shown in Fig. 1. The surface tension of the pure solvent (water) at $T = 298.15$ K is $\gamma_0 = 71.99$ mN/m [46]. The experimental values for the molecular area contained in Eqs. 17 and 20, and surface excess at saturation are $\overline{\omega} = 66.4$ Å^2 and $\Gamma_\infty = 2.5 \times 10^{-6}$ mol/m^2 [45], giving $RT/\overline{\omega} = 6.239$ mN/m. The free energy of adsorption, $\Delta\mu^0 = 42.00$ kJ/mol, for $C_{11}E_8$ was estimated from the surface free energy of the surfactant, which can be substituted into Eq. 19 to calculate the adsorption constant to give $b = 8282.540$ (mmol/L)$^{-1}$. The molecular volume, $v_m = 675$ Å^3, for $C_{11}E_8$ is determined from the volume of the ethylene group. Knowing the values for $RT/\overline{\omega}$ and b, surface tension of $C_{11}E_8$ can be calculated using the Szyszkowski–Langmuir equation (Eq. 20) and is shown in Fig. 1 by the thin line. Clearly, the Szyszkowski–Langmuir equation significantly under-predicts the surface tension of $C_{11}E_8$. The same trend is observed for the other surfactants of the octaethyleneglycol-n-alkyl ether series, which shows the deficiency of the Szyszkowski–Langmuir equation.

The Frumkin theory with Eqs. 17–18 presents the first improvement of the Szyszkowski–Langmuir theory and is shown in Fig. 1 by the thick line. The Frumkin theory requires input for the surfactant interaction parameter β, which is not known. In this paper, β has been obtained by the non-

---

[3] Mixtures of ionic-nonionic surfactants were recently examined in [44].

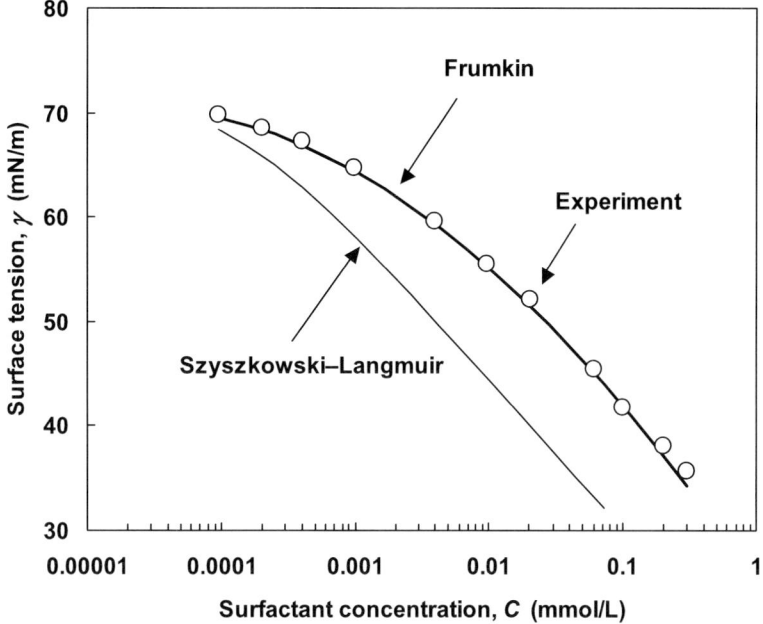

Fig. 1 Surface tension versus solution concentration of nonionic surfactant $C_{11}E_8$ as measured at $T = 298.15$ K (*data points*) [45], and as predicted by the Szyszkowski–Langmuir adsorption model (*thin line*) described by Eq. 20 and by the Frumkin adsorption model (*thick line*) described by Eqs. 17–18

linear least-square regression analysis. The regression analysis was carried out using a program written in Visual Basic for Application and Solver in Microsoft Excel. The non-linear Eq. 18 was solved for the surface coverage, θ, at given b, C, and β by applying the bisection method. The difference between a set of N experimental data points and the model for surface tension was minimized by the revised chi-square χ^2 defined as

$$\chi^2 = \sum_{i=1}^{N} \left[\left(\gamma_i^{\text{exp}} - \gamma_i^{\text{theo}} \right) \right]^2 .$$

It was assumed that the experimental uncertainties followed Gaussian statistics with equal standard deviation σ for all points. Then the standard deviation was determined as $\sigma = (\chi^2/\nu)^{1/2}$, where ν is the number of degrees of freedom in the fit. ν is equal to the number of the experimental points less the number of parameters used in the minimization. The best fit with the Frumkin prediction has one free parameter (β) and gives $\beta = -2.051$ and $\sigma = 0.41$ mN/m. The standard deviation in surface tension is small, indicating that the fit with the Frumkin prediction is statistically significant. Similar best fits are obtained for the other surfactants of the homologue series of

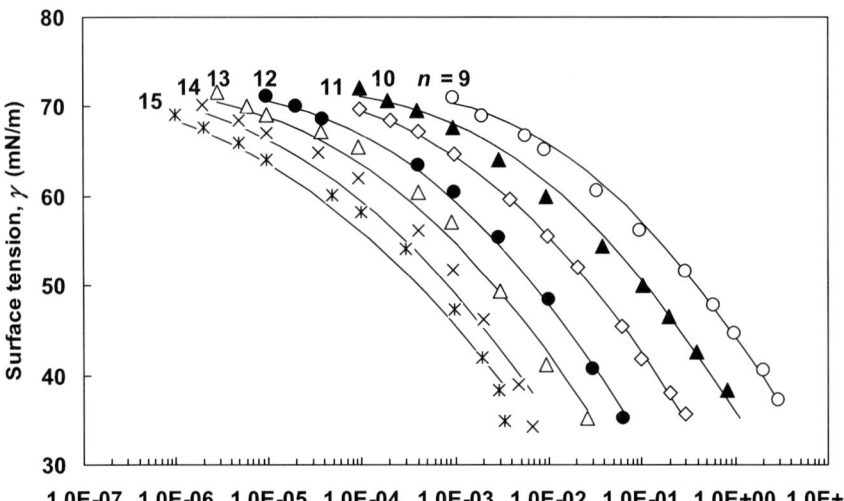

Fig. 2 Surface tension versus solution concentration for nonionic surfactants, octaethyleneglycol-n-alkyl ethers C_nE_8, as measured at $T = 298.15$ K (*points*) [45] and predicted by the Frumkin adsorption model (*lines*) described by Eqs. 17 and 18

Table 1 Experimental data for the surface excess Γ_∞ at saturation and adsorption free energy $\Delta\mu^0$ for C_nE_8 at $T = 298.15$ K [45], and the best fit results for β in the Frumkin model, Eq. 17, and the aggregation number n in the extended S–L model for the aggregation at the interface, Eq. 27

	$\Gamma_\infty \times 10^6$ (mol/m^2)[a]	$\Delta\mu^0$ (kJ/mol)[a]	Molecular interaction parameter, β	Aggregation number, n
C_9E_8	2.20	34.93	−1.085	3.002
$C_{10}E_8$	2.42	38.40	−1.705	3.002
$C_{11}E_8$	2.50	42.00	−2.051	3.003
$C_{12}E_8$	2.72	45.20	−2.423	3.001
$C_{13}E_8$	2.87	48.66	−3.286	3.001
$C_{14}E_8$	3.33	51.63	−4.386	2.998
$C_{15}E_8$	3.67	54.89	−6.213	2.998

[a] Experimental results [45].

the ethoxylated nonionics: from octaethyleneglycol nonyl ether (C_9E_8) to octaethyleneglycol pentadecyl ether ($C_{15}E_8$). The comparison is shown in Fig. 2. The results for β are presented in Table 1.

The data in Table 1 shows that the standard free energy of adsorption increases, on average, by 2.5 kJ/mol for each methylene group added to the hydrophobic tail. However, the intermolecular interaction parameter β in the Frumkin model obtained by the regression analysis is negative, indicating significant repulsion between the adsorbed molecules. The corresponding dimensional intermolecular interaction parameter ($= \beta RT/\Gamma_\infty$) is between -1×10^9 and -5×10^9 J m^2/mol^2. In the Frumkin theory [22], β has been considered to be positive [47, 48]. A negative sign of the interaction parameter has been reported in the literature for some nonionic surfactants [49, 50]. Fainerman et al. [17] consider that β is not related to the nature of the adsorption layer and is only a matching parameter. According to other authors [49, 50] the possible reason for the unexpected negative β should be sought in the dipole–dipole repulsion at the interface. However, detailed analysis [5] shows that the dipole–dipole hypothesis cannot be quantitatively justified.

To resolve the problem of negative β values obtained with the Frumkin theory, the improved Szyszkowski–Langmuir models which consider surfactant orientational states and aggregation at the interface have been considered [17]. For one-surfactant system with two orientational states at the interface, we have two balances, i.e., $\Gamma_t = \Gamma_1 + \Gamma_2$ and $\Gamma_t \overline{\omega} = \Gamma_1 \omega_1 + \Gamma_2 \omega_2$, which can be used in conjunction with Eq. 24 to derive two important equations for determining the total surface excess and averaged molecular area required in the calculation of surface tension, i.e.,

$$\Gamma_t \overline{\omega} = bC \sum_{i=1}^{2} (1 - \Gamma_t \overline{\omega})^{\frac{\omega_i}{\overline{\omega}}} \exp\left[-\frac{\omega_i}{\overline{\omega}} - 2\beta \Gamma_t \overline{\omega} + \beta \left\{1 - \frac{\omega_i}{\overline{\omega}}\right\} (\Gamma_t \overline{\omega})^2\right] \quad (45)$$

$$\Gamma_t \overline{\omega}^2 = bC \sum_{i=1}^{2} (1 - \Gamma_t \overline{\omega})^{\frac{\omega_i}{\overline{\omega}}} \exp\left[-\frac{\omega_i}{\overline{\omega}} - 2\beta \Gamma_t \overline{\omega} + \beta \left\{1 - \frac{\omega_i}{\overline{\omega}}\right\} (\Gamma_t \overline{\omega})^2\right] \omega_i .$$

(46)

Equations 45 and 46 can be numerically solved for Γ_t and $\overline{\omega}$ using a relaxation method. The surface tension can be predicted using Eq. 23 and directly compared with the experimental data for surface tension. The model parameters, β, ω_1 and ω_2, have been obtained by the non-linear least-square regression analysis described in the previous section. Typical results for surfactant $C_{11}E_8$ are shown in Fig. 3. Clearly, the improvement of the Szyszkowski–Langmuir model through consideration of the molecular orientational states at the surface, as described by Eqs. 21 and 22, does not help much in describing the experimental data. The improved Frumkin adsorption model described by Eqs. 23 and 24 agrees very well with the experimental data for surface tension but the intermolecular interaction parameter β is negative, as in the case of the best fit with the original Frumkin model. The same trend has been observed for the other surfactants of the series $C_n E_8$.

The same regression analysis methodology has been applied for analyzing the model for aggregation of surfactants at the interface. The non-linear Eq. 28 has been numerically solved by the bisection method. The surface tension predicted by Eq. 27 has been fitted to the experimental data by min-

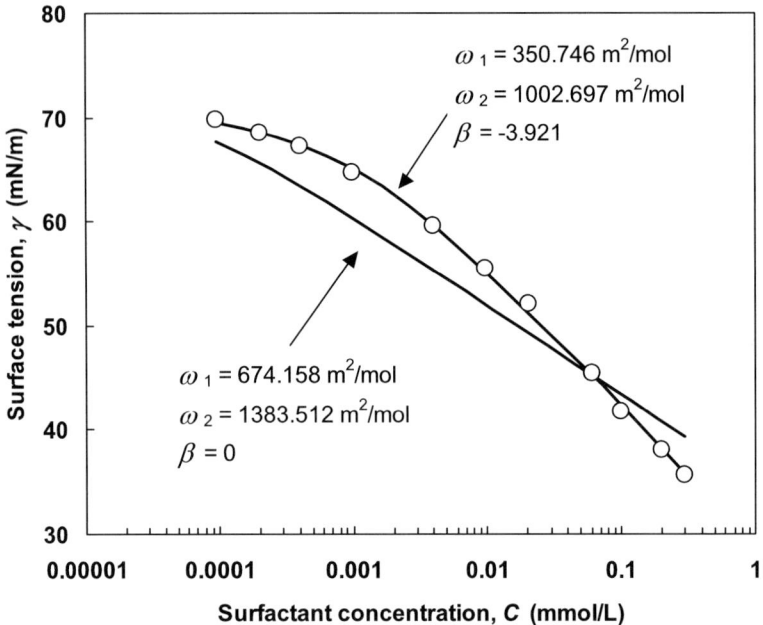

Fig. 3 Comparison of the surface tension for nonionic surfactant $C_{11}E_8$ as measured at $T = 298.15$ K, *data points* [45], with improved models considering orientational states of surfactant molecules at the surface. The data shown are obtained by regression analysis minimizing the revised chi-square χ^2. The calculation with $\beta = 0$ represents the best fit of the improved Szyszkowski–Langmuir model described by Eqs. 21 and 22. The other calculated curve with $\beta = -3.921$ shows the best fit of the improved Frumkin adsorption model described by Eqs. 23 and 24

imizing the revised chi-square. The results obtained with $C_{11}E_8$ surfactant are shown in Fig. 4. The obtained aggregation number is about $n = 3$, which is acceptable. The results $\omega_1 = 557.691$ m^2/mol and $\Gamma_c = 1004.471$ m^2/mol are also within the expected range. The surface aggregation model has produced physically consistent results. The last column in Table 1 shows the aggregation numbers obtained by the best fit. Fainerman et al. [51] have reported the same magnitude for the aggregation number of a similar surfactant series.

It is noted that the investigation of a mixed adsorption layer of $C_{10}E_8$ and TPeAB (tetrapentyl ammonium bromide) [35] shows evidence for attractive forces ($\beta > 0$), which suggests that the presence of the ionic surfactant can prevent aggregation in the extended S–L adsorption layer. Therefore, the main question of interest concerns how the Frumkin model and the aggregation model are related. One can find from Eq. 29 that the size of the elementary adsorption cell increases with the aggregation number resulting in a reduction in the number of cells. Negative β has the same effect of de-

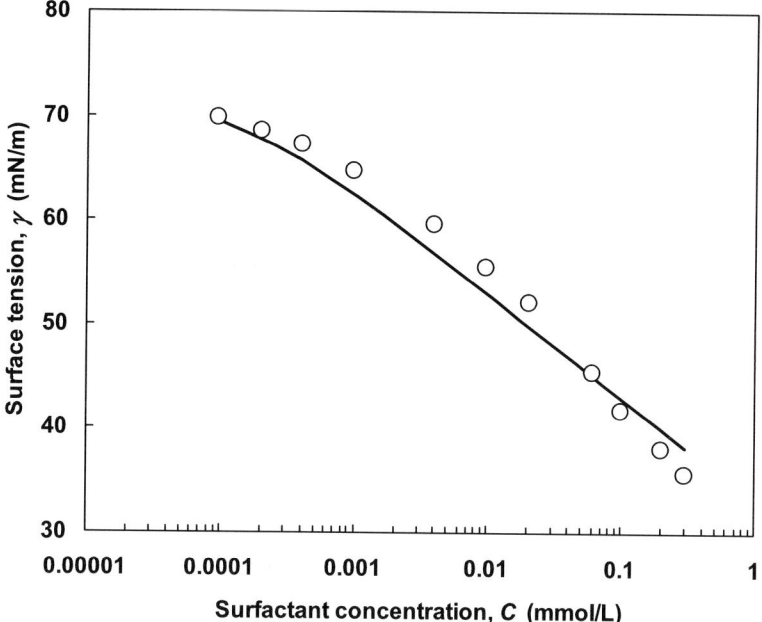

Fig. 4 Comparison between the experimental surface tension data for $C_{11}E_8$ surfactant, *data points* [45], and the extended S–L adsorption model (*line*) by the aggregation of surfactant molecules at the surface described by Eq. 27–29. The best fit gives $\omega_1 = 557.691$ m^2/mol, $\Gamma_c = 1004.471$ m^2/mol, and $n = 3.003$

creasing adsorption. Subsequently, the existence of negative β supports the hypothesis of aggregation in the adsorption layer.

5.2
Ionic Surfactants

Regression analysis developed on the basis of the proposed model described by Eqs. 36 and 38–40 has been applied for the adsorption of ionic surfactants. The regression analysis minimizes the revised chi-square,

$$\chi^2 = \sum_{i=1}^{N} \left[(\gamma_i^{\exp} - \gamma_i^{\text{theo}}) \right]^2 .$$

The standard deviation has been determined as $\sigma = (\chi^2/\nu)^{1/2}$, where ν is the number of degrees of freedom in the fit. The parameters for the molecular interaction β, the maximum adsorption Γ_∞, the equilibrium constant for adsorption of surfactant ions K_1, and the equilibrium constant for adsorption of counterions K_2, are thus obtained. The non-linear equations for the Frumkin adsorption isotherm have been numerically solved by the bisection method.

The following relations have also been used in the analysis:

$$K_1 = \frac{\delta_1}{\Gamma_\infty} \exp\left(\frac{\Delta\mu_1^0}{k_B T}\right) \qquad (47)$$

$$\frac{K_2}{K_1} = \frac{\delta_2}{\Gamma_\infty} \exp\left(\frac{\Delta\mu_2^0}{k_B T}\right), \qquad (48)$$

where δ_1 is the thickness of the adsorption layer, δ_2 is the thickness of the Stern layer, $\Delta\mu_1^0$ is the standard free energy of adsorption of surfactant ions, $\Delta\mu_2^0$ is the standard free energy of adsorption of counterions in the Stern layer, and $k_B T$ is the thermal energy of the surfactant molecule. For sodium dodecyl sulfate (SDS), δ_1 is ca. 2 nm. Here, δ_1/Γ_∞ presents the molar volume; the volume of one $-CH_2-$ group is ca. 25 Å^3 [52].

The experimental data for surface tension used in the regression analysis were obtained with the following surfactants:

1. A series of sodium hexadecyl n-sulfates with the sulfate group located at the 2nd, 3rd, 4th, 5th, 6th, 7th, and 8th positions [53].
2. A homologue series of sodium octyl, decyl, dodecyl, and tetradecyl sulfate [54].
3. A series of alkali dodecylsulfates, MeDS, where Me = Li, Na, K, Rb and Cs [55].

The results of the regression analysis for K_1, K_2, β, and Γ_∞ together with the standard deviation σ, and the occupation θ_2, of the adsorbed counterions in the Stern layer are summarized in Tables 2 to 4.

Table 2 shows a decrease in Γ_∞ when the location of the sulfate group in the surfactant molecular structure is shifted towards the middle of the hydrocarbon tail. The adsorbed species actually acquires two (shorter) hydrophobic tails thereby increasing the area per molecule at the interface as pictorially shown in Fig. 5. The decrease in Γ_∞ is explained by the increase in the area

Table 2 Results of the regression analysis for adsorption and surface tension of isomeric sodium hexadecyl n-sulfate (n describes the location of the sulfate group in the hydrocarbon chain)

$C_{16}H_{33}\,n-(SO_4)^-$	$\Gamma_\infty \times 10^6$ (mol/m^2)	$K_1 \times 10^6$ (m^3/mol)	$K_2 \times 10^4$ (m^3/mol)	β	σ (mN/m)	θ_2
$n=2$	3.21	3.0	8.73	−1.66	0.31	0.68
$n=3$	1.79	1.3	31.95	0.00198	0.25	0.71
$n=4$	1.51	4.8	18.15	0.00514	0.18	0.83
$n=5$	1.51	2.1	6.67	0.00223	0.22	0.50
$n=6$	1.56	2.2	4.05	0.68	0.05	0.45
$n=7$	1.37	2.5	11.80	1.63	0.18	0.55
$n=8$	1.35	2.3	4.96	1.97	0.17	0.43

Table 3 Results of the regression analysis for adsorption and surface tension of sodium alkyl sulfate homologues

	$\Gamma_\infty \times 10^6$ (mol/m^2)	$K_1 \times 10^6$ (m^3/mol)	$K_2 \times 10^4$ (m^3/mol)	β	σ (mN/m)	θ_2
Octyl	2.5	1.52	2.57	0.092	0.29	0.76
Decyl	3.0	15.1	6.61	−0.45	0.24	0.83
Dodecyl	3.4	60.6	7.72	0.28	0.43	0.84

Table 4 Results of the regression analysis for adsorption and surface tension of sodium alkyl sulfate homologues

Me	R_h (Å)[a]	$\Gamma_\infty \times 10^6$ (mol/m^2)	$K_1 \times 10^6$ (m^3/mol)	$K_2 \times 10^4$ (m^3/mol)	β	σ (mN/m)	θ_2
Li$^+$	3.4	3.03	13.6	0.038	−0.27	0.1	0.89
Na$^+$	2.8	3.80	20.4	0.029	−0.11	0.31	0.90
K$^+$	2.3	4.58	15	0.029	−0.93	0.07	0.91
Rb$^+$	2.3	4.00	10.4	0.058	0.30	0.12	0.93
Cs$^+$	2.3	3.80	19.8	0.03	0.47	0.14	0.93

[a] Radius of the hydrated counter-ions.

of contact between the hydrophobic tails of neighboring molecules. The average free energy of adsorption in Eq. 47 is $\Delta\mu_1^0 = 3.5$ KJ/mol of CH$_2$. The mean value of the standard free energy $\Delta\mu_1^0$ of adsorption of the surfactant ions is ~ 55.7 kJ/mol. The adsorption of counterions into the Stern layer is quite high, although no clear trend can be established when changing the position of the polar head.

Table 3 presents the results for the analysis of the homologue series of the alkyl sulfate surfactants. The maximum adsorption, Γ_∞, increases, together with the increasing number of carbon atoms in the hydrophobic tail. Consequently, there is an increase in the attraction forces: the stronger attractions lead to smaller areas occupied by the surfactant ions. This increases the number of the counterion bindings (except the last homologue–tetradecyl sulfate). The model has not been able to best fit the data for tetradecyl sulfate in the presence as well as in the absence ($\Delta\mu_2^0 = 0$) of a Stern layer.

Table 4 shows the results for the regression analysis of dodecylsulfate surfactants with different alkali counterions. The degree of surfactant ion/counterion association in the adsorption layer is evidently high (from 89.9% to 92.6% counterion coverage). There is also a correlation between the hydrated radius (volume) of the counterions and Γ_∞. The decrease in the hydrated volume of the counterions results in the higher value of Γ_∞, and increases the attractive force between the molecules. A pictorial presentation

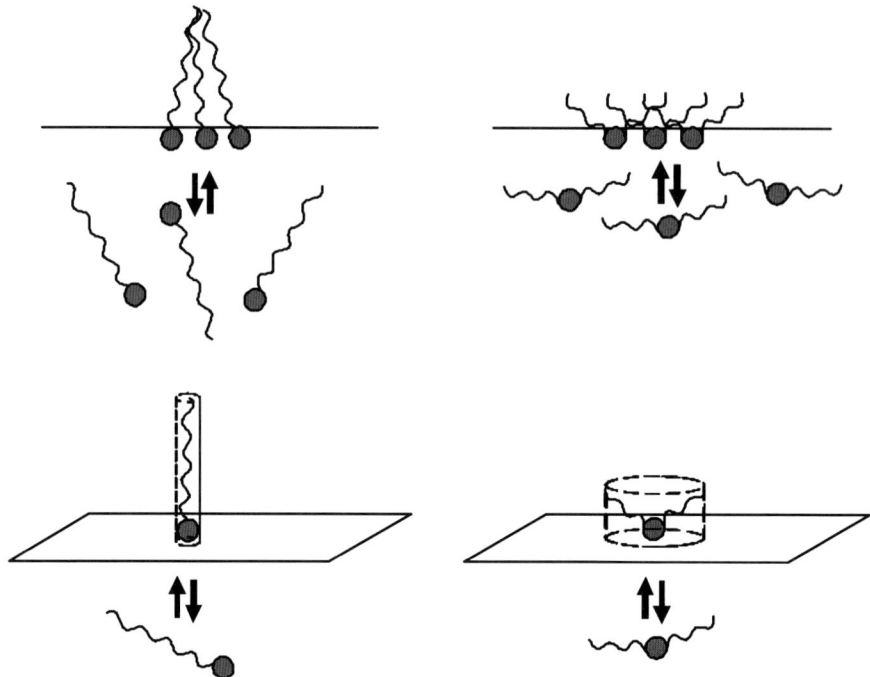

Fig. 5 Increase in the contact area between molecules with two hydrocarbon tails adsorbed at the interface (*top diagrams*) and expansion of the molecular area occupied by the adsorbed molecules (*bottom diagrams*)

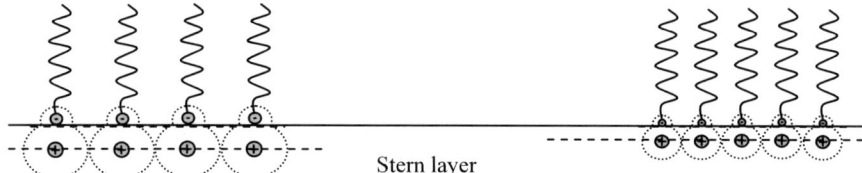

Fig. 6 Compression of the Stern layer due to the decrease in the hydration radius of the counterions

of the effect is shown in Fig. 6. An increase in the association in the adsorption layer can also be noted: the smaller the hydrated ion, the greater the degree of association. This can be possibly explained by the reduction in the electrostatic repulsion forces due to the stronger association.

5.3
Mixtures of Nonionic and Ionic Surfactants

The theories for the adsorption of mixtures of nonionic and ionic surfactants have been validated with tetraethylene glycol mono-*n*-octyl ether (C_8E_4) and

Table 5 Results of the regression analysis of adsorption and surface tension for C_8E_4, TPeAB and their mixture, TPeAB + C_8E_4

Surfactant	C_8E_4	TPeAB	C_8E_4 + TPeAB
$\Gamma_{\text{non}\infty} \times 10^6$ (mol/m^2)	3.42	–	4
$\Gamma_\infty \times 10^6$ (mol/m^2)	–	2.1	2.9
K_1 (m^3/mol)	–	6100	19.4
$\Delta\mu_1^0$ (kJ/mol)	–	41.3	26.98
K_2 (m^3/mol)	–	2.46×10^{-3}	2.06×10^{-4}
$\Delta\mu_2^0$ (kJ/mol)	–	5.79	0.42
K_{non} (m^3/mol)	24.64	–	15.5
$\Delta\mu_{\text{non}}^0$ (kJ/mol)	29.78	–	28.68
β	0.0354	– 6.9	0.35

tetrapentyl ammonium bromide (TPeAB). The adsorption state and isotherm of the individual surfactants and their mixtures have been analyzed using Eqs. 17 and 18 for C_8E_4 Eqs. 38 to 40 for TPeAB, and Eqs. 41 to 44 for their mixtures TPeAB + C_8E_4, together with experimental data for surface tension [5]. The regression analysis has been carried out to best fit the predictions for surface tension with the experimental data by minimizing the reduced chi-square as described in Sect. 5.1. The best-fitted results are summarized in Table 5.

It is noted that C_8E_4 is more surface-active than TPeAB, although the value of surface free energy of adsorption of TPeAB is greater then that of C_8E_4 because of the electrostatic repulsion between the TPeAB ions. In the surfactant mixture the attraction among molecules on the surface is stronger. The strong attraction increases the compression in the adsorption layer of the mixed surfactants. Similar conclusions have been reported in the literature [56, 57]. Concurrently, one can notice a reduction in the free energy of adsorption of the TPeAB species. The strong attraction in the mixed adsorption layer of TPeA$^+$ and C_8E_4 is due to the interaction between ionic and nonionic species. The ion–dipole attraction force is stronger than the attraction of both the dipole–dipole and ion–ion forces. Our regression analysis has confirmed the validity and applicability of the models developed for the adsorption state and isotherm for nonionic and ionic surfactant mixtures.

6
New Perspectives and Further Developments of the Adsorption Theories

The available modeling approaches are entirely thermodynamic. The models usually contain four or more free parameters that are obtained by regression analysis. The improvements of the available models are to reduce the number of free parameters by the fundamental modeling of the adsorption processes and the molecular interactions in the adsorption layers. The Poisson–Boltzmann equation used in the modeling of ionic surfactant adsorption is based on point charges, which could introduce some uncertainties in describing the counterion binding. For example, the Graham equation can be improved using the relationship between the surface charge density and the potential of the inner Helmholtz plane. In the following, the Frumkin adsorption isotherm to describe the adsorption of surface-active ions and the counterions is chosen for simplicity.

The adsorption of ionic surfactants creates an adsorption layer of surfactant ions, a Stern layer of counterions and a diffusive layer distributed by the electric field of the charged surface. Every layer has its own contribution to surface tension. For example, the adsorption of dodecyl sulfate (DS^-) ions from the sodium dodecyl sulfate solution is described by the modified Frumkin isotherm as

$$K_1 a_\infty = \frac{\Gamma_1 - \Gamma_2}{\Gamma_\infty - \Gamma_1} \exp\left(-\frac{2\beta_1 \Gamma_1 + F\Psi_s}{RT}\right), \tag{49}$$

where K_1 is the equilibrium constant of DS^- adsorption, $a_\infty = c_\infty \exp(-0.037\sqrt{C_\infty})$ is the activity of the surfactant with the bulk concentration C_∞, Γ_∞ is the surfactant adsorption at saturation, Γ_1 and Γ_2 are the adsorptions of the surfactant and counterions, respectively, β_1 is a parameter of interaction between DS^- ions, Ψ_s is the surface potential and F is the Faraday constant. In Eq. 49, the effect of counterions on the surfactant adsorption is accounted for via Γ_2. The adsorption of counterions is also described by the Frumkin isotherm as

$$K_2 a_\infty = \frac{\Gamma_2}{\Gamma_1 - \Gamma_2} \exp\left(-\frac{2\beta_2 \Gamma_2 - F\Psi_s}{RT}\right), \tag{50}$$

where K_2 is the equilibrium constant of counterion adsorption and β_2 is a parameter of the lateral interaction between the counterions in the Stern layer. The molecular interaction parameters are the second virial coefficients from the state equations of the adsorption layers, which can be described as [47]

$$\beta_i = -\pi N_A^2 \int_{d_i}^{\infty} U_i r \, dr, \tag{51}$$

where U_i is the interaction energy between two neighboring ions and d_i is the ion diameter. Counterions experience repulsion and surfactant ions undergo attraction between the hydrophobic tails and repulsion between the heads. The interaction between two counterions in the Stern layer can be described by the Debye–Hückel theory, which gives $U_2 = e^2 \exp(-\kappa r)/(4\pi\varepsilon_0\varepsilon r)$. The potential energy of lateral interaction for surfactant ions is described as $U_1 = e^2 \exp(-\kappa r)/(4\pi\varepsilon_0 \varepsilon r) - 9\pi C/(2\delta r^5)$, where $C = 5 \times 10^{-78}$ J/m^6 is the London constant, $\delta = 1.27$ Å is the diameter of a methylene group, and $L = 12\delta$ is the length of the dodecyl chain. These parameters correspond to the vertical orientation of the surfactant chains since the lateral interaction is important at relatively high Γ_1. The term $3\pi CL/8\delta^2 r^5$, describes the attraction energy between two parallel hydrophobic chains [58]. Inserting the available expressions for the interaction energies into Eq. 51 and integrating gives

$$\beta_1 = \frac{3N_A^2 \pi^2 C \sqrt{(\pi N_A \Gamma_\infty)^3}}{2\delta} - \frac{F^2}{4\varepsilon_0 \varepsilon \kappa} \tag{52}$$

$$\beta_2 = \frac{F^2}{4\varepsilon_0 \varepsilon \kappa} . \tag{53}$$

The condition: $\kappa d_2 \ll 1$ and the virial relationship: $1/\Gamma_\infty = N_A \pi d_1^2$ are used in simplifying the above equations.

The equilibrium constant for counterions adsorption in Eq. 50 can be determined as

$$K_2 = N_A v_2 \exp\left(\frac{\Delta\mu_2^0}{RT}\right), \tag{54}$$

where v_2 is the volume of a hydrated counterion. Since the driving force for the adsorption of counterions on the adsorbed surfactant ions is the electrostatic interaction promoted by a lower value of the dielectric permittivity at the interface, the free energy of adsorption of counterions on the surfactant ions can be calculated as $\Delta\mu_2^0 = eF/(4\pi\varepsilon_0\varepsilon_{12}d_{12})$, where ε_{12} and d_{12} are the dielectric permittivity and the distance between the adsorption and Stern layers, respectively. The distance d_{12} is equal to the average of the hydration diameters of a sulfate ion and of a counterion (the available data are given in Table 6).

The double-layer electrostatics gives

$$\Gamma_1 = 4\left(\frac{a_\infty}{\kappa}\right)\sinh\left(\frac{F\Psi_s}{2RT}\right), \tag{55}$$

where $\kappa = F\sqrt{2a_\infty/\varepsilon_0\varepsilon RT}$ is the Debye constant.

Applying Eqs. 49–55 and the Gibbs adsorption isotherm gives

$$\gamma = \gamma_0 + \Gamma_\infty RT \ln\left(1 - \frac{\Gamma_1}{\Gamma_\infty}\right) + \beta_1 \Gamma_1^2 + \beta_2 \Gamma_2^2$$
$$- 8RT\left(\frac{a_\infty}{\kappa}\right)\left[\cosh\frac{F\Psi_s}{2RT} - 1\right] . \tag{56}$$

Equation 56 presents an improvement of Eq. 40 for predicting surface tension of ionic surfactants. For adsorption of alkali dodecyl sulfates, experimental data are available for the adsorption at saturation Γ_∞ and for the equilibrium constant K_1 [55]. Table 6 summarizes the available data for alkali dodecyl sulfates. β_1, β_2, and K_2 are calculated using Eqs. 52–54. The value for the di-

Table 6 Experimental data, calculated data, and best-fitted results for ε_{12} obtained for the adsorption and surface tension of alkali dodecyl sulfates

Surfactant	LiDS	NaDS	KDS	RbDS	CsDS
v_2 (Å^3)	165.2	88.3	52.5	49.8	49.8
K_1 (m^3/mol)[a]	137.2	137.2	137.2	137.2	137.2
$\Delta\mu_1^0$ (kJ/mol)[a]	33.98	33.98	33.98	33.98	33.98
Γ_∞ (μmol/m^2)[a]	3.2	3.3	3.9	4.5	4.3
$\beta_1 \times 10^{-10}$ (J m^2/mol)[b]	1.99	2.28	3.27	4.73	4.32
$\beta_2 \times 10^{-10}$ (J m^2/mol)[b]	−3.05	−3	−3.52	−3.68	−3.53
d_{12} (Å)	5.8	5.2	4.8	4.7	4.7
ε_{12} (fitted)	5.7	5.6	5.3	5.2	5.1
$K_2 \times 10^{-3}$ (m^3/mol)	1.25	6.77	9.32	1.26	2.36
$\Delta\mu_2^0$ (kJ/mol)	41.6	47.5	55.5	56.4	58

[a] Experimentally determined by Eq. 50; [b] average data

electric permittivity of the first water layer ε_{12} is known to be around 6 [48]. However, since it is not an exact value, the surface tension isotherms of all alkali dodecyl sulfates have been best fitted to precisely determine ε_{12}. Indeed, the best-fitted results are within the reported range. Comparison between Eq. 56 and available data for surface tension is shown in Fig. 7. The improved prediction by Eq. 56 agrees with the experimental data better than the conventional theory described by Eq. 40, which has been best fitted to the data in the literature [10]. The results of the regression analysis of Eq. 56 for surface potential of the air-water interface with the adsorption of alkali dedecyl sulfate molecules as a function of the surfactant concentration in the bulk solution are shown in Fig. 8. The surface potentials are within the range of the experimental data [59–61] and are significantly lower than the values predicted by the conventional theory described by Eq. 40, which are between −100 and −150 mV [10].

It is noted that the molecular interaction parameter described by Eq. 52 of the improved model is a function of the surfactant concentration. Surprisingly, the dependence is rather significant (Fig. 9) and has been neglected in the conventional theories that use β_1 as a fitting parameter independent of the surfactant concentration. Obviously, the resultant force acting in the inner Helmholtz plane of the double layer is attractive and strongly influences the adsorption of the surfactants and binding of the counterions. Note that surface potential Ψ_s is the contribution due to the adsorption only, while the experimentally measured surface potential also includes the surface potential of the solvent (water). The effect of the electrical potential of the solvent on adsorption is included in the adsorption constants K_1 and K_2.

The improved treatment of the counterions presented in this section separates the interaction in the adsorption layer into two parts, namely, the

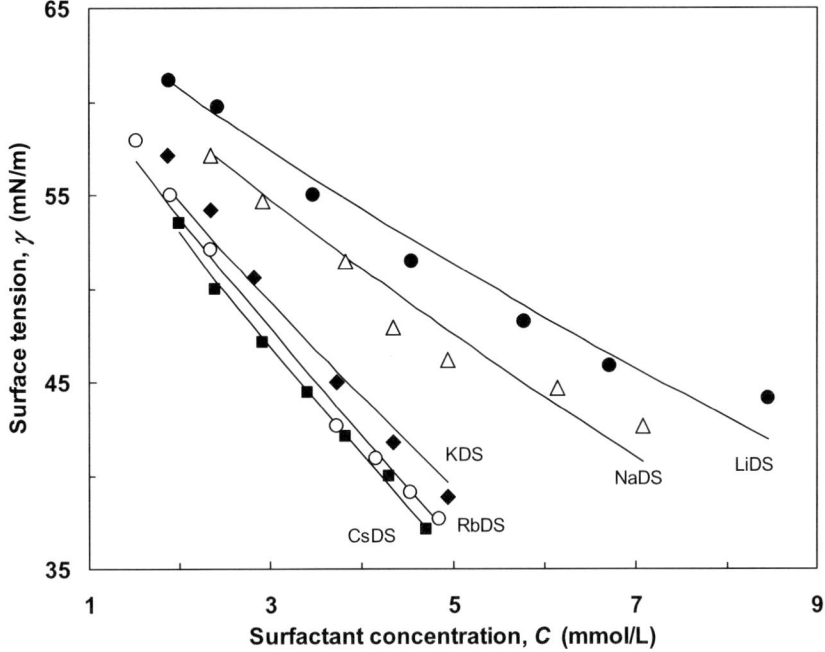

Fig. 7 Comparison between the full prediction (*lines*) by Eq. 56 and experimental results (*data points*) [55] for surface tension of alkali dodecyl sulfate solutions

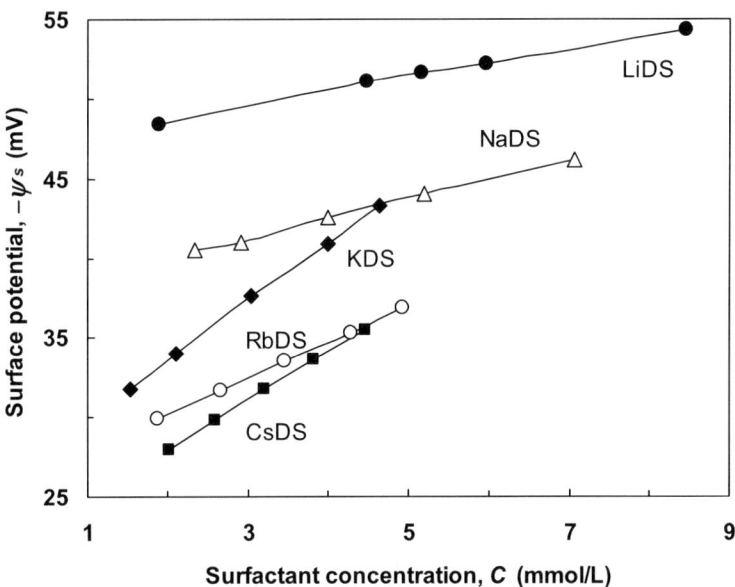

Fig. 8 Results of the regression analysis of Eq. 56 for surface potential of the air–water interface with the adsorption of alkali dodecyl sulfate molecules as a function of the surfactant concentration in the bulk solution

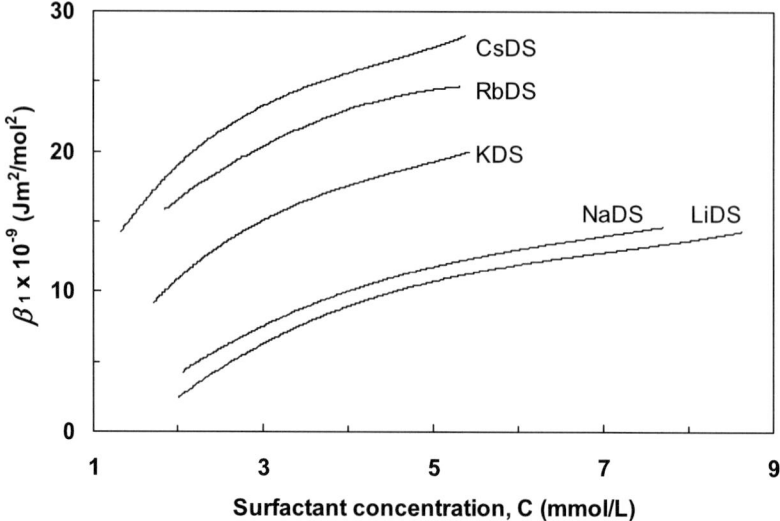

Fig. 9 Dependence of the molecular interaction β_1 between adsorbed dodecyl sulfate ions on the concentration of alkali dodecyl sulfate surfactants

interaction in the inner Helmholtz plane and the interaction in the outer Helmholtz plane. The theory based on such an assumption predicts values of the free energy of counterions binding in the range of the ionic bond (Table 6). In contrast to this consideration, the conventional model described by Eq. 40 only uses one averaged parameter for the two interactions and predicts lower values of the free energy of counterion binding. Neglecting the electrostatic repulsion between the counterions in the outer Helmholtz plane leads to error, producing significant decrease in the values of the equilibrium adsorption constant of the counterions and negative values of the molecular interaction parameter. In addition, the higher magnitude of the surface potential (100–150 mV) predicted by the conventional theories, as discussed earlier, is also due to neglecting the electrostatic repulsion between the counterions in the outer Helmholtz plane.[4]

7
Summary and Conclusion

Theoretical models for equilibrium adsorption of nonionic and ionic surfactants at the gas–liquid interface have been developed, critically reviewed, and evaluated using experimental data for surface tension. The thermodynamic

---

[4] The dispersion interaction between the surface active ions and the water–air interface was recently considered in the modeling of the equilibrium adsorption [62]. The molecular dynamic simulations are used in the recent years to describe the surfactant adsorption at the air–water interface [63–65].

approach of Butler and the Lucassen-Reynders dividing surface have successfully been used to describe the adsorption of nonionic surfactants. The obtained equations for the adsorption layer state and adsorption isotherm as a function of partial molar area are capable of describing many complicated adsorption phenomena, including the surfactant orientational states and aggregation at the interface. The traditional approach with the Gibbs dividing surface and Gibbs adsorption isotherm, and the Gouy–Chapman electrical double layer electrostatics have been used to model the adsorption of ionic surfactants and ionic-nonionic surfactant mixtures. The developed theories have been validated by comparison with experimental data for surface tension. Regression analysis by minimizing the reduced chi-square has been used to best fit the models to the experimental data to obtain the free parameters of the models. For the surfactant homologous series of octaethyleneglycol-n-alkyl ethers $C_nH_{2n+1}O(CH_2CH_2)_8H$, the negative sign of the intermolecular interaction parameter obtained in the regression analysis of surface tension has not been resolved by the model for the surfactant orientational state at the interface. For this surfactant series, the surface aggregation model gives physically consistent fitting and parameters. The models for adsorption of ionic surfactants have been validated using the surface tension of a series of sodium n-hexadecylsulfate with the sulfate group located at different positions in the hydrocarbon chain, a homologue series of sodium alkyl sulfate, and a series of alkali dodecylsulfate. Improved adsorption models for ionic surfactants have been developed by fundamental modeling of the adsorption processes and the molecular interactions in the adsorption layers. The improved predictions reduce the required number of free parameters and agree with the surface tension and surface potential data better than conventional models.

Acknowledgements The authors gratefully acknowledge the financial support provided by the Australian Research Council and the U.S. National Science Foundation (Grant No. INT 0227583) via their international collaboration programs. Also, the financial support provided by the DOE program (Grant No. DE-FG02-93ER14315) is certainly appreciated.

References

1. Shah DO (ed) (1998) Micelles, Microemulsions, and Monolayers. Marcel Dekker, New York
2. Levich VG (1962) Physicochemical Hydrodynamics. Prentice-Hall, Englewood Cliffs
3. Adamson AW, Gast AP (1997) Physical Chemistry of Surfaces, 6th ed. John Wiley & Sons Inc., New York
4. Karakashev SI, Manev ED, Nguyen AV (2004) Adv Colloid Interface Sci 112:31–36
5. Karakashev SI, Manev ED (2002) J Colloid Interface Sci 248:477
6. Fainerman VB, Miller R, Wuestneck R (1997) J Phys Chem B 101:6479
7. Lucassen-Reynders EH, Van den Tempel M (1967) Chem Phys Appl Surf Active Subst 2:779

8. Davies JT, Rideal EK (1963) Interfacial Phenomena. Academic Press, New York
9. Borwankar RP, Wasan DT (1988) Chem Eng Sci 43:1323
10. Kralchevsky PA, Danov KD, Broze G, Mehreteab A (1999) Langmuir 15:2351
11. Karakashev S, Tsekov R, Manev E (2001) Langmuir 17:5403
12. Fainerman VB, Miller R, Aksenenko EV, Makievski AV, Kragel J, Loglio G, Liggieri L (2000) Adv Colloid Interface Sci 86:83
13. Ruckenstein E, Bhakta A (1994) Langmuir 10:2694
14. Butler J (1932) Proc R Soc Ser A 138:348
15. Lucassen-Reynders EH (1964) J Coll Sci 19:584
16. Gibbs JW (1961) The Scientific Papers of J. Willard Gibbs. Dover, New York
17. Fainerman VB, Lucassen-Reynders EH, Miller R (1998) Colloids Surfaces A 143:141
18. Prigogine I (1968) The molecular theory of solutions. North-Holland, Amsterdam
19. Guggenheim EA (1952) Mixtures. Clarendon Press, Oxford
20. Read RC, Prausnitz JM, Sherwood TK (1977) The properties of gases and liquids, 3rd ed. McGraw-Hill, New York
21. Lucassen-Reynders EH (1994) Colloids Surfaces A 91:79
22. Frumkin AZ (1925) Physik Chem 116:501
23. Helfand E, Fish H, Lebowitz J (1961) J Chem Phys 34:1037
24. Parsons R (1964) J Electroanal Chem 7:136
25. Lucassen-Reynders EH (1966) J Phys Chem 70:1777
26. Lucassen-Reynders EH (1981) Anionic Surfactants. In: Lucassen-Reynders EH (ed) Physical Chemistry of Surfactant Action. Marcel Dekker, New York
27. Joos P (1967) Bull Soc Chim Belg 76:591
28. Damaskin BB, Frumkin AN, Dyatkina SL (1967) Izvestiya Akademii Nauk SSSR, Seriya Khimicheskaya, p 2171
29. Karolczak M, Mohilner DM (1982) J Phys Chem 86:2845
30. Hua XY, Rosen MJ (1982) J Colloid Interface Sci 90:212
31. Fainerman VB, Lylyk SV (1983) Kolloidnyi Zhurnal 45:500
32. Rodakiewitz-Nowak J (1982) J Colloid Interface Sci 85:586
33. Krotov VV (1985) Kolloidnyi Zhurnal 47:1075
34. Hamdi M, Schuhmann D, Vanel P, Tronel-Peyroz E (1986) Langmuir 2:342
35. Karakashev SI, Manev ED (2003) J Colloid Interface Sci 259:171
36. Bevington PR, Robinson DK (2003) Data reduction and error analysis for the physical sciences. McGraw Hill, Boston
37. Maldonado-Valderrama J, Martin-Molina A, Martin-Rodriguez A, Cabrerizo-Vilchez MA, Galvez-Ruiz MJ, Langevin D (2007) J Phys Chem C 111(6):2715–2723
38. Alahverdjieva VS, Grigoriev DO, Fainerman VB, Aksenenko EV, Miller R, Moehwald H (2008) J Phys Chem B 112(7):2148–2155
39. Robinson RA, Stokes RH (1959) Electrolyte Solutions, 2nd ed. Butterworths Scientific Publications, London
40. Hachisu S (1970) J Colloid Interface Sci 33:445
41. Hunter RJ (1981) Zeta Potential in Colloid Science. Academic Press, London
42. Ivanov IB, Ananthapadmanabhan KP, Lips A (2006) Adv Colloid Interface Sci 123–126:189-212
43. Para G, Warszynski P (2007) Colloids Surface A 300(3):346–352
44. Jarek E, Wydro P, Warszynski P, Paluch M (2006) J Colloid Interface Sci 293(1):194–202
45. Ueno M, Takasawa Y, Miyashige H, Tabata Y, Meguro K (1981) Colloid Polym Sci 259:761
46. Vargaftik NB, Volkov BN (1983) J Phys Chem Ref. Data 12(3):817–820

47. Hill TL (1960) An Introduction to Statistical Thermodynamics. Addison-Wesley, Reading, MA
48. Adamson AW (1982) Physical Chemistry of Surfaces, 4th ed. John Wiley and Sons Inc., New York
49. Pethica BA, Glasser ML, Mingins J (1981) J Colloid Interface Sci 81:41
50. Gurkov TD, Kralchevsky PA, Nagayama K (1996) Colloid Polym Sci 274:227
51. Fainerman VB, Miller R, Aksenenko EV, Makievski AV, Kragel J, Loglio G, Liggieri L (2000) Adv Colloid Interface Sci 86:83
52. Barneveld PA, Scheutjens JMHM, Lyklema J (1991) Colloids Surfaces 52:107
53. Pushel F (1966) Tenside 3:71
54. Tajima K (1973) Nippon Kagaku Kaishi, 883
55. Lu JR, Marrocco A, Su T, Thomas RK, Penfold J (1993) J Colloid Interface Sci 158:303
56. Ingram BT (1980) Colloid Polym Sci 258:191
57. Savchik J, Chang B, Rabitz J (1983) J Phys Chem 87:1990
58. Israelachvili JN (1992) Intermolecular and Surface Forces. Academic Press, London
59. Yoon RH, Yordan JL (1986) J Colloid Interface Sci 113:430
60. Okada K, Akagi Y, Kogure M, Yoshioka N (1990) Can J Chem Eng 68:393
61. Paruchuri VK, Nguyen AV, Miller JD (2004) Colloids Surfaces A 250:519
62. Ivanov IB, Marinova KG, Danov KD, Dimitrova D, Ananthapadmanabhan KP, Lips A (2007) Adv Colloid Interface Sci 134–135:105–124
63. Vacha R, Jungwirth P, Chen J, Valsaraj K (2006) Phys Chem Chem Phys 8(38):4461–4467
64. Howes AJ, Radke CJ (2007) Langmuir 23(4):1835–1844
65. Howes AJ, Radke CJ (2007) Langmuir 23(23):11580–11586

Surfactants Containing Hydrolyzable Bonds

Dan Lundberg · Maria Stjerndahl · Krister Holmberg (✉)

Applied Surface Chemistry, Chalmers University of Technology, 412 96 Göteborg, Sweden
kh@chem.chalmers.se

1	Introduction	58
2	Background	58
3	Environmental Regulations	60
4	Biodegradation of Surfactants	61
5	Technical Incentives for Hydrolysable Surfactants	63
6	Surfactants Containing Normal Ester Bonds	64
7	Ester Quats	68
8	Betaine Ester Surfactants	70
9	Surfactants Containing Carbonate Bonds	73
10	Surfactants Containing Amide Bonds	74
11	Cyclic Acetals	75
12	Cyclic Ketals	77
13	Ortho Ester Surfactants	78
14	Summary	80
	References	80

Abstract There is a growing demand for hydrolyzable surfactants, i.e., surfactants that break down in a controlled way by changing the pH. Environmental concern is the main driving force behind current interest in these surfactants, but they are also of interest in applications where surfactants are needed in one stage but later undesirable at another stage of a process. This chapter summarizes the field of hydrolyzable surfactants with an emphasis on their more recent development. Surfactants that break down either on the acid or on the alkaline side are described. It is shown that the susceptibility to hydrolysis for many surfactants depends on whether or not the surfactant is in the form of micelles or as free unimers in solution. It is shown that whereas nonionic ester surfactants are more stable above the CMC (micellar retardation), cationic ester surfactants break down more readily when aggregated than when present as unimers (micellar catalysis).

1
Introduction

Hydrolyzable surfactants are, as the name implies, surfactants that break down by hydrolysis. Implicit in the phrase is that the surfactant is designed in such a way that hydrolysis will be simplified. Hydrolyzable surfactants are sometimes seen as synonymous with cleavable surfactants, which is a phrase commonly seen in the literature. In a strict sense there is a difference between the two, however, since surfactants may be cleaved by means other than hydrolysis. Hydrolyzable surfactants are therefore just a fraction, although a major fraction, of cleavable surfactants.

The interest in cleavable surfactants has increased rapidly in recent years and the topic has been covered in review papers during the last decade [1–4]. This chapter begins with a relatively thorough discussion about the incentive for hydrolyzable surfactants, continues with a discussion about biodegradation of surfactants, which is important for understanding the concept of hydrolyzable surfactants, and then gives an account of the development of hydrolyzable surfactants with an emphasis on recent results.

2
Background

Until about a decade ago there was a general belief that surfactants should be as chemically stable as possible. Depending on the application, surfactants are used under different conditions, in particular at different pH values, and, in order to be able to use a specific product for as broad a range of applications as possible, there should not be labile bonds in the surfactant. In cases where weak bonds were present in the molecule, such as in the widely used anionic surfactant sodium dodecyl sulfate (SDS), this was regarded as a nuisance. SDS, which is a monoester of sulfuric acid, is readily cleaved when exposed to acid and the cleavage generates sulfuric acid. The acid formed gradually brings down the pH, and is thereby accelerating the breakdown. This autocatalytic degradation of SDS is well-known by those who formulate surfactant-containing products and has been regarded as a substantial disadvantage for this specific surfactant.

SDS was an exception. Among the traditional surfactant work-horses, including anionics (such as alkylbenzene sulfonates and alkyl sulfates), nonionics (such as alcohol ethoxylates and alkylphenol ethoxylates), and cationics (such as alkyl quats and dialkyl quats), only alkyl sulfates are not chemically stable under normal conditions.

The general attitude towards surfactants that are readily hydrolyzable has changed in recent years. Environmental concern has become one of the main driving forces for the development of new surfactants and the rate of biodegradation has turned into a major issue. Many of the established surfactants have gradually been phased out, due to insufficient rate of biodegradation, and have been replaced by surfactants that are found to be more benign to the environment. The first major example of a widely used surfactant that was taken out of the market solely because of poor biodegradability was the branched alkylbenzene sulfonates. These used to be the major surfactant constituent in laundry detergent formulations, and, indeed, they were very efficient in removing stains from fabrics. They were virtually non-degradable in the environment, however, and the famous persistent foam that a discharge of a formulation based on this surfactant gave rise to in the British river Lea in the 1960s was one of the events that started the debate in Europe on the fate of chemicals in the environment. That debate lead to the regulations that were gradually enforced in many European countries during the 1980s and 1990s. Branched alkylbenzene sulfonates, the alkyl chain of which is tetrapropylene, have now been almost completely replaced by linear alkylbenzene sulfonates (LAS or LABS), which have a linear alkyl chain, usually dodecyl, in p-position to the sulfonate group.

A second example of a surfactant class that is not much used today due to poor biodegradability (and also considerable aquatic toxicity) is the alkylphenol ethoxylates. Octyl- and nonylphenol ethoxylates were widely used as nonionic surfactants until the 1980s, but they have now largely been replaced by fatty alcohol ethoxylates. Replacing an alkylbenzene ethoxylate by an alcohol ethoxylate looks straight-forward on paper but is sometimes not easy in practice. It is probably true to say that for some applications, in particular when it comes to stabilization of dispersed systems, no other type of nonionic surfactant gives the same performance as the alkylphenol ethoxylates. This is a typical example where environmental concerns are seen as more important than technical quality. The sometimes better performance of an alkylphenol ethoxylates over an alcohol ethoxylate of the same hydrophilic-lipophilic balance has nothing to do with differences in biodegradability. The good performance of alkylphenol ethoxylates in applications such as pigment dispersants is probably related to the bulky hydrophobe, leading to a higher so-called packing parameter [5], which is generally beneficial, and to the fact that the benzene ring contains π-electrons that may form electron donor-acceptor complexes with unsaturated moieties on the pigment surface [6].

A third and more recent example of where readily biodegradable surfactants have replaced surfactants that are more long-lived in the environment is the taking over of the market for textile softener surfactants by the ester quats from the stable quats. This transition is still in progress today but for several years the "big soapers" all base their softener formulations for the US and Western Europe markets on ester quats instead of traditional, stable quats. The change from stable quats to ester quats is probably the best example of the concept of introducing a cleavable bond in a given surfactant structure because the two types of products are very similar in structure and physical chemical charac-

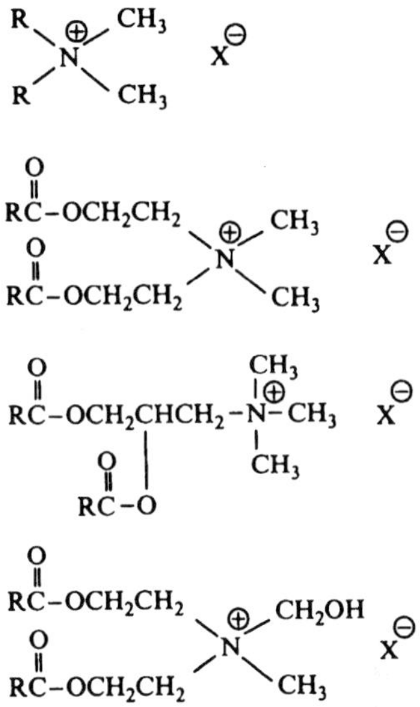

Fig. 1 A stable quat (*top*) and three ester quats of the type that is used for textile softening

teristics, as well as in performance. The stable quats used as fabric softeners almost invariably contain a nitrogen atom substituted with two long and two very short alkyl chains, with a halide, acetate, or methylsulfate as counterion. The ester quats have similar structure although their synthesis is very different. Figure 1 shows the structure of a stable quat and of three commercially available ester quats. Ester quats are discussed in more detail below.

3
Environmental Regulations

The three transitions mentioned above, replacement of branched alkylbenzene sulfonates by linear alkylbenzene sulfonates, replacement of alkylphenol ethoxylates by alcohol ethoxylates and substitution of ester quats for stable quats are all driven by environmental concerns. Rate of biodegradation and aquatic toxicity are the major parameters taken into account, and, in order to pass the regulations that the European Union has adopted, a surfactant must pass the criteria of showing:

- at least 60% degradation during 28 days incubation with active sludge under aerobic conditions
- a value of aquatic toxicity (LC_{50} for fish or EC_{50} for daphnia or algae, where LC stands for lethal concentration and EC for effective concentration) of above 1 mg/l

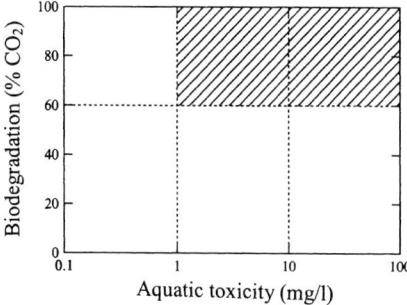

Fig. 2 Environmental assessment of surfactants is based on values of biodegradation and aquatic toxicity. A surfactant must lie within the *shaded areas* in order to meet the OECD regulatory directives

It is common practice to put the experimentally obtained values for biodegradation and aquatic toxicity into a kind of matrix, as is illustrated by Fig. 2. The shaded areas in the matrix are the "approved areas". Out of the six product types discussed above linear alkylbenzene sulfonates, alcohol ethoxylates, and ester quats lie within the shaded areas, while branched alkylbenzene sulfonates, alkylphenol ethoxylates and stable quats do not.

The discussion above is meant to illustrate that one of the ways, although not the only way, to design readily biodegradable surfactants is to build a bond with limited stability into the structure. For practical reasons the weak bond is usually the bridging unit between the polar head group and the hydrophobic tail of the surfactant, which means that degradation immediately leads to destruction of the surface activity of the molecule, an event usually referred to as the primary degradation of the surfactant. The biodegradation then proceeds along various routes depending on the type of primary degradation products. The ultimate decomposition of the surfactant, often expressed as the amount of carbon dioxide evolved during four weeks exposure to appropriate microorganisms (counted as percent of the amount of carbon dioxide that could theoretically be produced), is the most important measure of biodegradation. It seems that for most surfactants containing easily cleavable bonds the values for ultimate degradation are also higher than for the corresponding surfactants lacking the weak bond. Thus, the strong trend towards more environmentally benign products favors the cleavable surfactant approach on two accounts.

4
Biodegradation of Surfactants

As mentioned above, surfactants are usually toxic to aquatic organisms, even at very low concentrations, and should therefore be removed from wastewater before entering receiving waters. The generally employed method to decompose surfactants in sewage treatment plants is biodegradation, i.e., removal or destruction of chemical compounds through the biological action of living organisms [7, 8]. The microorganisms carrying out the biodegradation process are able to metabolize organic chemicals and convert them into less complex chemicals by a series of enzymatic reactions. Under aerobic conditions microorganisms convert organic substances into carbon dioxide, water, and biomass, resulting in ultimate degradation. A product that does not undergo natural biodegradation is stable and persists in the environment.

For surfactants the main route to the environment is via a sewage plant. In sewage treatment plants chemicals are exposed to aerobic biodegradation, which also applies when chemicals are released into natural waters [9]. Therefore, most biodegradation test methods for surfactants are based on aerobic conditions. Certain terms are used in connection to biodegradability that usually relate to a specific test method: *ultimate biodegradation* is the complete breakdown of the surfactant into water, carbon dioxide, inorganic materials (also called mineralization), and cellular matter [10]. In *ready biodegradation* the term *ready* has a very specific meaning, originally defined by the Organisation for Co-operation and Development (OECD). Readily biodegradable means that the substance reaches a specific level of ultimate biodegradability within a specified time frame using a ready biodegradability test method [10]. One of the most common of these methods is the Closed Bottle Test in which 60% biodegradation should be reached within 28 days for the substance to be defined as readily biodegradable [11]. In addition, the rules say that the time needed to go from 10 to 60% degradation should not exceed 10 days. This requirement is being questioned from a relevance point, however, as is discussed below. Another term, *inherently biodegradable* is used for substances that do not fulfill the demands for ready biodegradation but are degraded to a certain extent within a specified time frame. It is based on the philosophy that if a substance is degraded to, say, 20% within the test period there is no reason to believe that it should not break down completely given sufficient time [10].

Complete biodegradation of surfactants normally requires the concerted action of at least two microorganisms, as a single organism usually lacks the full complement of enzymatic capabilities. This results in a sequential degradation of the hydrophobic tail and the polar head group [12, 13]. The demand that the surfactant should go from 10 to 60% degradation within ten days is, however, based on the assumption that only one microorganism is needed to bring about total degradation. Thus, the "10-day window" for classification of surfactants is not entirely adequate, which has been recognized by the Scientific Committee on Toxicity, Ecotoxicity and the Environment (CSTEE) of the EU. Nevertheless, it is still used in the OECD classification of surfactants.

The biodegradation pathway of alcohol ethoxylates has been thoroughly investigated [14–16]. The metabolism can be initiated by an attack on the ter-

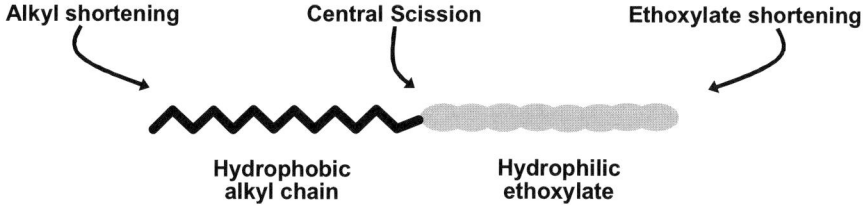

Fig. 3 Possible pathways for biodegradation of alcohol ethoxylates

minal hydroxyl group, by terminal oxidation of the alkyl chain, or by cleavage of the ether bond adjacent to the hydrophobic group (central scission), see Fig. 3. In surfactants with straight alkyl chains the central scission dominates. If the alkyl carbon next to the ether bond is substituted, this mechanism is more complicated and the attack from the surfactant ends may dominate.

When an alcohol ethoxylate is subjected to central scission, the ether bond is often first converted into an ester. The separation of the hydrophobe and the hydrophile is obtained through hydrolysis and results in the liberation of an alkyl chain (as alcohol, aldehyde or acid) and an oxyethylene chain. The alkyl moiety is then oxidized through so-called β-oxidation, removing two carbons at a time by the action of coenzyme A. The alkyl chain can also be metabolized while attached to the polar head group. However, before the β-oxidation can begin the terminal end (the ω-position) of the alkyl chain must be oxidized by ω-oxidation to give a carboxyl group that can be activated by coenzyme A. The process of β-oxidation can cope with limited amounts of branching in the alkyl chain, but when tetrasubstituted carbons are present alternative routes have to be employed. The polar head group of the alcohol ethoxylate is degraded by the successive removal of glycol units from the terminus.

The biodegradation of the hydrophobic tail, if linear, is considerably faster than that of the polyoxyethylene chain of alcohol ethoxylates. The length of the hydrophilic moiety has been shown to significantly influence the biodegradation rate; longer chains show increased bioresistance [14, 17]. The reason for the slow biodegradation of long polyoxyethylene moieties is that the pronounced hydrophilicity and the large molecular dimensions limit the rate of transport through the cell walls.

5
Technical Incentives for Hydrolysable Surfactants

In addition to environmental concerns, there are other driving forces for the development of cleavable surfactants. One such incentive is to avoid complications such as foaming or formation of unwanted, stable emulsions after the use of a surfactant formulation. Use of a cleavable surfactant instead of a stable surfactant may offer a way to circumvent such problems. If the weak bond is present between the polar and the nonpolar part of the molecule, cleavage will lead to one water-soluble and one water-insoluble product. Both these moieties can usually be removed by standard work-up procedures. This approach has been of particular interest for surfactants used in preparative organic chemistry and in various biochemical applications.

Yet another use of surfactants with limited stability is to have the cleavage product impart a new function. For instance, a surfactant used in personal care formulations may decompose on application to form products beneficial to the skin. Surfactants that impart a new function after cleavage are sometimes referred to as functional surfactants.

Finally, surfactants that break down into non-surface active products in a controlled way may find use in specialized applications, such as in the biomedical field. For instance, cleavable surfactants that form vesicles or microemulsions can be of interest for drug delivery, provided the metabolites are nontoxic.

Most cleavable surfactants contain a hydrolyzable bond. Chemical hydrolysis is either acid- or alkali-catalyzed and many papers discuss the surfactant breakdown in terms of either of these mechanisms. In the environment, bonds susceptible to hydrolysis are often degraded by enzymatic catalysis, but only few papers dealing with cleavable surfactants have dealt with such processes in vitro. Other approaches that have been taken include incorporation of a bond that can be destroyed by UV irradiation or use of a bond which is cleaved when exposed to ozone.

6
Surfactants Containing Normal Ester Bonds

In order to study the effect of substituents near the hydrolyzable bond, four fatty acid ethoxylates with different degrees of steric hindrance near the ester bond, see Fig. 4, have been synthesized [18]. The homologue pure surfactants were prepared by reacting the appropriate acid chlorides with a large excess of tetra(ethylene glycol) in the presence of pyridine.

To elucidate how the hydrolytic stability of the ester bond is influenced by substitution at the α-carbon of the acyl chain, the base-catalyzed hydrolysis of the ester surfactants was investigated (at concentrations well below the CMC). The half-life of the methyl-substituted surfactant 2 was almost the same as that of the unsubstituted surfactant 1, indicating that one methyl substituent in α-position does not influence the hydrolysis to any great extent. However, the surfactant with two methyl substituents, 4, underwent a much more sluggish hydrolysis, with the half-life being about two orders of magnitudes longer than that of the unsubstituted surfactant. Somewhat surprisingly the ethyl-substituted surfactant 3 had almost the same half-life

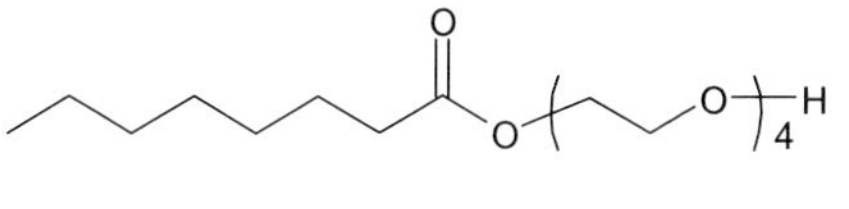

Fig. 4 Four fatty acid ethoxylates with different substituents on the acyl carbon next to the ester bond, i.e., the α-carbon

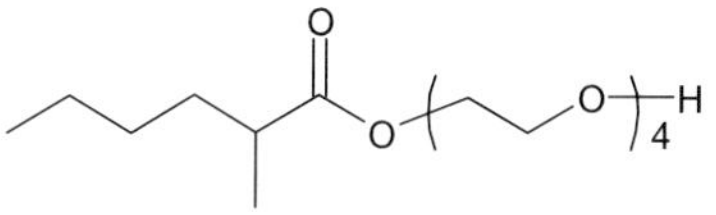

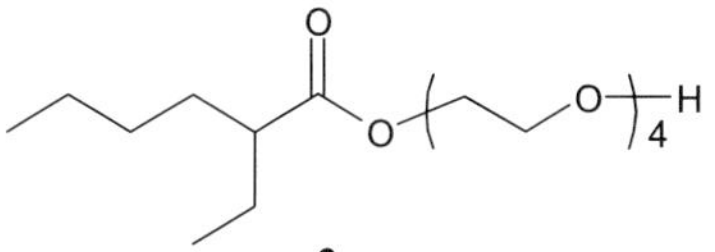

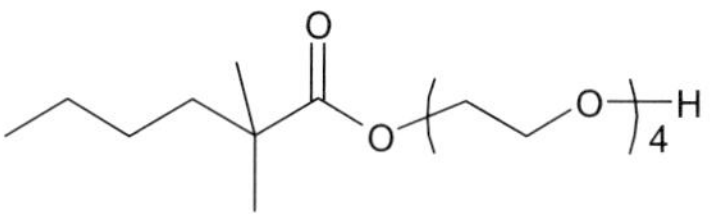

as the surfactant with two methyl groups in α-position. Evidently, an ethyl substituent gives rise to much more severe steric hindrance than a methyl substituent.

A comparison of the hydrolysis behavior above and below the CMC revealed that whereas the half-life of the surfactant was constant at concentrations below the CMC it increased linearly with increasing surfactant concentrations above the CMC [19]. Fitting the data from the hydrolysis to rate equations showed that while below the CMC the values were in accordance with the pseudo first-order rate equation, above the CMC, they fit a zero-order rate equation. For zero-order reactions the rate is independent of the concentration of the reacting species. This implies that only surfactant molecules present as unimers are cleaved, whereas those residing in micelles are protected from hydrolysis. It is known that hydroxide ions are depleted from polymers containing polyoxyethylene groups; hence, the stabilizing effect exerted by micellization of the ester surfactant is probably due to the fact that the hydroxide ions are depleted from the oxyethylene chains surrounding the micelle.

The possibility to control the hydrolysis rate of ester surfactants by addition of an ionic surfactant was investigated by studying the alkaline hydrolysis of the linear ester surfactant **1** in mixtures with a stable cationic surfactant, dodecyltrimethylammonium chloride, or a stable anionic surfactant (sodium decylsulfate) using hydrolysis of the ester surfactant alone as reference [20]. It was found that in the mixed aggregates with stable ionic surfactants the hydrolysis rate of the ester was significantly affected. Whereas the hydrolysis rate was increased in co-aggregates with the cationic surfactant, the hydrolysis was retarded when mixed with the anionic surfactant. The former effect is attributed to the formation of positively charged micelles that attract hydroxyl ions. The latter is probably a consequence of the decrease in CMC accompanying mixed micelle formation, which in turn increases the fraction of aggregated surfactant.

The stability of the ester surfactants against enzymatic hydrolysis by two different microbial lipases, *Mucor miehei lipase* (MML) and *Candida antarctica lipase B* (CALB) added separately to the surfactant solutions, was also investigated, see Fig. 5 [19]. It is obvious that hydrolysis of the unsubstituted surfactant is much faster with both CALB and MML than that of the substituted surfactants, i.e., increased steric hindrance near the ester bond leads to decreased hydrolysis rate. Since the specificity of the enzyme against its substrate is determined by the structure of the active site, it can be concluded, as expected, that the straight chain surfactant most easily fits into the active site of both enzymes.

Lipases generally show low hydrolytic activity when their ester substrates are dissolved in aqueous media and present in unimeric form. A pronounced increase in activity is observed when the substrate concentration reaches the solubility limit and a separate phase is formed. In the case of surfactants this implies that a possible increase in activity can be expected above the CMC. Attempts to investigate how the hydrolysis is affected by micellization were made for the linear surfactant 1 of Fig. 4. The CMC of this surfactant is 10 mM, and a marked change in the activity of the MML is indeed observed when this concentration is exceeded, see Fig. 6. The initial reaction is faster (steeper slope) above the CMC. When CALB was used to catalyze the reaction, no increase of the reaction rate was observed above the CMC. It was also found that the rate, expressed in moles of surfactant consumed per minute, was independent of the start concentration (same slope). A tentative explanation to the fact that the MML but not the CALB-catalyzed hydrolysis is accelerated by the presence of micelles may be that MML but not CALB is able

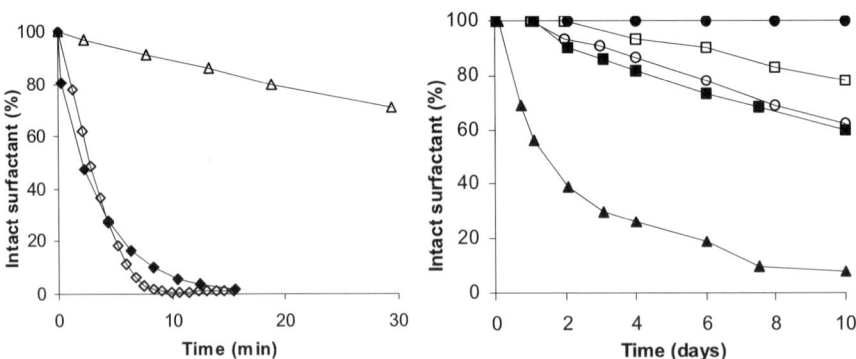

Fig. 5 Hydrolysis of the esters surfactants in the presence of lipase at 20 °C. *Left*: (♦) Linear ester + MML, (◊) linear ester + CALB, and (△) methyl substituted ester + CALB. *Right*: (▲) Methyl substituted ester + MML, (■) ethyl substituted ester + MML, (●) dimethyl substituted ester + MML, (□) ethyl substituted ester + CALB, and (○) dimethyl substituted ester + CALB. MML and CALB stand for *Mucor miehei* and *Candida antarctica B*, respectively. (Redrawn from [19])

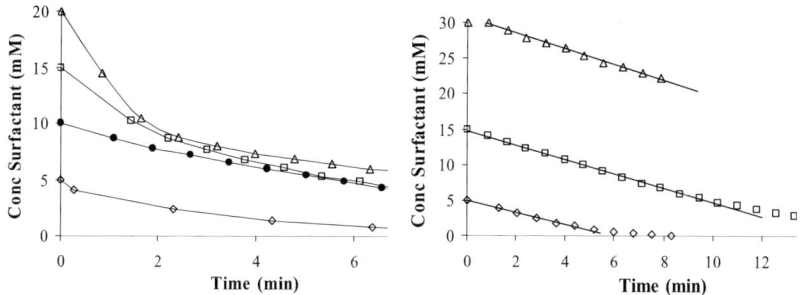

Fig. 6 Hydrolysis of the linear ester surfactant in the presence of lipase at 20 °C. *Left*: 0.5 g/l MML. (◊) 5 mM, (•) 10 mM, (□) 15 mM, and (△) 20 mM. *Right*: 0.5 g/l CALB. (♦) 5 mM, (□) 15 mM, and (△) 30 mM. MML and CALB stand for *Mucor miehei* and *Candida antarctica B*, respectively. (Redrawn from [19])

to interact with the hydrophobic core of the micelle and thereby become activated. It is known that MML, but not CALB, has a "lid" covering its active site and that this lid needs to open to allow the substrate to enter. The bottom of the lid has a hydrophobic character and the lid-opening process is driven by hydrophobic interactions, which in this case may be interactions with the hydrophobic core of the surfactant micelles.

Biodegradation tests of the four ester surfactants have shown that the linear, the methyl and the ethyl substituted ester surfactants (1, 2 and 3, respectively, of Fig. 4) biodegrade by almost the same path in a plot of biodegradation versus time. All three had reached 60% biodegradation at day 28; hence, these substances meet the main criterion for ready biodegradability. The disubstituted ester surfactant (4 of Fig. 4) had reached only 31% biodegradation after 28 days. The sluggish rate of biodegradation of the disubstituted surfactant was due to slow hydrolysis of the ester bond, as proved by the fact that the biodegradation of a 1 : 1 mixture of the two hydrolysis products, 2,2-dimethylhexanoic acid and tetra(ethylene glycol), was much faster than the biodegradation of the intact surfactant. Hence, it could be concluded that central scission was the rate determining step of the biodegradation of the disubstituted ester surfactant.

By comparing the results from the biodegradation tests with those obtained from the study of chemical hydrolysis, it is obvious that a simple correlation between the two does not exist.

The physical chemical properties of the surfactants that contain an ester bond between the hydrophobic tail and the polar head group are very similar to those of alcohol ethoxylates of the same alkyl chain length and the same number of oxyethylene units. The CMC and the cloud point values of the linear ester surfactant **1** of Fig. 4 are approximately the same as those of the straight chained alcohol ethoxylate tetra(ethylene glycol)monooctyl ether (C8E4), i.e., around 10 mM and 40 °C, respectively. Thus it appears that the carbonyl group of the ester bond gives approximately the same driving force for aggregation as does a methylene group, when situated in-between the hydrophobe and the hydrophile.

7
Ester Quats

The term ester quat is commonly used to describe a family of cationic surfactants in which the long chain alkyl group or groups are fatty acid ester(s) with the ester linkage situated close to the quaternary ammonium moiety. Typical ester quats are shown in Fig. 1. Ester quats exhibit a very good biodegradation profile, which is the expected behavior considering their structure. In-depth studies on one commercial ester quat, di(tallow fatty acid)ester of dihydroxyethyldimethylammonium chloride have shown that the biodegradation starts with a biologically mediated cleavage of the two ester bonds, which produces fatty acid salts and a small, highly water soluble quaternary ammonium species [21]. Fatty acid salts are known to degrade rapidly in the environment. Ester quats also degrade rapidly under anaerobic conditions [22]. Furthermore, ecotoxicity studies both on the parent ester quat and on the primary degradation products, i.e., the monoester and the polyalcohol quaternary ammonium salt, show very low toxicity values. In summary, ester quats must be regarded as benign to the environment.

The other side of the coin is that ester quats may hydrolyze in the formulation during storage. All esters are susceptible to hydrolysis, in particular under alkaline conditions, but ester quats are more unstable than normal esters because of the quaternary ammonium group adjacent to the ester group. The positively charged nitrogen attracts electrons from the neighboring atoms, which leads to an electron deficiency of the carbonyl carbon, which, in turn, favours nucleophilic attack by hydroxyl ions. The rate of hydrolysis becomes strongly pH dependent and the maximum stability is obtained far down on the acidic side, as can be seen from Fig. 7. As a result, all aqueous formulations containing ester quats must have a pH between 2 and 5.5.

When present in micelles, ester quats hydrolyze faster than free unimers in the bulk phase. This is due to an increased hydroxyl ion concentration around the micelle, i.e., the local pH in the vicinity of the micelle surface is higher than in the bulk. The phenomenon is referred to as micellar catalysis and is further discussed in the Betaine esters section.

Ester quats with one, two, or three fatty acid ester moieties exist in the molecule, but the dominating type are the diester quats, i.e., products that contain two fatty acid ester chains and two short alkyl groups. These are synthesized by reaction of two moles of fatty acid with one mole of alkanolamine using an acid catalyst under conditions where water is stripped off. The fatty

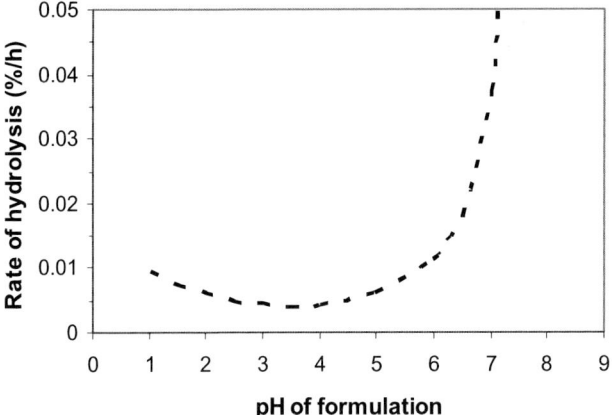

Fig. 7 Influence of pH on the hydrolytic stability of dicetylester of bis-2-hydroxyethylammonium chloride at 25 °C. (Redrawn from [22])

acids are usually tallow or palm based, i.e., they are essentially C16–C18 mixtures. Methyldiethanolamine is the most commonly used alkanolamine, but triethanolamine is also often used, in which case a mixture of mono-, di-, and triesters are formed. Monoesteramine and unreacted fatty acid are the main byproducts when methyldiethanolamine is used. The esterification step is followed by quaternization, usually with either methyl chloride or dimethyl sulphate as the quaternizing agent.

The dominating use of ester quats is as fabric softener and the transition from the traditional di(hydrogenated tallow)dimethylammonium chloride, which had been the work-horse softener for more than three decades, to ester quats was made in Europe in the early 1990s as a response to pressure from environmental authorities in the European Union [23]. The transition has been less rapid in the United States, but ester quats have gradually replaced the traditional quats in most of the major softener products there as well. The same trend is ongoing in the rest of the world.

Fabric softeners can be applied in different ways. The dominating types of products are the rinse cycle softeners, which are liquid aqueous dispersions. The softener may also be introduced as tumble dryer sheets, a practice that is relatively common in the United States but not in Europe. A smaller but growing use of the softener product is as softener-containing detergents, sometimes referred to as softergents.

The diester quats used for fabric softening are hydrophobic species with very limited water solubility. In water they form lamellar phases and liposomes at low concentration. The low water solubility in combination with the hydrolytic instability makes formulation of softener products a demanding task, and a plethora of patents and patent applications related to ester quat formulations have been published during the last decade [21]. The diester quats used as rinse cycle softeners are usually formulated together with additives such as alcohol ethoxylates with relatively high HLB value and/or water miscible solvents, such as isopropanol [21, 23, 24].

The same type of diester quats that is now the dominating active ingredient in softener products is presently finding use as hair conditioner. Ester quats are also used or explored in other traditional applications of quaternary ammonium compounds, both in the personal care sector and for industrial use. Cationics are established as additives in a range of industrial applications, such as mineral ore flotation, textile finishing, fiber processing, and organoclays. Cationic surfactants also have an established use as disinfectants in many types of products. In most of these applications ester quats, mono-, di-, and sometimes even triesters, are seen as environmental-friendly alternatives to the stable quats. In several cases a simple replacement of the stable quat by an ester quat is not possible, however, due to the limited chemical stability of the latter.

8
Betaine Ester Surfactants

The hydrolysis of esters of the amino acid betaine (trimethylglycine), see Fig. 8, shows extraordinary strong pH dependence. In an alkaline environment the rate of hydrolysis is much higher for these substances than for esters in general, whereas they are more stable in an acidic environment [25, 26]. In fact, betaine esters are generally hydrolyzed at a significant rate by the base-catalyzed mechanism even at neutral pH.

The special hydrolysis characteristics of betaine esters can be explained by two effects caused by the presence of the strongly electron-withdrawing, positively charged quaternary ammonium group in close proximity to the ester bond. Firstly, this will, in the same way as described for the ester quats above, give rise to a decreased electron density at the carbonyl carbon and thereby make it more prone to nucleophilic attack by hydroxyl ions. The other effect is an inherent destabilization of the ground state, caused by repulsion between the partial positive charge at the carbonyl carbon and the positive charge on the nitrogen atom. This repulsion is relieved by attack of a hydroxide ion, but augmented by protonation.

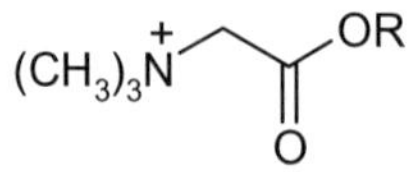

Fig. 8 General structure of a betaine ester. In a surface active betaine ester the R group makes up the hydrophobic tail

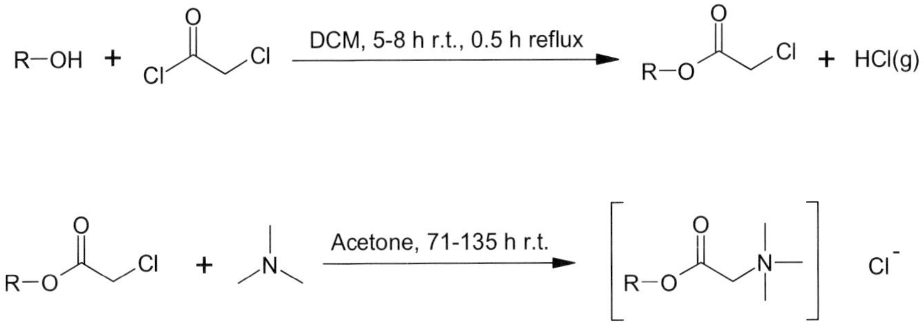

Fig. 9 A synthesis route used for the preparation of betaine ester surfactants. DCM is dichloromethane

Most work on surface active betaine esters have been performed on substances made from saturated long-chain alcohols [20, 27–30, 32–34]. These compounds can be prepared by the simple two-step synthesis route shown in Fig. 9.

For a surface active betaine ester the rate of alkaline hydrolysis shows significant concentration dependence. Due to a locally elevated concentration of hydroxyl ions at the cationic micellar surface, i.e., a locally increased pH in the micellar pseudophase, the reaction rate can be substantially higher when the substance is present at a concentration above the critical micelle concentration compared to the rate observed for a unimeric surfactant or a non-surface active betaine ester under the same conditions. This behavior, which is illustrated in Fig. 10, is an example of micellar catalysis. The decrease in reaction rate observed at higher concentrations for the C12–C18 : 1 compounds is a consequence of competition between the reactive hydroxyl ions and the inert surfactant counterions at the micellar surface. This effect is in line with the essential features of the pseudophase ion-exchange model of micellar catalysis [29, 31].

The physical chemical behavior of betaine esters of long-chain alcohols shows strong similarities to the common, closely related alkyltrimethylammonium surfactants both in dilute and concentrated aqueous systems. In consistence with the findings about CMC:s of surfactants containing normal ester bonds (see above) it has been found that the CMC for a betaine ester with a hydrocarbon chain of n carbons is very close to the value for an alkyltrimethylammonium chloride surfactant with a hydrocarbon chain of $n + 2$ carbons [32]. The binary phase diagram of dodecyl betainate-water has an appearance very similar to that of an alkyltrimethylammonium surfactant with a hydrophobic tail of a similar size [30].

The alcohol formed on hydrolysis of a betaine ester surfactant has strong effects on the shape of its aggregates and its phase behavior. In a micellar solution of dodecyl betainate, addition of 10 to 20% of dodecanol causes a significant aggregate growth (higher additions cause phase separation) [29]. In

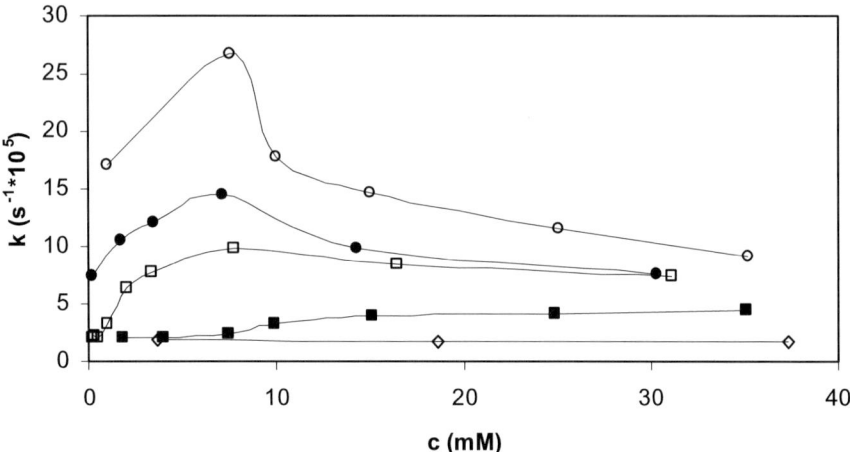

Fig. 10 Concentration dependence of the pseudo first-order rate constants in a 100 mM phosphate buffer at pH 7.5 and 37 °C for a number of surface active betaine esters with hydrophobic tails of different sizes. For comparison, the rate constant for a non-surface active compound (ethyl betainate) is included. (○) Oleyl betainate, (●) tetradecyl betainate, (□) dodecyl betainate, (■) decyl betainate, (◊) ethyl betainate. All compounds have chloride counterions

more concentrated samples of the same system, addition of just a few percent of dodecanol turns a hexagonal phase (present at surfactant concentrations between 35 and 75 wt. % in the binary aqueous system) into a lamellar phase. In this context it deserves to be noted that when a betaine ester is present in an unbuffered aqueous solution, the production of betaine during the initial hydrolysis leads to a lowering of the pH to values at which the compound is stable.

The effects of "dilution" of the micellar surface charge on the rate of alkaline hydrolysis of a betaine ester surfactant have been investigated for a mixture of decyl betainate and a nonionic surfactant with a similar CMC. It was shown that the relation between micellar composition and the hydrolysis rate essentially parallels the relation between micellar composition and counterion binding to mixed micelles made up of ionic and nonionic surfactants [20].

Another interesting system containing a surface active betaine ester is the dilute aqueous mixture of dodecyl betainate and hydrophobically modified hydroxyethylcellulose (HM-HEC) that has been studied by Karlberg et al. [33]. It is well known that the viscosity of mixtures of HM polymers and surfactants is strongly dependent on the concentration of the amphiphile. By preparing a mixture of a surface active betaine ester and HM-HEC in a solution buffered at a pH where the surfactant is hydrolyzed, it is possible to make a gel with a time-dependent viscosity.

Since surface active betaine esters can be degraded under mild conditions and the hydrolysis products, i.e., the amino acid betaine and a long-chain alcohol can be expected to be less toxic than the intact surfactant, these amphiphiles are interesting candidates for use in applications where surfactant toxicity is an issue. Surface active betaine esters have been evaluated as temporary bactericides [34] and have been studied as potential candidates for use as pharmaceutical excipients (pharmaceutical helper molecules) [30].

9
Surfactants Containing Carbonate Bonds

Nonionic surfactants containing a carbonate bond have been synthesized by reaction between an alkyl chloroformate and tetra(ethylene glycol), using a large excess of the latter reactant, see Fig. 11 [35].

By comparing the rate of base-catalyzed hydrolysis of a linear carbonate surfactant and a linear ester surfactant it was found that the carbonate bond was more stable against alkaline hydrolysis than the ester bond [35]. The time needed to hydrolyze 50% of the carbonate surfactant was twice as long as the time needed to hydrolyze 50% of the ester surfactant. This result was somewhat unexpected since the carbonyl carbon of the carbonate, having electronegative atoms on both sides, would at first sight be expected to be strongly electrophilic and thus readily attacked by the nucleophilic hydroxyl ion. A possible contributing factor for the relative stability of the carbonate bond is that it is stabilized by resonance. Delocalization of the electrons over the three oxygen atoms of the carbonate bond is likely to reduce the electrophilicity of the carbonyl carbon.

The activity of three ester splitting enzymes, *Candida antarctica lipase B* (CALB), *Mucor miehei lipase* (MML) and esterase, towards the carbonate surfactant was studied. While CALB and esterase were found to catalyze the hydrolysis of the carbonate bond, MML showed no activity.

Biodegradation tests have shown that surfactants containing a carbonate bond between the hydrophobic tail and the polar head group are readily biodegradable. In comparative tests such carbonate surfactants biodegrade somewhat faster than the corresponding surfactants containing an ester bond [35]. The carbonate bond is not only susceptible to alkaline hydrolysis

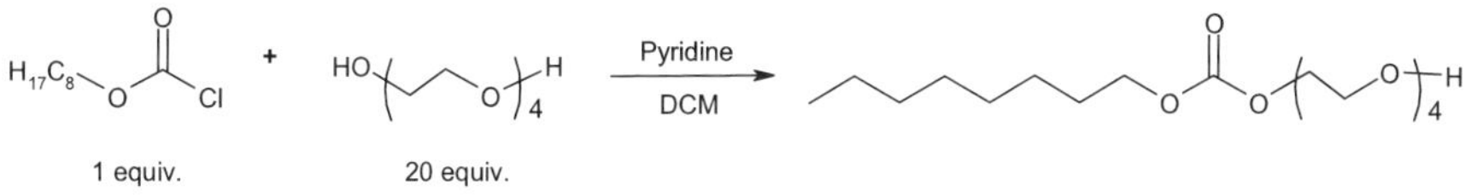

Fig. 11 Synthesis route for preparation of carbonate surfactants. DCM stands for dichloromethane

Table 1 Effective diffusion coefficients in the various phases for pure nonionic surfactants

	CMC (mM)	γ_{CMC} (mN/m)	CP (°C)	A_{CMC} (NM^2)
Linear carbonate	3.2	30	24	0.44
Branched carbonate	–	–	< 0	–
Linear ester	10	31	42	0.4 [a]
Branched ester	40	31	28	0.41
Linear amide	65	36	> 100	0.61
Linear ether [37, 38]	9	30	40	0.49

[a] Uncertain due to the non-distinct slope of the surface tension isotherm

but also to heat, which can be used as an alternative trigger for surfactant cleavage.

As mentioned above, the carbonyl group of the ester bond can be treated as an integral part of the hydrophobic tail, giving the same apparent hydrophobic contribution as a methylene group. The CMC of the linear carbonate surfactant with the same number of carbon atoms in the hydrophobic tail is a factor of three smaller than the CMC of the linear ether and the linear ester, see Table 1. There is an established relation between the length of the alkyl chain and the CMC, which says that adding an extra methylene group decreases the CMC of a nonionic surfactant by a factor of three [36]. Thus by comparing the CMC value of the linear carbonate with those of the ether and ester surfactants, one may conclude that, in a formal sense, the –O(C=O)– part of the carbonate bond also gives the same hydrophobic contribution as a methylene group (when situated between the hydrophobic tail and the polar head group of a surfactant).

10
Surfactants Containing Amide Bonds

One synthesis pathway for preparation of nonionic surfactants containing an amide bond from a fatty acid chloride and tetra(ethylene glycol) is illustrated in Fig. 12 [39].

The chemical stability of the amide bond is high. When the surfactant containing an amide bond was subjected to 1 M sodium hydroxide during five days at room temperature, only 5% of the amide surfactant was cleaved. The corresponding experiment performed in 1 M HCl resulted in no hydrolysis. The amide bond was, however, found to be slowly hydrolyzed when lipase from *Candida antarctica* or peptidase was used as catalyst. Amidase and lipase from *Mucor miehei* was found to be ineffective. Despite the very high chemical stability, the amide surfactant biodegrades by a similar path in the

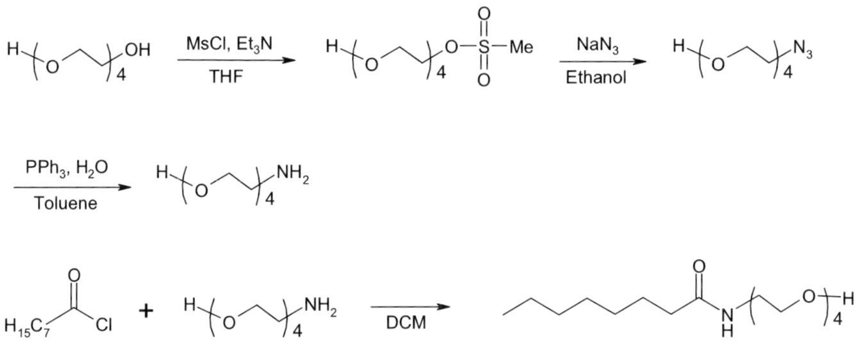

Fig. 12 Synthesis route for preparation of amide surfactants. MsCl, Et₃N, THF and DCM stand for mesyl chloride, triethylamine, tetrahydrofuran and dichloromethane, respectively

plot of biodegradation versus time as the corresponding ester surfactant, and, hence, it reaches 60% biodegradation within 28 days.

Values of surface tension at the CMC, γ_{CMC}, and area per molecule, A_{CMC}, indicate that the amide surfactants pack less densely than the corresponding ester, ether, and carbonate surfactants. The amide-containing surfactant also has markedly higher CMC and cloud point values (see Table 1 above). The greater hydrophilic character of amide surfactants over the corresponding ether, ester, and carbonate surfactants is due to the fact that the amide bond is highly polar, giving rise to higher water solubility of the amide surfactant and also higher CMC and cloud point.

11
Cyclic Acetals

The groups of Burczyk, Takeda, and others have made thorough studies of cyclic acetals, such as 1,3-dioxolane (five-membered ring) and 1,3-dioxane (six-membered ring) compounds, illustrated in Fig. 13. They are typically synthesized from a long-chain aldehyde by reaction with a diol or a higher polyol. Reaction with a vicinal diol gives the dioxolane [40–42] and 1,3-diols yield dioxanes [43, 44].

A hydroxyl acetal may form if the diol contains an extra hydroxyl group, such as in glycerol. The remaining hydroxyl group can subsequently be derivatized to give anionic or cationic surfactants, as illustrated in Fig. 14. It has been claimed that glycerol gives ring closure to dioxolane, yielding a free, primary hydroxyl group, but it is likely that some dioxane with a free, secondary hydroxyl group is formed as well. The free hydroxyl group can be treated with SO₃ and then neutralized to give the sulfate [45]; it can be reacted with propane sultone to give the sulfonate [46], or it can be substi-

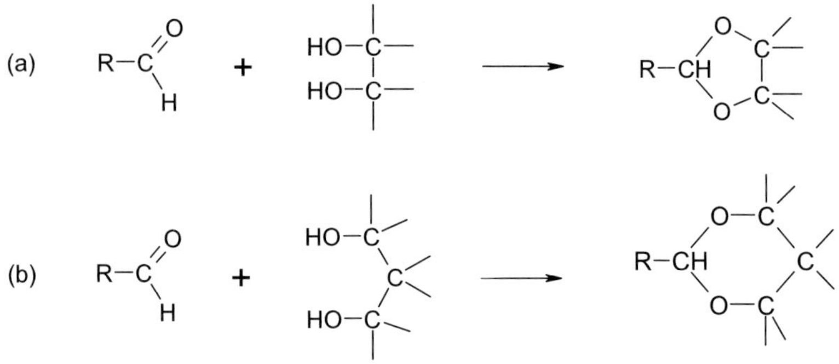

Fig. 13 Preparation of 1,3-dioxolane surfactant (**a**) and 1,3-dioxane surfactant (**b**) from a long-chain aldehyde and a 1,2-and a 1,3-diol, respectively

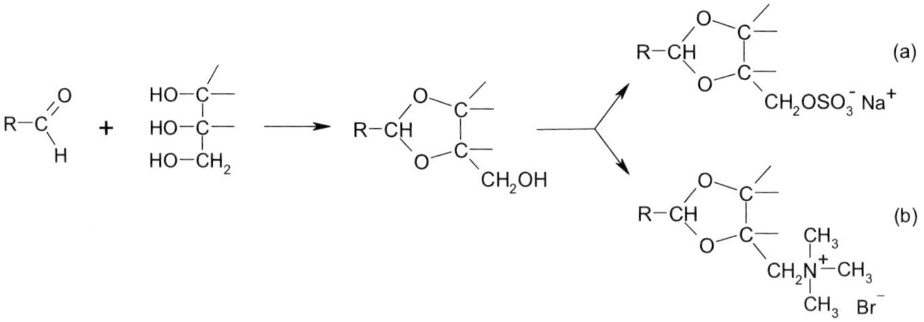

Fig. 14 Examples of anionic (**a**) and cationic (**b**) 1,3-dioxolane surfactants

tuted by bromine or chloride and then reacted with dimethylamine to give a tertiary amine as polar group. Quaternization of the amine can be made in the usual manner, e.g., with methyl bromide [47]. An analogous reaction with pentaerythritol as the diol yielded a 1,3-dioxane with two unreacted hydroxymethyl groups, which can be reacted further, e.g., to give a dianionic surfactant [46]. The remaining hydroxyl group may also be ethoxylated. These acetal surfactants have been commercialized [48]. The rate of decomposition in sewage plants of this class of nonionic surfactants is much higher than for normal ethoxylates [49].

Hydrolysis of acetals yields aldehydes, which are intermediates in the biochemical β-oxidation of hydrocarbon chains. Acid catalyzed hydrolysis of unsubstituted acetals is generally facile and occurs at a reasonable rate at pH 4–5 at room temperature. Electron-withdrawing substituents, such as hydroxyl, ether oxygen, and halogens, reduce the hydrolysis rate, however [50]. Anionic acetal surfactants are more labile than cationic [40], a fact that can be ascribed to the locally high oxonium ion activity around such micelles. The same effect can also be seen for surfactants forming vesicular aggregates, again undoubtly due to differences in the oxonium ion activity in the pseudophase surrounding the vesicle. Acetal surfactants are stable at neutral and high pH.

One advantage of using a cleavable acetal surfactant instead of a conventional amphiphile has been elegantly demonstrated in a work by Bieniecki and Wilk [51]. A cationic 1,3-dioxolane derivative was used as surfactant in a microemulsion formulation that was employed as a reaction medium for an organic synthesis. When the reaction was complete, the surfactant was decomposed by addition of acid and the reaction product easily recovered from the resulting two phase system. Through this procedure the problems of foaming and emulsion formation, frequently encountered with conventional surfactants, could be avoided.

It has been found that the 1,3-dioxolane ring corresponds to approximately two oxyethylene units with regard to effect on the critical micelle concentration and adsorption characteristics [42]. Thus, surfactant type I in Fig. 14 should resemble ether sulfates of the general formula $R-(OCH_2CH_2)_2OSO_3Na$. This is interesting since the commercial alkyl ether sulfates contain two to three oxyethylene units.

12
Cyclic Ketals

Reaction between a long-chain ketone and a diol will lead to a surfactant that contains a ketal bond. The synthesis is analogous with those given in Figs. 13 and 14 for the preparation of acetal surfactants [52]. Ketal-based surfactants have also been prepared in good yields from esters of keto acids by either of two routes, as shown in Fig. 15 [53–55].

Figure 16 shows the biodegradation profiles of the dioxolane surfactants of Fig. 15 [54]. As expected, the degradation rate is strongly dependent on the alkyl chain length. The process is markedly faster for the labile surfactants (and particularly for structure I which contains an extra ether oxygen) than for the conventional carboxylate surfactant of the same alkyl chain length used as reference. Ketal surfactants are in general more labile than the corresponding acetal surfactants [56]. As an example, a ketal surfactant kept at pH 3.5 was cleaved to the same extent as an acetal surfactant of similar structure kept at pH 3.0 [57]. The relative lability of the ketal linkage is due to the greater stability of the carbocation formed during ketal hydrolysis compared to the carbocation formed during acetal hydrolysis. (It is noteworthy that biodegradation of an acetal surfactant has been found to be faster than for a ketal surfactant of very similar structure [54].) This is another example showing that there is no simple correlation between rate of biodegradation and susceptibility to chemical hydrolysis (compare the discussion above for the ester and amide surfactants).

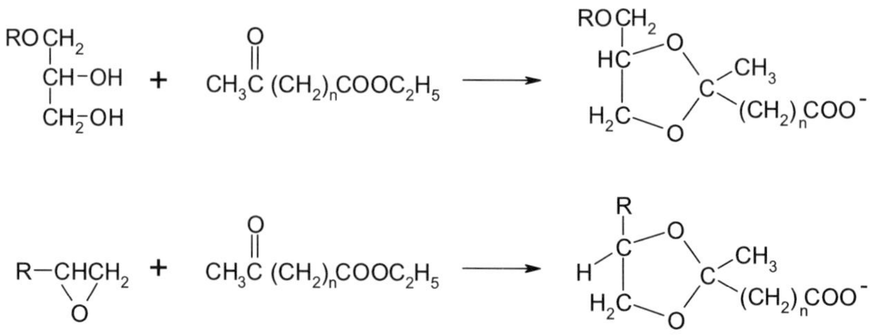

ROCH$_2$–CH(OH)–CH$_2$OH + CH$_3$C(O)(CH$_2$)$_n$COOC$_2$H$_5$ ⟶ [ROCH$_2$–HC–O, H$_2$C–O]C(CH$_3$)(CH$_2$)$_n$COO$^-$ (I)

R–CHCH$_2$ (epoxide, O) + CH$_3$C(O)(CH$_2$)$_n$COOC$_2$H$_5$ ⟶ [HC(R)–O, H$_2$C–O]C(CH$_3$)(CH$_2$)$_n$COO$^-$ (II)

Fig. 15 Preparation of anionic 1,3-dioxolane surfactant from ethyl esters of ketoacids

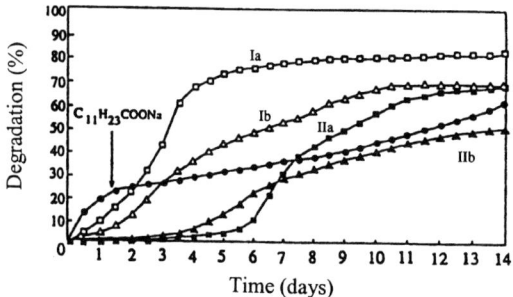

Fig. 16 Rate of biodegradation vs. time for four ketal surfactants and for sodium decanoate, which is used as a reference. I and II relate to the compounds of Fig. 15; a: R = $C_{12}H_{25}$, $n = 2$; b: R = $C_{16}H_{33}$, $n = 2$. (From [54])

13
Ortho Ester Surfactants

Surfactants containing the ortho ester bond have been explored in recent years [58, 59]. They are examples of acid-sensitive surfactants. Surface active ortho esters are easily prepared by transesterification of a low molecular weight ortho ester (usually triethyl orthoformate) with a fatty alcohol and a methyl-capped poly(ethylene glycol) (PEG). An example of a structure and a typical method of preparation are given in Fig. 17. Due to the trifunctionality of the ortho ester, a complex mixture of species is obtained, each of which can be identified by NMR using the chemical shift of the central methine proton as a marker [59]. Furthermore, if the reactant alcohol is difunctional, then cross-linking will occur, and a large network may be formed. Such compounds have been shown to be effective foam depressants (an example based on poly(propylene glycol) (PPG) and methyl-capped PEG is shown in Fig. 18). By varying the number and types of substituents (fatty alcohol, alkyleneoxy group, end blocking), the properties of the ortho ester-based surfactant or block copolymer can be tailor-made for a specific field of application.

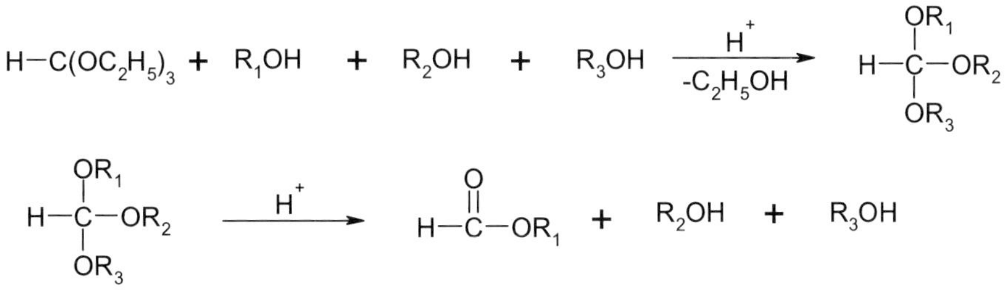

$$H-C(OC_2H_5)_3 + R_1OH + R_2OH + R_3OH \xrightarrow[-C_2H_5OH]{H^+} H-\underset{OR_3}{\overset{OR_1}{\underset{|}{\overset{|}{C}}}}-OR_2$$

$$H-\underset{OR_3}{\overset{OR_1}{\underset{|}{\overset{|}{C}}}}-OR_2 \xrightarrow{H^+} H-\overset{O}{\overset{\|}{C}}-OR_1 + R_2OH + R_3OH$$

Fig. 17 Synthesis and hydrolysis of ortho esters. R_1, R_2, and R_3 are alkyl groups

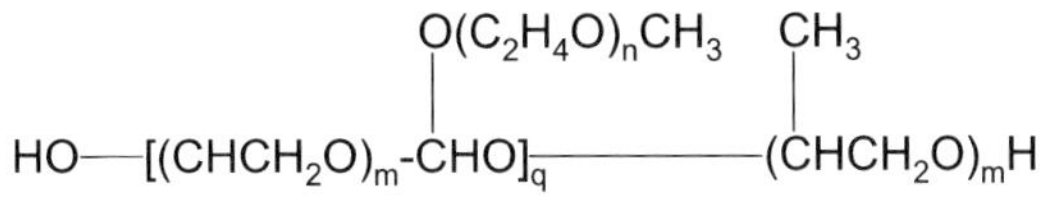

$$HO-[(CHCH_2O)_m-\underset{O(C_2H_4O)_nCH_3}{\overset{|}{CHO}}]_q-\underset{CH_3}{\overset{|}{(CHCH_2O)_m}}H$$

Fig. 18 An ortho ester-based block copolymer

Hydrolysis of ortho esters occurs by a mechanism analogous to that of acetals and ketals and gives rise to one mole of formate and two moles of alcohols [60, 61]. Both formates and alcohols can be regarded as non-toxic substances, and recent research by Bergh et al. has shown that surface active formates (similar to surface active alcohols and esters but in contrast to surface active aldehydes) have little or no dermatological effect, evaluated in terms of sensitizing capacity and irritancy [62]. Ortho ester-based surfactants undergo acid-catalyzed cleavage much more readily than acetal-based surfactants under the same conditions [63]. For instance, a water-soluble ortho ester based on octanol and monomethyl-PEG is hydrolyzed to 50% in 2 h at pH 5. The structure of the surfactant has been found to influence the hydrolysis rate and, in general, a more hydrophilic surfactant has a higher decomposition rate.

Ortho ester linkages can also be used to improve biodegradation properties in long chain ethoxylates or block copolymers. It has been shown that a conventional PEG-PPG copolymer with a molecular weight of 2200 biodegrades to only 3% in 28 days. However, if an equivalent molecule is built up from PEG 350 and PPG 400, connected by ortho ester links, it will reach 62% biodegradation within 28 days and thus be classified as "readily biodegradable" [64].

Recently, nonionic ortho ester surfactants have been used as emulsifiers for squalene, a polar oil [65]. In this case a polymer is used together with the surfactant. The emulsification is made under acidic conditions, and the surfactant breaks down rapidly after the emulsion is formed, leaving a surfactant-free, polymer stabilized emulsion with reasonable stability.

Cationic ortho ester surfactants have also been described [66]. They can be synthesized by a route analogous to that used for nonionic ortho ester surfactants (Fig. 17), using an amino alcohol, such as N,N-dimethylaminoethanol, instead of the methyl-capped PEG. Cationic ortho esters have been suggested for use as surfactant or as a bactericidal agent with a pH-dependent half-life [63, 66]. The surfactants are completely stable at pH 10 while they break down readily under mild acidic conditions. In this respect they are the reverse to the betaine esters discussed above.

14
Summary

A wide range of hydrolyzable surfactants exist that are suitable for different applications. The decisive factor for choosing the best hydrolyzable surfactant is the pH of the application; Fig. 19 illustrates the susceptibility of the different types of surfactants that have been discussed in this review to acid and alkaline hydrolysis. As can be seen from the figure, the formulator has a tool box of surfactants with different hydrolysis characteristics to choose from. The figure also illustrates that for some surfactants, e.g., betaine esters, the susceptibility to hydrolysis increases above the CMC, i.e., when the surfactant is present in aggregated form, while for other surfactants, e.g., normal esters, the opposite holds true. This behavior is practically important and must be taken into account in the formulation work.

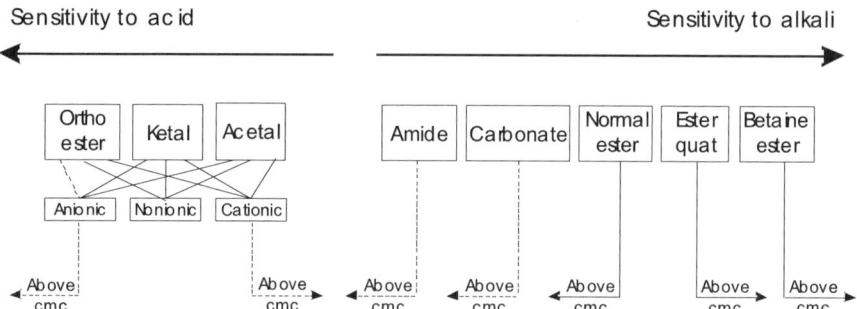

Fig. 19 Hydrolysis characteristics of different hydrolyzable surfactants

References

1. Jaeger DA (1995) Supramol Chem 5:27–30
2. Tehrani-Bagha AR, Holmberg K (2007) Curr Opin Colloid Interface Sci 12:81–91
3. Holmberg K (2001) Cleavable Surfactants. In: Texter J (ed) Reactions and Synthesis in Surfactant Systems. Marcel Dekker, New York
4. Stjerndahl M, Lundberg D, Holmberg K (2003) Cleavable surfactants. In: Holmberg K (ed) Novel Surfactants, 2nd ed. Marcel Dekker, New York

5. Holmberg K, Jönsson B, Kronberg B, Lindman B (2002) Surfactants and Polymers in Aqueous Solution, 2nd ed. Marcel Dekker, New York
6. Holmberg K (1993) Surf Coatings Int 76:481–448
7. Swisher RD (1987) Surfactant Biodegradation. Marcel Dekker, New York
8. White GF, Russel NJ (1995) What is biodegradation? In: Karsa DR (ed) Biodegradability of Surfactants. Blackie, Glasgow
9. Brown D (1995) Introduction to surfactant biodegradation. In: Karsa DR (ed) Biodegradability of Surfactants. Blackie, Glasgow
10. Painter HA (1995) Biodegradability testing. In: Karsa DR (ed) Biodegradability of Surfactants. Blackie, Glascow
11. Painter HA (1995) Testing strategy and legal requirements. In: Karsa DR (ed) Biodegradability of Surfactants. Blackie, Glascow
12. van Ginkel CG (1996) Biodegradation 7:151–164
13. van Ginkel CG, Stroo CA, Kroon AGM (1993) Tenside Surfact Deterg 30:213–216
14. Balson T, Felix MSB (1995) Biodegradability of non-ionic surfactants. In: Karsa DR (ed) Biodegradability of Surfactants. Blackie, Glasgow
15. Huber M, Meyer U, Rys P (2000) Environ Sci Technol 34:1737–1741
16. Marcomini A, Zanette M, Pojana G, Suter MJF (2000) Chem 19:549–554
17. Carvalho G, Novais JM, Pinheiro HM (2003) Environ Technol 24:109–114
18. Stjerndahl M, Holmberg K (2003) J Surfact Deterg 6:311–318
19. Stjerndahl M, van Ginkel CG, Holmberg K (2003) J Surfact Deterg 6:319–324
20. Lundberg D, Stjerndahl M, Holmberg K (2005) Langmuir 21:8658–8663
21. Overkempe C, Annerling A, van Ginkel CG, Thomas PC, Boltersdorf D, Speelman J (2003) Esterquats. In: Holmberg K (ed) Novel Surfactants. Marcel Dekker, New York
22. Garcia MT, Campos E, Sanchez-Leal J, Ribosa I (2000) Chemosphere 41:705–710
23. Steichen DS (2002) Cationic Surfactants. In: Holmberg K (ed) Handbook of Applied Surface and Colloid Chemistry, Vol. 1. Wiley, New York
24. Jacques A, Schramm CJ (1997) Fabric softeners. In: Lai K-Y (ed) Liquid Detergents. Marcel Dekker, New York
25. Wright MR (1968) J Chem Soc B 5:548–550
26. Wright MR (1969) J Chem Soc B 6:707–710
27. Thompson RA, Allenmark S (1989) Acta Chem Scand 43:690–693
28. Thompson RA, Allenmark S (1992) J Colloid Interface Sci 148:241–246
29. Lundberg D, Holmberg K (2004) J Surfact Deterg 7:239–246
30. Lundberg D, Ljusberg-Wahren H, Norlin A, Holmberg K (2004) J Colloid Interface Sci 278:478–487
31. Romsted LS (1977) A general kinetic theory of rate enhancements for reactions between organic substrates and hydrophilic ions in micellar systems. In: Mittal KL (ed) Micellization, Solubilization, Microemulsions. Plenum Press, New York
32. Rozycka-Roszak B, Przestalski S, Witek S (1988) J Colloid Interface Sci 125:80–85
33. Karlberg M, Stjerndahl M, Lundberg D, Piculell L (2005) Langmuir 21:9756–9763
34. Lindstedt M, Allenmark S, Thompson RA, Edebo L (1990) Agents Chemother 34:1949–1954
35. Stjerndahl M, Holmberg K (2005) J Colloid Interface Sci 291:570–576
36. Rosen M (1989) Surfactants and Interfacial Phenomena, 2nd ed. Wiley, New York
37. Corti M, Degiorgio V, Zulauf M (1982) Phys Rev Lett 48:1617–1620
38. Ohta A, Takiue T, Ikeda N, Aratono M (2001) J Solution Chem 30:335–350
39. Stjerndahl M, Holmberg K (2005) Surfact Deterg 8:1–6
40. Jaeger DA (1995) Supramol Chem 5:27–30
41. Bieniecki A, Wilk KA, Gapinski J (1997) J Phys Chem B 101:871–875

42. Sokolowski A, Bieniecki A, Wilk KA, Burczyk B (1995) Colloids Surfaces A 98:73–82
43. Wang G-W, Yuan X-Y, Liu Y-C, Lei X-G, Guo Q-X (1995) J Am Oil Chem Soc 72:83–87
44. Wang G-W, Liu Y-C, Yuan X-Y, Lei X-G, Guo Q-X (1995) J Colloid Interface Sci 173:49–54
45. Sokolowski A, Piasecki A, Burczyk B (1992) J Am Oil Chem Soc 69:633–638
46. Wang G-W, Yuan X-Y, Liu Y-C, Lei X-G (1994) J Am Oil Chem Soc 71:727–730
47. Wilk KA, Bieniecki A, Burczyk B, Sokolowski A (1994) J Am Oil Chem Soc 71:81–85
48. Galante DC, Hoy RC (1996) Eur Patent Appl EP 0 742 177 A1
49. Hoy RC, Joseph AF (1996) INFORM 7:428–429
50. Piasecki A (1985) J Prakt Chem 327:731–738
51. Bieniecki A, Wilk KA (1995) J Phys Org Chem 8:71–76
52. Yamamura S, Nakamura M, Kasai K, Sato H, Takeda T (1991) J Jpn Oil Chem Soc 40:1002–1006
53. Masuyama A, Ono D, Yamamoto A, Kida T, Nakatsuji Y, Takeda T (1995) J Jpn Oil Chem Soc 44:446–450
54. Ono D, Yamamura S, Nakamura M, Takeda T, Masuyama A, Nakatsuji Y (1995) J Am Oil Chem Soc 72:853–856
55. Ono D, Masuyama A, Nakatsuji Y, Okahara M, Yamamura S, Takeda T (1993) J Am Oil Chem Soc 70:29–36
56. Kida T, Morishima N, Masuyama A, Nakatsuji Y (1994) J Am Oil Chem Soc 71:705–710
57. Song B-K, Wolf K (1995) DWI Rep 114:549–554
58. Hellberg P-E, Bergström K (2002) An ortho ester based surfactant, its preparation and use. Eur Patent 1 042 266
59. Hellberg P-E, Bergström K, Juberg M (2000) J Surfact Deterg 3:369–379
60. Cordes EH, Bull HG (1974) Chem Rev 74:581–603
61. Potts RA, Schaller RA (1993) J Chem Ed 70:421–424
62. Bergh M, Shao LP, Magnusson K, Gäfvert E, Nilsson JLG, Karlberg A (1999) J Pharm Sci 88:483–488
63. Hellberg P-E (2002) Licentiate thesis. Chalmers University of Technology, Göteborg, Sweden
64. Bergström K, Hellberg P-E (1998) PCT Int Patent WO 98/00452
65. Mohlin K, Holmberg K (2006) Nonionic ortho esters used as emulsifiers. J Colloid Interface Sci 299:435–442
66. Hellberg P-E (2002) J Surfactants Deterg 5:217–227

Practical Surfactant Mixing Rules Based on the Attainment of Microemulsion–Oil–Water Three-Phase Behavior Systems

Raquel E. Antón · José M. Andérez · Carlos Bracho · Francia Vejar · Jean-Louis Salager (✉)

Lab. FIRP, Universidad de Los Andes, Mérida, Venezuela
salager@ula.ve

1	Introduction	84
2	The Way to Quantitatively Measure the Mixture Behavior	86
3	Collective Behavior and Ideal Mixing Rules	92
4	The Main Source of Non-Ideality: Mixture Partitioning	95
5	Deviations from Ideal Mixing Rules	101
5.1	Anionic-Nonionic Mixtures (Weakly non-Linear)	101
5.2	pH Sensitive System (Highly Non-linear because pH Scale is Logarithmical)	102
5.3	Anionic–Cationic Mixtures: Highly Non-Linear Synergy but Amenable to Linear	103
5.4	pH Sensitive Anionic–Cationic Mixtures – a Complex Situation	106
References		108

Abstract Surfactant mixture are generally used to fine-tune formulations to an exact property value, such as changing its hydrophilicity. To do so a precise characterization method has to be used. The presented technique consists of the attainment of a microemulsion–oil–water Winsor III three-phase behavior in a reference system. It allows one to classify surfactants in a hydrophilicity scale with an accuracy equivalent to one tenth of HLB unit. The characterization method is applied in different ways, including simple and double scans, to an unknown surfactant and to mixtures of two base surfactants. It is also used to test the ideality of the mixing rule expression, which is equivalent to a linear variation of the characteristic parameter versus the mixture composition. Conditions for linearity of the mixing rule are discussed. The selective partitioning of different species results in non-linear mixing rules, whose detection is discussed according to the aspect of the three-phase region in different diagrams. Typical mixing rules for pH sensitive systems containing fatty acids and fatty amines are shown. Anionic–nonionic mixtures are found to exhibit a slight deviation from ideality. The special case of antagonistic anionic–cationic mixture is shown to be easily linearized by introducing a virtual, catanionic species.

1
Introduction

All formulators use mixed surfactants, particularly since Griffin's HLB concept and its associated mixing rule was introduced 50 years ago as a yardstick for hydrophilicity. There are essentially two reasons: either the commercial products they use are already mixtures, or the proper surfactant property is not attainable with a commercial product, and it is attempted by mixing available species.

Most commercial products are mixtures because of the way they are manufactured. For instance many surfactant hydrophobes come from assorted products such as petroleum alkylate cuts or triglyceride oils, with a molecular weight distribution that could be narrow or wide. Usually, a purification and separation of single isomeric species would be too costly and, in most cases, pointless. Moreover, the synthesis reactions involved in the surfactant manufacturing might be the intrinsic reason of the production of a mixture, such as in the case of polycondensation of ethylene oxide which results in an often wide spread ethylene oxide number (EON) distribution. A residual content of some intermediates or by-products might also be a significant cause for mixture effects.

Alternatively, people mix surfactants in the fine-tuning operations of formulating products. In general, the primary goal of mixing is to attain a property that is intermediate between those of two available products, particularly an averaged degree of hydrophilicity. Other reasons may be to cummulate the properties characteristic of two substances, such as detergency and hard water tolerance, or to avoid precipitation, or to lower the critical micelle concentration (CMC). Finally, the mixing of two substances might lead to the formation of a new compound with properties completely different from the products that are mixed, because of some antagonism or synergy.

The most common concept of averaging is the so-called linear mixing rule, in which the property of the mixture is directely proportional to the amount of each species contained in the mixture, i.e.,

$$P_M = \sum_i x_i P_i, \qquad (1)$$

where P indicates the property and x is the (generally molar, sometimes weight) fraction of each species in the mixture. Of course this kind of arithmetic mean implies that the property contributions may be added, which might be the case if they are related with interaction energies, although many averaging processes at the molecular level are known to be better described by a London-type geometric mean. Therefore, the way the mixing rule is expressed should be deduced from experimental measurements.

The property of interest to characterize a surfactant or a mixture of surfactants is its hydrophilic–lipophilic tendency, which has been expressed in many different ways through a variety of concepts such as the hydrophilic-lipophilic balance (HLB), the phase inversion temperature (PIT), the cohesive energy ratio (CER), the surfactant affinity difference (SAD) or the hydrophilic–lipophilic deviation (HLD) [1], which were found to be more or less satisfactory depending on the case. In the next section, the quantification of the effects of the different compounds involved in the formulation of surfactant–oil–water systems will be discussed in details to extract the concept of characteristic parameter of the surfactant, as a way to quantify its hydrophilic–lipophilic property independently of the nature of the physicochemical environment.

When a single surfactant species is introduced in a surfactant–oil–water (SOW) system, its molecules distribute at the interface and in the bulk liquid phases in different amounts, but since there is only a single species, the nature of the substance present at interface and in the phases is the same.

When mixing two surfactants species in a SOW system, an equilibrium takes place between the oil and water phases and the interface for each species. Since the two species do not necessarily exhibit the same affinity for the interface and the oil and water bulk phases, the compositions of the surfactant mixtures at interface and in the phases might be different. For instance if a very hydrophilic species is mixed with a very lipophilic one, as often recommended in the old formulation literature, then the hydrophilic surfactant has a strong tendency to partition in water, whereas the lipophilic one would partition in the oil. In this case the surfactant mixture in water will contain a large majority of hydrophilic species, i.e., it will be very hydrophilic, whereas the oil phase will predominantly contain the lipophilic species, with the remaining adsorbing at interface. This situation in which each species actuates on its own, more or less independently of the other, has been called *non-collective* behavior. Since the surfactant mixture composition at interface is often the one that commands the actual property of the system, such as the interfacial tension or the stability of the emulsion, it is most important to know how to calculate or measure the characteristics of the mixture present at interface. Such methods will be discussed in the next section.

Conversely, instead of being completely independent from each other, the two surfactant species might interact to produce a compound with totally new properties, e. g., more tension lowering, insensible to temperature or to hardness, able to produce other aggregation structures, and so-forth. This transformation is often referred to as synergy and is often due to strong interactions of the antagonistic type.

2
The Way to Quantitatively Measure the Mixture Behavior

The quantification of the formulation of SOW systems was studied in detail during the 1970's when a considerable research drive was dedicated to enhanced oil recovery by surfactant flooding methods [2]. The basic concepts came from Winsor's work on the phase behavior of SOW systems and its R ratio of interactions between the surfactant molecules adsorbed at interface and oil and water [3], which has been presented thoroughly in a review book [4].

Winsor reported that the phase behavior of SOW systems at equilibrium could exhibit essentially three types, so called WI, WII and WIII, illustrated by the phase diagrams indicated in Fig. 1. In the WI (respectively, WII) case, the surfactant bears a stronger affinity for the water (respectively, oil) phase and most of it partitions into water (respectively, oil). As a consequence, the system exhibits a two-phase behavior in which a microemulsion is in equilibrium with excess oil (respectively, water).

In the WIII case, provided that there is enough surfactant but not too much, e.g., 1 wt. %, the system splits into three phases, i.e., a microemulsion in equilibrium with excess water and excess oil. At a higher surfactant concentration than the top vertex of the 3ϕ triangle, a single phase microemulsion often called WIV behavior is attained. However, this occurrence generally requires a large amount of surfactant, e.g., 20 wt. %, which is in most practical cases too much for cost reasons. At a very low surfactant concentration, around the CMC, only two phases are in equilibrium, and the tension is not necessarily very low. Hence, the convenient surfactant concentration to carry out a phase behavior study is in the range 0.5–3 wt. % for which three-phase behavior and a very low interfacial tension is exhibited in most WIII cases.

Three-phase behavior takes place exactly when the surfactant exhibits an equal affinity for both phases, which is indicated by a unit value of Winsor R ratio. This situation has been called *optimum formulation* because it is associated with an ultra low minimum of the interfacial tension, which en-

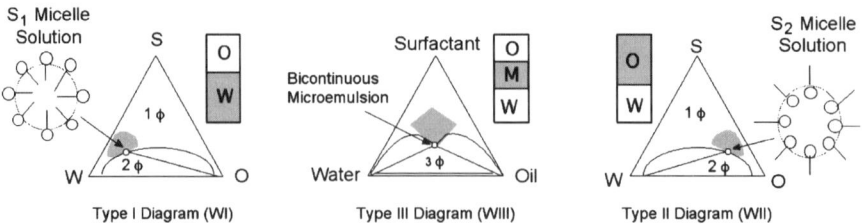

Fig. 1 Phase behavior types of surfactant–oil–water systems as Winsor Diagrams for diferent cases of the ratio R of interactions between the surfactant adsorbed at interface and the oil and water molecules

ables enhanced oil recovery by surfactant flooding [5]. The exact balance of affinity of the surfactant for the water and oil phases does not only depend on the surfactant, but also on the nature of the oil, electrolytes in water, the presence of alcohol cosurfactant, which is generally added to avoid the formation of mesophases, as well as temperature and in some instances even pressure. All these effects, which were treated qualitatively in Winsor's R ratio expression [3], were studied experimentally in the 1960s by Shinoda and collaborators through the phase inversion temperature (PIT) concept [6], and were numerically laid down in the late 1970's as the correlations for the attainment of three-phase behavior [7, 8].

Later these correlations were reported to be identical to the expression of the *surfactant affinity difference* (SAD) [9], i.e., the variation of Gibbs free energy when a surfactant molecule passes from oil to water:

$$\text{SAD} = \mu_w^* - \mu_o^* = \Delta \mu_{o \to w}^* = RT \ln X_o / X_w, \tag{2}$$

where the μ^* indicate the standard chemical potential at some reference concentration, and X is a dimensionless surfactant concentration, i.e., a concentration C, divided by some reference C_{ref}. With references taken as the concentration in the excess phases at equilibrium in a three-phase WIII system, a generalized dimensionless surfactant affinity difference, so-called *hydrophilic–lipophilic deviation* (HLD) has been defined [10]:

$$\text{HLD} = (\text{SAD} - \text{SAD}_{\text{ref}})/RT = \ln C_o / C_w - \ln C_{\text{oref}} / C_{\text{wref}} = \ln K - \ln K_{\text{ref}}, \tag{3}$$

where K_{ref} is the value of the partitioning coefficient between excess phases at optimum formulation. Since the partition coefficient can be experimentally measured, its variation can be established for all formulation variables [11–14].

The expression of *HLD* was found to match the empirical correlations found for the attainment of three-phase behavior for a SOW system containing a *n*-alkane, a NaCl brine and an anionic or cationic [7, 15]

$$\text{HLD} = \ln S - k\text{ACN} + f(A) + \sigma - a_T(T - 25), \tag{4}$$

or a polyethoxylated phenol or alcohol nonionic surfactant [8]

$$\text{HLD} = \alpha - \text{EON} - k\text{ACN} + bS + \phi(A) + c_T(T - 25), \tag{5}$$

where S is the salinity in wt.% NaCl, ACN is the alkane carbon number of the oil, T is the temperature, $f(A)$ and $\phi(A)$ are almost linear functions of the alcohol type and concentration. The characteristic parameter of the surfactant is σ for ionics, and α – EON for ethoxylated nonionics, EON being the average number of ethylene oxide group per surfactant molecule. Parameters a_T, b, c_T and k are constants.

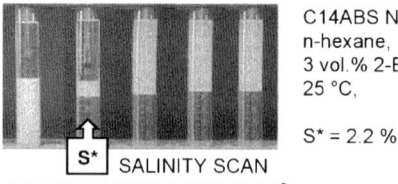

C14ABS Na
n-hexane,
3 vol.% 2-Butanol,
25 °C,

S* = 2.2 % NaCl

Fig. 2 Aspect of test tubes of a unidimensional salinity scan

Values of the constants and more sophisticated expressions have been reported for non-alkane oils, electrolytes other than NaCl, and a variety of alcohols and other conditions [5, 9, 16–26].

When HLD = 0 three-phase behavior takes place because of the exact compensation of the effect of the different formulation variables expressed in Eqs. 4 and 5. In practice, an unidimensional formulation *scan* is carried out to detect the occurrence of three-phase behavior. In such a scan, only one of the formulation variables (appearing in Eqs. 4 and 5) is changed, while all the others are held constant, as well as composition variables, i.e., surfactant concentration and water/oil ratio (WOR).

Figure 2 shows the test tube aspect of a salinity scan with an anionic surfactant at a concentration about 1 wt. % and for WOR = 1. In all test tubes the surfactant, oil, alcohol, and temperature are the same, i.e., in Eq. 4 all values are set but salinity. The test tube that exhibits three-phase behavior corresponds to the salinity $S^* = 2.2\%$ NaCl, so-called optimum salinity in this case, which satisfies HLD = 0 according to Eq. 4.

The absolute value of the characteristic parameter of the surfactant σ can be estimated from a single experiment by using Eq. 4 for HLD = 0 if all other variables values are known. For instance in the example of Fig. 2 scan: oil ACN = 6 for hexane, 3 vol % 2-butanol $f(A) = -0.16$, temperature 25 °C, and since the three-phase behavior is exhibited for the test tube with aqueous phase salinity $S = 2.2\%$ NaCl, then the surfactant parameter value is $\sigma = 0.32$.

This method requires knowledge of the characteristics values for the oil and alcohol effects, which is not always the case, in particular if some natural ill-defined product like a petroleum refinery cut is used. Alternatively, it might be impossible to attain three-phase behavior in the feasible experimental range, for instance the salinity that satisfies Eq. 4 might be too high to be attainable in practice. In such a case, another variable should be changed to keep the optimum value of the scan in the feasible range, for instance the introduction of another alcohol, which would alter the value of $f(A)$. However, this tends to introduce inaccuracies.

A second method, the so-called double scan technique, might be used instead. Its principle is to measure not an absolute value but a variation of the characteristic parameter of the surfactant. It should be noted first that if the surfactant is changed, then the value of σ in Eq. 4 is changed, and, for the

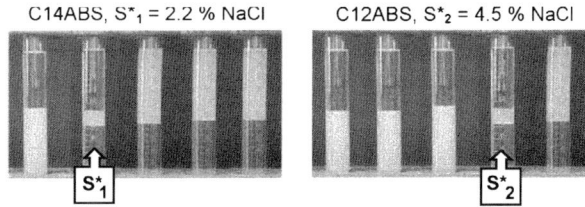

Fig. 3 Double scan technique (oil, alcohol and temperature are the same in both scans)

same oil, alcohol, and temperature, another value of salinity would satisfy Eq. 4 for the attainment of three-phase behavior at HLD = 0.

For instance, if the surfactant in Fig. 3 (left) scan is an alkyl-benzene sulfonate with a 14 carbon atom alkylate, and it is changed for a 12 carbon atom alkylate, then another scan with the same oil, alcohol, and temperature shows that the optimum salinity becomes $S_2 = 4.5$ wt. % NaCl (Fig. 3 right). It follows that $\ln S$ has increased by 0.72 unit, hence σ should have decreased by 0.72 unit according to

$$\ln 2.2 - k\text{ACN} + f(A) + \sigma_{(C_{14}S)} - a_T(T - 25) = 0 \tag{6}$$
$$\ln 4.5 - k\text{ACN} + f(A) + \sigma_{(C_{12}S)} - a_T(T - 25) = 0, \tag{7}$$

by substraction $\ln(4.5/2.2) = 0.72 = \sigma_{(C_{14}S)} - \sigma_{(C_{12}S)}$ and $\sigma_{(C_{12}S)} = -0.40$.

This double scan technique allows to experimentally estimate the characteristic parameter of an unknown surfactant from the value of a known one, which should preferably be close enough to make the change not too large. The advantage of the technique is that the contributions of the other variables (oil, alcohol, and temperature) do not need to be known accurately, because they are canceled out. This technique is particularly handy if the formulator has a series of known surfactants on hand that possess a wide range of characteristic values, from very hydrophilic to very lipophilic. It is recommended that the selected base surfactant have a characterisitic parameter close to the one of the unknown surfactant, so that the variation of optimum formulation does not go outside the feasible scanned variable range.

As expected from the relation of HLD with the free energy of transfer from oil to water, the removal of one carbon atom to the alkyl tail of the surfactant would make it easier for the surfactant to be transferred, and would hence tend to reduce the value of HLD, through a reduction of the characteristic parameter σ or α. The experience shows in effect that [23]:

$$\sigma/k = \sigma_0/k + 2.25 \text{ SACN} \tag{8}$$
$$\alpha/k = \alpha_0/k + 2.25 \text{ SACN}, \tag{9}$$

where SACN is the surfactant alkyl carbon number, i.e., the number of carbon atoms in the surfactant alkyl tail; the 0 susbcript indicates the extrapolation at SACN = 0 and is thus a characteristic of the surfactant head group and overall structure; the term k is introduced because it slightly depends on the head group of the surfactant, while Eq. 4 and Eq. 5 divided by the corresponding k values are expressed in ACN units, which are absolutely the same in all systems and allow a numerical comparison of the hydrophilicity of all kinds of surfactants. Table 1 lists the numerical values of surfactant characteristic parameters and k's. The linearity of Eqs. 8 and 9 is consistent with the energy nature of the terms in Eqs. 4 and 5 and the additivity of the different contributions.

The single and double scan methods allow estimation of the value of the characteristic parameters for both ionic and nonionic surfactants (see Table 1). The extrapolation of Eqs. 8 and 9 to SACN = 0 allows the classification of the head groups in some hydrophilicity scale. Inspection of Table 1 data indicates that the branching and isomeric structure does have quite an influence on the characteristic parameter, as it has been reported in the literature [12, 27–32].

This means that when isomerically ill-defined compounds are to be used, which is essentially the case with many commercial products, it is quite recommendable to estimate their characteristic parameter from some experience rather than to try to deduce it from an assumed or approximate structure.

It is worth noting that the scanned techniques presented to estimate the value of the characteristic parameter of a surfactant, can be used equally to determine the characteristic parameter of the oil phase, particularly if it is not an alkane, or of other effects. However, this is out of the scope of the present chapter.

The same principle may be extended to characterize mixtures of surfactants. If the case of anionic surfactants is still taken as an example, and if

Table 1 Experimentally determined values of parameters σ, α, k, a_T, c_T for some typical surfactants. (C_X represents a linear alkyl chain with X carbon atoms and iso-C_X an alkylate with a branched chain)

Surfactant	σ or α	k	σ/k or $[\alpha - \text{EON}]/k$	a_T or c_T
iso-C_{12} benzene sodium sulfonate	– 0.6	0.16	– 3.5	– 0.01
iso-C_{12} o-xylene sodium sulfonate	+ 0.5	0.16	+ 3.5	– 0.01
C_{12} sodium carboxylate	– 2.5	0.10	– 24	– 0.01
C_{12} sodium sulfonate	– 3.0	0.10	– 30	– 0.01
C_{12} ammonium chloride (at pH 3)	– 1.0	0.17	– 5	– 0.02
C_{12} trimethylammonium chloride	– 3.5	0.19	– 19	– 0.02
iso-C_{12} phenol EON = 5	+ 7.5	0.15	16.6	0.06
C_{12} alcohol EON = 5	+ 6.0	0.15	6.6	0.06
iso-C_9 phenol EON = 5	+ 6.5	0.15	10	0.05

the value of k is the same for the two base surfactants in the ionic mixture, then the optimum salinities for three-phase behavior of the scans with surfactant 1, surfactant 2 and a mixture M, would satisfy the following relationships:

$$\text{HLD} = \ln S_1 - k\text{ACN} + f(A) + \sigma_1 - a_T(T - 25) = 0 \tag{10}$$
$$\text{HLD} = \ln S_2 - k\text{ACN} + f(A) + \sigma_2 - a_T(T - 25) = 0 \tag{11}$$
$$\text{HLD} = \ln S_M - k\text{ACN} + f(A) + \sigma_M - a_T(T - 25) = 0, \tag{12}$$

in which σ_1 and σ_2 are known, and S_1, S_2, S_M are experimentally determined. The combination of Eqs. 10–12 with any of the previous ones allows one to determine the characteristic parameter value of the mixture σ_M. If the composition of the mixture is for instance given by the (mole) fraction of each species, i.e., x_1 and x_2 (with $x_1 + x_2 = 1$ for a binary surfactant mixture), then $\ln S_M$ or σ_M can be plotted against the mixture composition represented as x_M in Fig. 4.

If the relationship is linear, i.e., if

$$\ln S_M = x_1 \ln S_1 + x_2 \ln S_2, \tag{13}$$

it means that the characteristic parameter σ also follows a linear relationship, i.e.,

$$\sigma_M = x_1 \sigma_1 + x_2 \sigma_2. \tag{14}$$

If the experimental data indicate a linear relationship, such as Eq. 1, which will be considered as an "ideal" case, then it means that the mixing in the system produces exactly the same mixing at interface as in the whole system, and a so-called *collective* behavior of the surfactant species takes place.

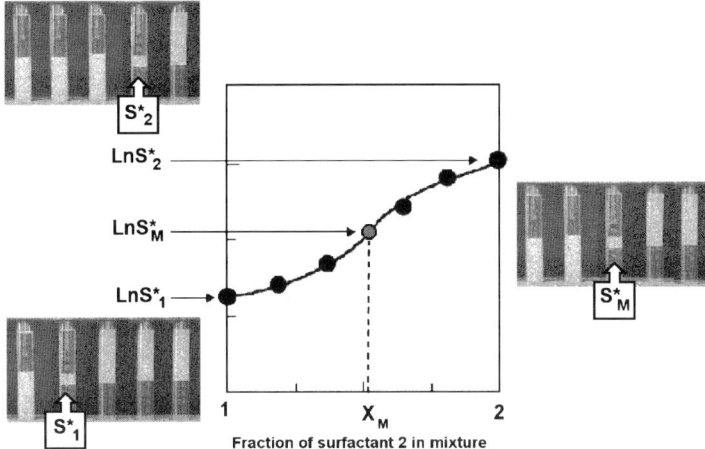

Fig. 4 Formulation scan for surfactant binary mixtures

It is worth noting that if the two surfactants which are mixed do not share the same value of k, as for instance an alkylbenzene sulfonate ($k = 0.16$) and an alkylsulfate ($k = 0.10$), then Eqs. 10–12 have to be divided by k to be expressed in ACN units which have the same meaning in all cases, whatever the k's [33]. The value σ/k has been called EPACNUS (Extrapolated Preferred Alkane Carbon Number at Unit Salinity and no-alcohol), since it is the value of ACN when the conditions $S = 1$ wt % NaCl, $f(A) = 0$ and $T = 25\,°C$ are satisfied. If the optimum formulation is determined experimentally, not from the occurrence of three-phase behavior but from the minimum of interfacial tension, as is often the case at low surfactant concentration or when the oil phase is not transparent, it is called $EACN_{min}$ or n_{min} [17, 34, 35].

This technique has been used to evaluate mixing rules [33] and, provided that the mixing rule is assumed to be linear, it is used to characterize surfactants which do not exhibit three-phase behavior alone, as will be discussed later [36, 37].

Ethoxylated nonionic surfactants approximately obey a linear mixing rule expressed as Eq. 1 when the characteristic property is the averaged number of ethylene oxide groups per molecules (EON) [35]. The goodness of the fit depends on the partitioning phenomena, which will be discussed later, in Sect. 4.

3
Collective Behavior and Ideal Mixing Rules

The collective behavior corresponds to the same mixture proportions in bulk phases as at interface, so that the mixture may be treated as a pseudocomponent. This implies the same partitioning for the different species [38].

As far as the formulation is concerned the linearity exhibited in Eq. 13 is fully equivalent to the linearity in characteristic parameter (Eq. 14), and is the easiest proof of "ideal" behavior. Figure 5 (left) shows the variation of $\ln S^*$ in a case of perfect linearity for binary mixtures of dodecylbenzene sulfonate and a petroleum sulfonate. When the oil phase is changed the relationship stays linear, and the line remains parallel to itself, since the change is just a change of ACN value in Eq. 12. Figure 5 (right), which shows the same plot for mixtures of two relatively pure surfactants, i.e., nonyl and pentadecyl-orthoxylene-sulfonates, exhibits a slight deviation to linearity that indicates a slight predominance of the short chain species. Both species are actually mixtures and the pentadecyl species contain a small amount of the next alkylates as propylene polymer isomers (C12 and C18). The C18 isomer is likely to partition into oil, hence disappearing from the interface, which thus results in a slightly more hydrophilic mixture and a higher $\ln S^*$. This deviation is observed at the current low concentration (0.02 M) and essentially disappears at higher concentration.

Surfactant Mixing Rules 93

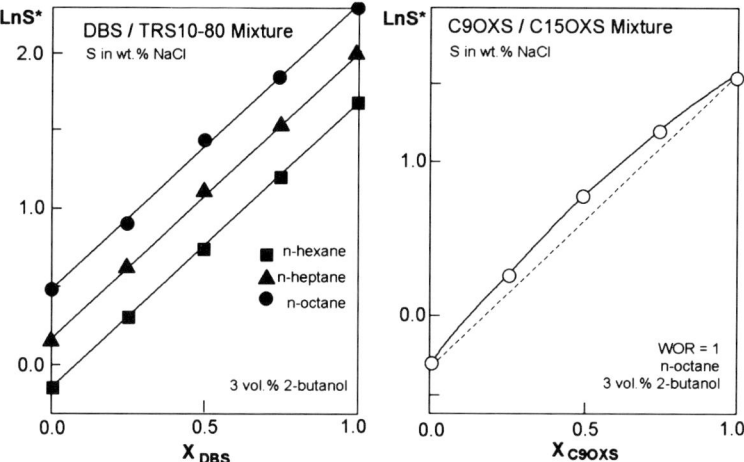

Fig. 5 Mixing rules for alkyl benzene sulfonate surfactants with same k values. After [39]

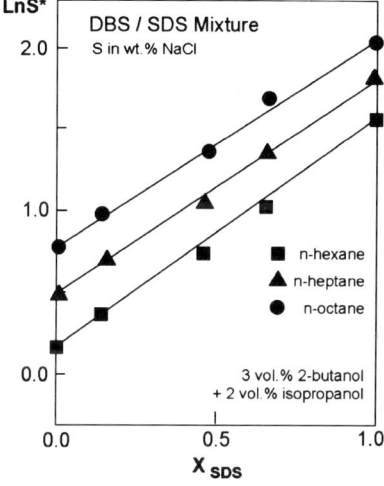

Fig. 6 Mixing rule for anionic surfactants with different k values. After [39]

Figure 6 indicates the variation of different mixtures of a dodecylbenzene sulfonate ($k = 0.16$) and a dodecyl sulfate ($k = 0.10$). The variations are linear, which indicates an ideal mixture as in Eq. 13 and Eq. 14. However, this time the lines expressing the mixing rule exhibit a slope which depends on the oil phase. This is due to the fact that the k's are not the same in Eqs. 10–12, which have to be combined after dividing them by their respective k. The k_M value for each mixture can be calculated as:

$$k_M = x_1 k_1 + x_2 k_2 \,. \tag{15}$$

These equations may be generalized to multicomponent systems with a linear mixing rule as in Eq. 1. It is worth remarking that the goodness of the fit tends to improve when the number of components of the mixture increases. For instance, the deviation shown in Fig. 5 (right) for the mixture of C9 and C15 ortho-xylene sulfonates, disappears if 20% of C12 ortho-xylene sulfonate is added. It is conjectured that the presence of intermediate species improves the collective behavior.

Nonionic surfactants can be mixed as well, but in practice most of them, at least the ethoxylated ones, are already a mixture because of the polycondensation mechanism in ethylene oxide adduction. Isomerically pure ethoxylates are extremely expensive and are exclusively reserved for research work. Little work has been carried out with mixture of isomerically pure nonionics and the bulk of the work on mixture deals with mixtures of commercial products, i.e., mixture of mixtures, which obey a linear mixing rule on EON, provided that the base mixtures are not too different [8, 35].

The main problem in mixing ethoxylated nonionics is that they often contain extremely different surfactant species that are likely to behave noncollectively, much more than in the case of ionic surfactants. This problem will be addressed in the next section. Again it seems that the presence of intermediate species tends to favor a collective behavior.

Figure 7 indicates the phase behavior of SOW systems containing ternary nonionic surfactant mixtures that in turn contain a very hydrophilic surfactant (Tween 60: Sorbitan + 20 EO stearate), a very lipophilic surfactant (Span 20: Sorbitan monolaurate), and an intermediate (Tween 85: Sorbitan 20 EO trioleate or Nonylphenol with an average of 5 EO groups). The two intermediate surfactants correspond exactly to an optimum formulation in the physicochemical conditions, i.e., they exhibit three-phase behavior with the system 1 wt. % NaCl brine–heptane–2-butanol. As the intermediate hydrophilicity surfactant is replaced by an equivalent mixture of the "extreme"

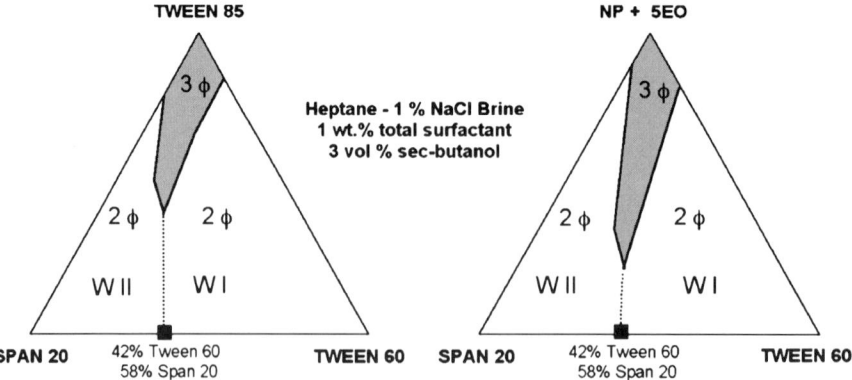

Fig. 7 Phase behavior of SOW systems containing ternary surfactant mixtures. After [40]

surfactants (downward path from the upper vertex in the diagrams), three-phase behavior still takes place at optimum formulation but its range diminishes. When there is less than a certain proportion of intermediate surfactant (40% for Tween 85 and 26% for NP + 5 EO), three-phase behavior no longer occurs at optimum formulation and the transition takes place directly from Winsor I to Winsor II phase behavior, at 42% Tween 60 and 58% Span 20. The vanishing of the three-phase behavior region indicates that the quality of the system is becoming too low, i.e., the interactions between the surfactant mixture and the oil and water phases are no longer sufficient to result in the formation of a microemulsion. From Fig. 7 data it can be said that the intermediate NP + 5 EO performs better than Tween 85, since a lesser amount is required to produce three-phase behavior. This concept of quality has been recently discussed in a review paper in terms of solubilization and interfacial tension [41].

It may be conjectured that collective behavior implies that the surfactants that make up the mixture are not too different, the presence of an intermediate being a way to reduce the discrepancy. When the activity coefficient is calculated from non-ideal models it is often taken to be proportional to the difference in solubility parameters [42, 43], which in case of a binary is the difference: $(\delta_1 - \delta_2)^2$; if the system is multicomponent, then the difference is $\sum(\delta_i - \delta_m)^2$, which is often less, because the mean value exhibits an average lower deviation. In other terms, it means that for a ternary in which the third term δ_3 is close to the average of the two first terms, then the introduction of the third component reduces the nonideality because $(\delta_1 - \delta_3)^2 + (\delta_2 - \delta_3)^2 < (\delta_1 - \delta_2)^2$.

4
The Main Source of Non-Ideality: Mixture Partitioning

Commercial ethoxylated nonionic surfactants, e.g., phenol or alcohol ethoxylates, generally exhibit a wide oligomer distribution, because of the mechanism of polycondensation of ethylene oxide. It means that a typical ethoxylated nonyl phenol with an average of 5 EO group per molecule might have oligomer species ranging from 1 EO to 12 EO as seen in Fig. 8, which illustrates the liquid chromatographic EO distribution analysis of such a commercial product.

Oligomer properties drastically depend on their ethoxylation degree. For instance, oligomers may be compared from the value of their partition coefficient in a typical water–oil system, which is the ratio of the oligomer concentrations in water and oil at ambient temperature and in absence of micelle ($K = C_w/C_o$). It is found that for oligomers with EON = 3 and EON = 7 the partition coefficients between water and heptane at 25 °C are respectively 0.003 and 0.17 [11].

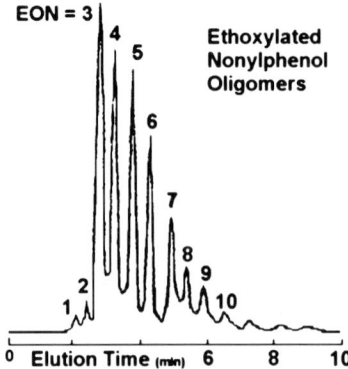

Fig. 8 Chromatogram showing the EON distribution of a commercial ethoxylated nonylphenol surfactant with an average EON of 5

Taking as a reference the EON = 5 oligomer, which is close to optimum formulation in a system with n-heptane and water at ambient temperature, and has a partition coefficient of 0.02, it means that the EON = 3 oligomer is 6 times less hydrophilic, and the EON = 7 oligomer 8 times more hydrophilic.

If the HLD scale is used to carry out comparisons in order to evaluate the change when passing from oligomer EON = 7 to oligomer 3, the surfactant parameter (α – EON) increases by 4 units, which, according to Eq. 9, is equivalent to adding 12 carbon atoms to the tail of the surfactant, or to adding 25 carbon atoms to the n-alkane oil. This indicates that there is a considerable difference in hydrophilicity between these two oligomers, and that they are likely to behave independently from each other, as far as their partitioning is concerned. Actually a portion of the distribution indicated in Fig. 8, i.e., the low EON range up to EON = 4, tends to partition into the oil phase, hence leaving more hydrophilic oligomers to adsorb at the interface with little left in water because of the very low CMC of these surfactants. As a consequence of the migration of low ethoxylation oligomers into the oil phase, the actual oligomer mixture at interface may exhibit an average EON one or two units above the average EON of the mixture that was put in the system in the first place. This phenomenon is out of the scope of the present report, but a detailed account may be found elsewhere [44–46].

Since this phenomenon takes place with all ethoxylated surfactants, the mixture of two commercial products would exhibit a similar loss of lipophilic oligomers. As a consequence, a mixture of two (commercial) mixtures, would approximately exhibit an apparent linear averaging rule which may be stated as:

$$\text{EON}_M = x_1 \text{EON}_1 + x_2 \text{EON}_2 , \qquad (16)$$

where EON now refers to the average EON value of each commercial product, which is itself a mixture. Equation 16 is fairly well satisfied if the two commercial base products are relatively close, as far as their degree of ethoxylation is concerned. On the contrary the mixing rule is not linear if the mixed products are extremely different, e.g., with EON = 2 and EON = 10. Figure 9 right plots illustrate the reason, which is that the two commercial products are not affected similarly by the partitioning phenomenon.

For each base commercial product the left (lipophilic) part of the distribution partitions into the oil phase and is thus lost as far as the adsorption at the interface is concerned. If the two products are not very different, then the part which is lost is not very different. On the contrary, in the case of a product with EON = 2 (as the left mode in Fig. 9 right plots), most of the this product distribution is lost to the oil, whereas very little is lost to the oil for the EON = 10 product (right mode in Fig. 9 right plots). Whatever the proportion of the EON = 2 and EON = 10 products in the mixture, the contribution of EON = 2 product to the interfacial mixture is always weak, because it essentially disappears into the oil. For instance if the mixture contains $x_1 = 50\%$ of EON = 2 product and $x_2 = 50\%$ of EON = 10 product, the actual average EON at interface will not be 6 but rather 8.

A way to reduce the impact of the selective partitioning of nonionics is to avoid mixing products that are too different, and if it is unavoidable, to introduce an intermediate product in-between. For instance if 20–30% of a EON = 5 commercial product (with a distribution indicated in Fig. 9 left plot) is added to the mixture of EON = 2 and EON = 10 (with a distribution indicated in Fig. 9 right plots), adjusting formulation becomes much easier. This is probably why the old literature [47], particularly in times of HLB pioneering use, advised to mix products with widely different characteristics. If this may help in attaining the right formulation, it is worth remembering that it favors the partitioning of the lipophilic species into the oil phase, and thus implies the use of more surfactant.

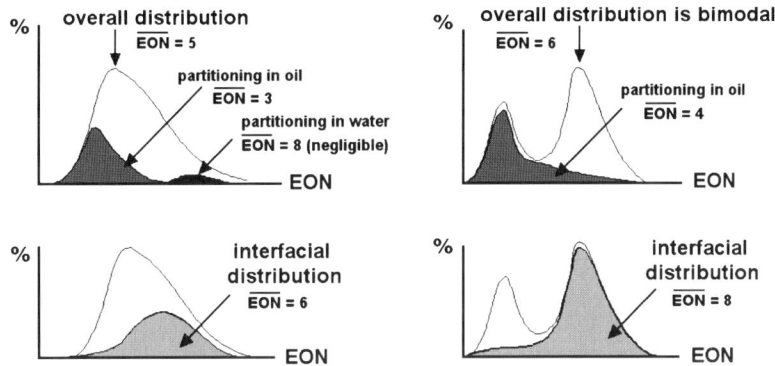

Fig. 9 Partitioning of lipophilic oligomers in the oil phase for two different products. *Left*: A monomodal mixture. *Right*: A bimodal mixture made up of very different commercial products, for instance with average EONs of 2 and 10

The partitioning can be revealed by different patterns that are found in experimental data. The first one is a reduction in solubilization because a part of the surfactant mixture is no longer at interface. For the same amount of surfactant in the system at optimum formulation, the volume of microemulson middle phase is lower, just because a certain proportion of the surfactant is no longer in the microemulsion, but has partitioned into one of the excess phases. However, it is worth noting that there are other reasons for the solubilization to decrease, hence this is only a hint.

Since partitioning is altered by the total concentration of surfactant [44], the optimum formulation for three-phase behavior or for minimum interfacial tension is likely to change with concentration [8, 48]. Thus, optimum salinity, optimum ACN, or optimum EON change with the concentration of surfactant if it is a mixture, whereas it is independent of the concentration for an isomerically pure surfactant, as shown in Fig. 10 left and center plots.

Reducing the surfactant mixture total concentration generally tends to increase partitioning. Consequently the change in formulation with concentration tends to be more severe as the concentration decreases, and on the other hand generally vanishes above 5–6% of surfactant [8]. This is also evidenced in the so-called fish or gamma phase behavior diagram, in which the optimum formulation line at the center of the one-phase or three-phase behavior zone tends to become more twisted at low concentration [49–51]. Figure 11 indicates that the optimum formulation (dashed) line departs more and more from a constant formulation or constant temperature as surfactant concentration decreases. Such twisting is not found to be significant in fish diagrams corresponding to isomerically pure surfactants of the ethoxylated alcohol type [52, 53], even when the concentration range is extremely wide.

The reduction of surfactant mixture concentration produces opposite formulation drifts for ionic and nonionic surfactants. In effect, lipophilic nonionic species tend to partition into oil, and very little is left in water because of

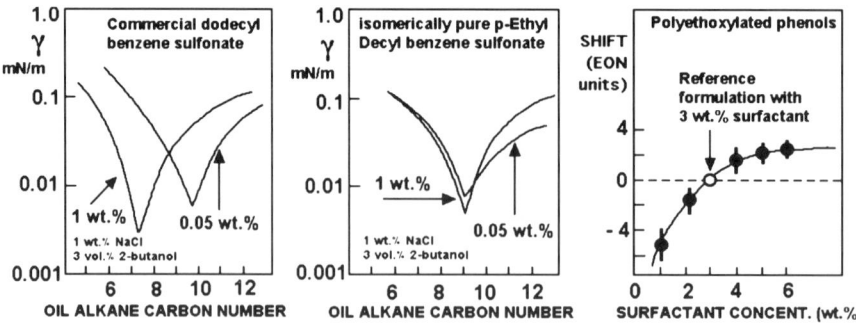

Fig. 10 Effect of surfactant concentration on the optimum formulation (minimum tension position) for anionic mixtures (*left*), pure anionic surfactant (*center*), and ethoxylated nonionic mixtures (*right*)

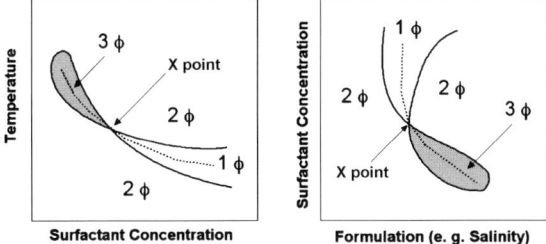

Fig. 11 Fish and gamma phase behavior diagrams exhibiting a severe twisting at low surfactant concentration

a low CMC, hence the interface becomes more hydrophilic. On the contrary, hydrophilic ionics (for instance alkylbenzene disulfonates) tend to partition into water while very little goes to oil because of a low CMC in oil, and as a result the interface becomes more lipophilic. As a consequence, an increase in partitioning tends to turn the interfacial nonionic surfactant mixture more hydrophilic and the ionic one more lipophilic. These opposite tendencies when surfactant concentration changes have been advantageously used to reduce the sensitivity to concentration by mixing the two kinds of surfactants, so that the effects cancel out [54].

The water–oil ratio (WOR) also has an effect on the surfactant partitioning. If the system contains more oil, the partitioning into the oil phase of the lipophilic species of a polyethoxylated commercial surfactant tends to increase, hence the surfactant mixture left to go to adsorb at the interface has a higher EON average than the surfactant mixture that was put in the system in the first place. As a consequence, it is necessary to reduce the EON of the whole product to maintain the interfacial EON constant. This means that the required average EON for optimum formulation tends to be reduced when the system contains more oil. Accordingly, the formulation (EON)–WOR or Temperature–WOR maps exhibit a three-phase band which is not at constant overall EON, but whose overall EON increases as WOR increases, as a way to keep the interfacial EON constant [51, 55–59]. The strong slanting of this band is a tell-tale sign that partitioning is severe in systems, as seen in Fig. 12, for polyethoxylated nonionics (center and right maps) or mixtures of a hydrophilic anionic surfactant such as sodium dodecyl sulfate and a lipophilic cosurfactant, such as *n*-pentanol (left map).

It is worth noting that this phenomenon might be important in applications in which the WOR inevitably changes, such as in rinsing after washing, or injecting a surfactant solution in a petroleum reservoir. The formulation changes with the dilution, and the effect has to be controlled or at least predicted so that the change takes place the right way.

Extreme partitioning has been found when the oil phase exhibits a particularly good interaction with the nonionic polar groups, as is the case with aromatic oils. A continuous change in oil mixture from hexane (ACN = 6)

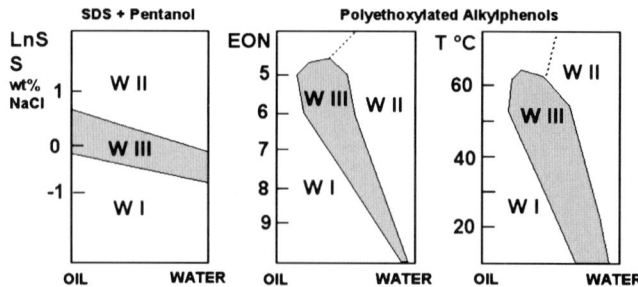

Fig. 12 Phase behavior in a formulation-WOR map. Slanting of the three-phase behavior zone indicates partitioning. After [56–58]

to benzene (EACN = 0), is expected to produce a reduction of the oil EACN, and thus to result in an increase in interaction between the surfactant and the oil, that results in a transition WI → WIII → WII. However, a second phenomenon takes place when the oil becomes more aromatic, which is an increased partitioning of the lipophilic oligomers of the polyethoxylated species into the oil phase, and hence an increase in the average EON of the remaining surfactant that adsorbs at interface. This second phenomenon just tends to favor the opposite transition, i.e., WII → WIII → WI. If the two transitions take place exactly at the same formulation, i.e., if the two WIII states coincide, then a WI → WIII → WI transition is observed, with the monotonous change in oil mixture from hexane to benzene. This curious phenomenon, which has been called *retrograde transition*, has been observed in different kinds of scans (EACN, alcohol, and temperature) with ethoxylated nonionic surfactant systems [60, 61]. It results in an apparent increase in hydrophilicity, though the opposite is expected from the general trends.

This phenomenon often result in the vanishing of the three-phase behavior region as in Fig. 12 when the oil content is high at low EON or high temperature [57, 58]. This is essentially the same as in Fig. 7, in which partitioning is reduced as a third intermediate hydrophilicity surfactant is introduced in the system. Moreover, a large partitioning of the surfactant into the oil phase, particularly at high surfactant concentration, could also modify the oil phase itself, turning it more polar, since the lipophilic surfactant species might be viewed as polar oils, which accumulate close to the interface as explained next.

It is worth noting here that this difference between the interface and in the bulk is not specific to surfactant mixtures. While oil mixtures of very similar substances, such as n-alkanes, exhibit a linear mixing rule written in terms of equivalent alkane carbon number or EACN [62–64], mixtures of oils containing substances with very different polarities behave in a non-ideal way and exhibit a segregation near the interface, which results in an accumulation most polar oil components close to the interface [65].

This effect can be of great importance, because it is susceptible to considerable alteration of the surfactant interaction between oil and water, and the solubilization in microemulsion (as in the so-called lipophilic and hydrophilic *linker* mechanisms). The role of the linker molecules is to extend the reach of the surfactant in the bulk phase and in practice to somehow modify the oil and water phases close to the interface, so that their characteristic parameters are altered [66–69].

The segregation of the different molecules according to their polarity is just the opposite of random mixing and is thus a clear sign of uncollective behavior characteristic of non-ideal mixing.

5
Deviations from Ideal Mixing Rules

5.1
Anionic-Nonionic Mixtures (Weakly non-Linear)

Ionic surfactants are sensitive to water hardness whereas polyethoxylated surfactant are not. Hence the mixing of both types often result in formulations that are salt tolerant for applications such as detergency or enhanced oil recovery.

If the mixing is followed (as for ionic surfactants) by carrying out a salinity scan, the aspect of the $\ln S^*$-mixture composition rule is often found to be non-linear as indicated in Fig. 13 left plot.

The optimum salinity variation does not follow a straight line on a $\ln S$ scale (nor with a S scale), but exhibits a downward deviation, i.e., it displays a value of optimum salinity which is systematically lower than expected from the use of Eqs. 4 and 5 [33]. This indicates that the anionic-nonionic association tends to reduce the hydrophilicity of the mixture, and it has been

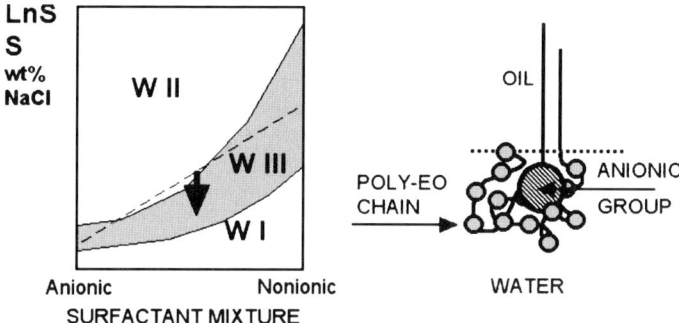

Fig. 13 Plot of optimum formulation (as optimum salinity) versus nonionic-ionic mixture composition. After [33]

suggested than the polyether chain wraps around the ionic head group and reduces its interaction with the water phase, as suggested by the illustration on the right of Fig. 13.

There is no available modeling of this effect, and only some qualitative trends can be provided by the reports on mixed surfactant micellization.

It has been suggested that this deviation to linear mixing rules comes from the fact that the nonionic species partitions to a much larger extent than the ionic one [51, 70].

It is worth noting that the effect of temperature on ionic and polyethoxylated nonionic surfactants is just opposite. As temperature increases, the nonionics become more lipophilic whereas the ionics turn more hydrophilic. By mixing the two types of surfactants in a proper proportion, these effects could cancel each other out, and the mixture is said to be insensitive to temperature. This interesting feature of ionic-nonionic surfactant mixtures may be considered as a synergy, since it could be very important in practice. Analysis of this feature is not included here, because plenty of information may be found in the literature on applications of such mixtures to equilibrated and emulsified systems [10, 71–74].

5.2
pH Sensitive System (Highly Non-linear because pH Scale is Logarithmical)

The phase behavior of systems containing pH-sensitive surfactants is another example of non-linearity of the mixing rule. If an oil phase containing an amphiphilic molecule, such as an organic acid, as in the case of naphtenic acids in crude oils, is put into contact with an alkaline water phase, the neutralization takes place at interface and results in a mixture of unneutralized acid (the lipophilic component) and its dissociated alkaline salt (the hydrophilic component). Hence, the interface contains a mixture of two surfactants whose relative proportion depends on the ionization (in the water and at interface), and thus of the pH [75].

If the organic acid is symbolized by AH and its ionized salt by A^-, then the equilibrium in the aqueous phase is given by

$$K_a = \frac{[A^-][H^+]}{[AH]}, \qquad (17)$$

where K_a is the dissociation constant in water (10^{-5} to 10^{-6} for fatty acids with more than 10 carbons atoms). The higher the pH, the larger the amount of hydrophilic ionic soap in water, and accordingly at interface. However, since the A^- and AH species compete for the interface, the most likely to adsorb is the lipophilic AH species. Consequently, a 90% "advantageous" proportion of the A^- species in water (attained at pH = pK_a + 1), might not be enough to attain a 50% proportion of A^- at interface. As a matter of fact the shift between the water composition and the interface composition has been related

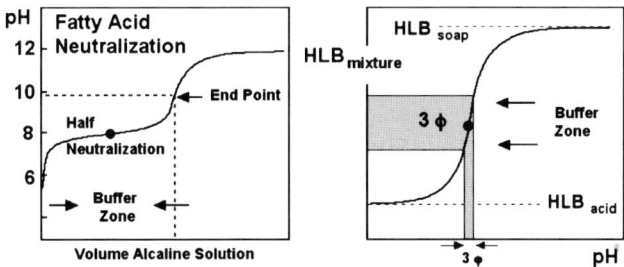

Fig. 14 Neutralization of a fatty acid (*left*) and resulting interfacial HLB in a SOW system as a funcion of pH

to the partition coefficient of the AH species between oil and water. Since this partition coefficient $P_a = [AH_o]/[AH_w]$ is at least 1000, the pH at which the composition of the interface is roughly 50% of each specie is about $pK_a + 3$ or higher [76].

The problem in the present case is that the variation of the pH with the amount of alkaline molecules to produce neutralization is the typical S-shape shown in Fig. 14.

The point at which, supposedly, 50% of the acid species is transformed in salt corresponds to the half-neutralization, i.e., when half the alkaline required to reach the equivalence point has been added. This position corresponds to a buffer zone in which the variation of pH is small with respect to the amount of added neutralization solution (Fig. 14 left plot). Hence, in this region a very slight variation of pH can produce a very large variation of neutralization (Fig. 14 right plot), i.e., a considerable alteration of the relative proportion of AH and A^-. Far away from this pH, the opposite occurs. Consequently, the pH could be used to carry out a formulation scan, but the scale is far from linear and the variation of pH does not render the variation of the characteristic parameter of the actual surfactant mixture that is at interface [77, 78]. The appropriate understanding of the behavior of this kind of acid–salt mixture is particularly important in enhanced oil recovery by alkaline flooding [79, 80] and emulsification processes that make use of the acids contained in the crude oils [81–83].

5.3
Anionic–Cationic Mixtures: Highly Non-Linear Synergy but Amenable to Linear

When anionic and cationic surfactants are mixed they strongly interact, and it could even be said that they almost react to produce a *catanionic* species, which is essentially a new surfactant, in which the two hydrophilic groups have merged into a nonionic or a somehow amphoteric head attached to a double tail. The first consequence is that the new daughter species has

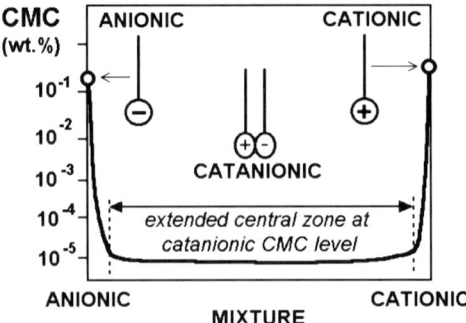

Fig. 15 CMC of anionic–cationic surfactant mixtures as a function of composition of the mixture

completely different properties than its parents: it is much more lipophilic, often non-water soluble, and less sensitive to electrolytes. When both parent species have straight chains with approximately the same length, crystals form readily and a precipitation takes place even at very low concentration.

The CMC of this new surfactant is several orders of magnitude lower than the CMC of its parent species. Figure 15 indicates a typical CMC plot versus the composition of the anionic–cationic (e.g., dodecyl sulfate-tetradecyl trimethyl ammomnium chloride) mixture in water. It can be seen that the CMCs of the anionic and cationic species are quite high, e.g., around 0.1 wt. %. As soon as a very small percentage of cationic is added to an anionic solution, the CMC falls several orders of magnitude. The same happens when a very small amount of anionic is added to a cationic solution. In both cases it seems that an equimolar catanionic species forms, and that its very low CMC dominates the mixing rule [84].

The phase behavior of anionic–cationic surfactant mixture/alcohol/oil/water systems exhibit a similar effect. First of all, it should be mentioned that because of the low solubility of the catanionic compound, it tends to precipitate in absence of co-surfactant, such as a short alcohol. When a small amount of cationic surfactant is added to a SOW system containing an anionic surfactant and alcohol (A), three-phase behavior is exhibited at the proper formulation, and the effect of the added cationic surfactant may be deduced from the variation of the optimum salinity (S^*) for three-phase behavior as in Figs. 5–6 plots. Figure 16 (left) shows that when some cationic surfactant is added to a SOWA system containing mostly an anionic surfactant, the value of $\ln S^*$ decreases strongly, which is an indication of a reduction in hydrophilicity of the surfactant mixture. The same happens when a small amount of anionic surfactant is added to a SOWA system containing mostly a cationic surfactant. As seen in Fig. 16 (left), the values of $\ln S^*$ at which the parent anionic and cationic surfactant systems exhibit three-phase behavior are quite high, which means that both base surfactants, e.g., dodecyl sulfate

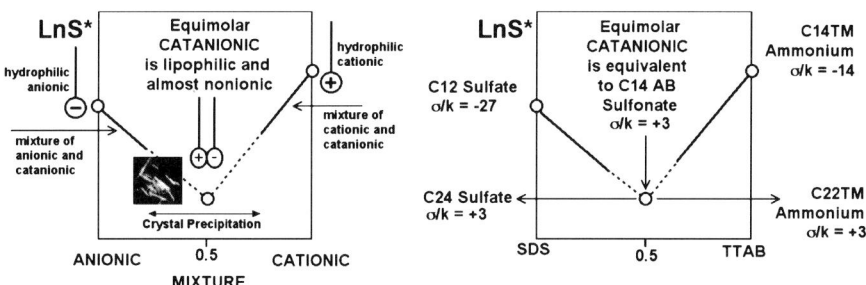

Fig. 16 ln S^* vs. composition plot for an anionic–cationic mixture. After [87]

and tetradecyl trimethyl ammonium, are quite hydrophilic. When they are mixed, ln S^* decreases and a precipitation takes place in the central region of the plot in spite of the presence of alcohol, even in high concentration.

However, if the trend of variation of ln S^* with composition is extrapolated (dashed lines in Fig. 16), it is found that the two lines starting from pure anionic and pure cationic, cross exactly at an equimolar mixture composition. It may be said that on the left side of the diagram (more anionic than cationic) there is a linear mixing rule between the pure anionic and the equimolar compound, whereas in the right part of the diagram the linear mixing rule is between the cationic and the equimolar compound [85, 86].

The equimolar compound is somehow virtual since it cannot produce a microemulsion alone, but instead a crystal, even with a high concentration of alcohol. It might not exist as an isolated species, but only in some kind of interfacial tesselation. However, for the sake of simplicity of the mixing rule, it may be considered to exist in mixtures with a "large" amount or either anionic or cationic species. The meaning of "large" is up to now about 70% or more, though there is no clear-cut limit to it. For instance, it may be conjectured that unequal or branched tail parent surfactants might produce an equimolar species able to produce a microemulsion alone.

Thanks to this linearity, the mixtures of anionic and cationic surfactants could be handled the same way in practice as other mixtures, provided that the collected data allow the positioning of the equimolar compound and that the rule is expressed as a mixture between the surfactant in excess and the equimolar compound. Since the value of ln S^* allows the calculation of the surfactant parameter σ/k, the catanionic species can be characterized by a σ/k parameter value, which is evidently quite higher, i.e., more lipophilic, according to the data shown in Fig. 16 (right). For a mixture of C12 anionic and C14 cationic, the catanionic equimolar compound is from its σ/k parameter value equivalent to a C24 anionic with same sulfate head group, or to a C22 corresponding cationic. The loss of hydrophilicity is thus equivalent to adding approximately 10 carbon atoms to the tail, which is roughly what is added to the first surfactant when the second sticks with it. Consequently, it may be said in a first aproximation that the head group of the catanionic species is roughly as hydrophilic as one of the parent head groups. Since the hydrophilicity is mostly due to the ionic character of the head group, it means that about half the ionization of the head groups has been lost in the association. Accordingly, a better view of the catanionic species is as an amphoteric surfactant with roughly one half positive charge and one half negative charge.

5.4
pH Sensitive Anionic–Cationic Mixtures – a Complex Situation

This last case is a combination of the two previous ones, in which the pH has an opposite effect on two surfactants. As shown in Fig. 17 (upper part) and discussed in Sect. 5.2 an increase in pH increases the ionization of the fatty acid, i.e., the proportion of the ionized hydrophilic soap, and hence the hydrophilicity of the acid–soap mixture in the water phase and consequently at the interface. At low pH, the acid, i.e., a lipophilic nonionic surfactant, prevails, whereas at high pH, it is the hydrophilic soap that dominates the formulation. The pH at which about half of the interfacial mixture is acid and half soap, i.e., the pH at which the interfacial mixture is at optimum formulation (and three-phase behavior is exhibited), is called pH* in Fig. 17.

Fatty amine are nonionic at high pH, but they get protonated at low pH and become the ionizated amine salt, which is a hydrophilic cationic surfactant. Figure 17 (lower part) indicates that the phenomenology with fatty amines is absolutely identical to the previous one with fatty acid if pOH substitutes the pH [76].

Hence there is a pH above which the acid–soap is mostly anionic, and another pH below which the amine–salt is mostly cationic. If both ionizations happen at a same pH the combination of the two ionic species will result in the formation of the catanionic compound.

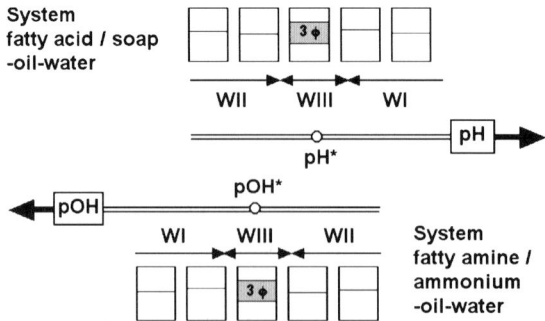

Fig. 17 Effect of a pH scan in both faty acid and fatty amine/oil/water systems

This happens when the pH* of the acid–soap system is lower than the pH* of the amine–salt systems. This is the case of Fig. 17, if the two pH scales are assumed to be coincident. For some intermediate pH both systems exhibit a WI phase behavior, that is to say that they contain a high percentage of the corresponding ionized species. If both the acid and amine are placed in the same system at such a pH, the two ionic species would combine to produce a catanionic one, in equilibrium with the nonionic acid and amine species. Therefore, five surfactants would be present in the system, with relative proportions directly linked to the pH through the dissociation equilibria, the partitioning equilibria, and the catanionic association equilibrium. How the phase behavior is altered by the pH through this complex scheme, does not seem easy to deduce and an experimental approach is surely the safer one.

Figure 18 shows a two-dimensional map in which the phase behavior is plotted for a SOWA system containing mixtures of octyl acid and octyl amine for which the pH* of the acid–soap transition (pH* = 7) is lower than the pH* for the amine–ammonium transition (pH* = 9). Hence, in the 7–9 range of pH, both the acid–soap and amine–ammonium systems exhibits a WI phase behavior, which means a predominance of the ionic species.

At pH = 8 with the amine–ammonium system (upper part of the diagram) the phase behavior is WI. As some acid is added, it is mostly ionized and the soap combines with the ammonium to produce the catanionic compound, which is less hydrophilic. The combination of the ammonium with the catanionic species results in a less and less hydrophilic mixture as more acid is added, up to the point where the formulation becomes balanced and three-phase behavior occurs. A further addition of acid results in a WII phase behavior and some precipitation of the excess of catanionic species whose proportion keeps increasing. The same happens by starting at pH = 8 on the lower part of the diagram with pure acid–soap system, and moving upward

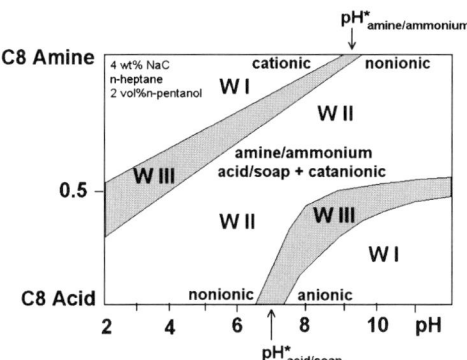

Fig. 18 Phase behavior map for a mixture of two pH sensitive surfactants with antagonistic dissociation behavior. After [87]

by adding some amine. Thus it can be said that at pH = 8 the change in acid–amine proportion produces a sequence of two phase behavior transitions, i.e., WI → WIII → WII → WIII → WI. Because the direction of change of HLD with pH is opposite for the two base cases, and since the pH* for three-phase behavior of the amine–ammonium and acid–soap are not the same, the two WIII bands are slanted as indicated in Fig. 18.

It is worth noting that a full pH range scan for an equimolar proportion of acid and amine, i.e., a horizontal cut in the middle of the map, would result in a similar double transition, with an extremely wide central region with WII phase behavior and precipitated catanionic species, with two WIII bands at high and low pH, and eventually a small WI region at extreme high and low pH values.

The other cases, i.e., when the pH* of both systems are equal or in opposite order, may be deduced easily from Fig. 18 map, since it is essentially the same case with a central WII region which covers all acid-amine proportions in the mid-pH range.

Acknowledgements The authors would like to acknowledge the financial support of their University Research Council CDCHT-ULA through grant I-834-05-08-AA.

References

1. Salager JL (1996) Quantifying the Concept of Physico-Chemical Formulation in Surfactant–Oil–Water Systems. Prog Colloid Polym Sci 100:137–142
2. Shah DO, Schechter RS (eds) (1977) Improved Oil Recovery by Surfactant and Polymer Flooding. Academic Press, New York
3. Winsor P (1954) Solvent Properties of Amphiphilic Compounds. Butterworth, London
4. Bourrel M, Schechter RS (1988) Microemulsions and Related Systems. Marcel Dekker, New York
5. Reed RL, Healy RN (1977) Some physicochemical aspects of microemulsion flooding: a review. In: Shah DO, Schechter RS (eds) Improved Oil Recovery by Surfactant and Polymer Flooding. Academic Press, New York, pp 383–347
6. Shinoda K, Kunieda H (1983) Phase Properties of Emulsions: PIT and HLB. In: Becher P (ed) Encyclopedia of Emulsion Technology, vol 1: Basic Theory, Chap. 5. Marcel Dekker, NY, pp 337–367
7. Salager JL, Vasquez E, Morgan J, Schechter RS, Wade WH (1979) Optimum formulation of surfactant–water–oil systems for minimum interfacial tension and phase behavior. Soc Petrol Eng J 19:107–115
8. Bourrel M, Salager JL, Schechter RS, Wade WH (1980) A Correlation for Phase Behavior of Nonionic Surfactants. J Colloid Interface Sci 75:451–461
9. Salager JL (1999) Microemulsions. In: Broze G (ed) Handbook of Detergents – Part A: Properties. Surfactant Sci Ser, vol 82, Chap 8. Marcel Dekker, New York, pp 253–302
10. Salager JL, Márquez N, Graciaa A, Lachaise J (2000) Partitioning of ethoxylated octylphenol Surfactants in Microemulsion–oil–water Systems. Influence of Temperature and relation between Partitioning Coefficient and Physicochemical Formulation. Langmuir 16:5534–5539

11. Márquez N, Antón RE, Graciaa A, Lachaise J, Salager JL (1995) Partitioning of Ethoxylated Alkyl Phenol Surfactants in Microemulsion–oil–water systems. Colloid Surface A 100:225–231
12. Márquez N, Antón RE, Graciaa A, Lachaise J, Salager JL (1998) Partitioning of Ethoxylated Alkyl Phenol Surfactants in Microemulsion–oil–water systems. Part II. Influence of Hydrophobe Branching. Colloid Surface A 131:45–49
13. Márquez N, Graciaa A, Lachaise J, Salager JL (2002) Partitioning of ethoxylated alkylphenol surfactants in microemulsion–oil–water systems: Influence of physicochemical formulation variables. Langmuir 18:6021–6024
14. Márquez N, Bravo B, Ysambertt F, Chávez G, Subero N, Salager JL (2003) Analysis of Polyethoxylated Surfactants in Microemulsion–oil–water Systemas. Part III. Fractionation and Partitioning of Polyethoxylated Alcohol Surfactants. Anal Chim Acta 477:293–303
15. Antón RE, Garcés N, Yajure A (1997) A correlation for three-phase behavior of cationic surfactant–oil–water systems. J Dispers Sci Technol 18:539–555
16. Marzdall L (1977) The effect of alcohols on the hydrophilic–lipophilic balance of nonionic surfactants. J Colloid Interface Sci 60:570–573
17. Hayes M, El-Emary M, Schechter RS, Wade WH (1979) The Relation between the EACNmin concept and Surfactant HLB. J Colloid Interface Sci 68(3):591–592
18. Antón RE, Salager JL (1990) Effect of the electrolyte anion on the salinity contribution to optimum formulation of anionic surfactant microemulsions. J Colloid Interface Sci 140:75–81
19. Bavière M, Schechter RS, Wade WH (1981) The influence of alcohols on microemulsion composition. J Colloid Interface Sci 81:266–299
20. Fotland P, Skauge A (1986) Ultralow interfacial tension as a function of pressure. J Dispers Sci Technol 7:563–579
21. Skauge A, Fotland P (1990) Effect of Pressure and Temperature on the Phase Behavior of Microemulsions. SPE Reserv Engin 5:601–608
22. Salager JL, Antón RE (1999) Ionic Microemulsions. In: Kumar P, Mittal K (eds) Handbook of Microemulsions Science and Technology, Chap 8. Marcel Dekker, New York, pp 247–280
23. Salager JL, Antón RE, Andérez JM, Aubry JM (2001) Formulation des microémulsions par la méthode HLD. In: Techniques de l'Ingénieur. Thème: Chimie et Bio. Base: Formulation. Dossier J2157. Editions T.I. Paris
24. Nardello V, Chailloux N, Poprawski J, Salager JL, Aubry JM (2003) HLD concept as a tool for the characterization of cosmetic hydrocarbon oils. Polym Int 52:602–609
25. Pierlot C, Poprawski J, Catté M, Salager JL, Aubry JM (2003) Experimental design for the determination of the physicochemical parameters of optimum water–oil surfactant systems. Polym Int 52:614–618
26. Poprawski J, Catté M, Marquez L, Marti MJ, Salager JL, Aubry JM (2003) Application of hydrophilic–lipophilic deviation formulation concept to microemulsions containing pine oil and nonionic surfactants. Polym Int 52:629–632
27. Doe PH, Wade WH, Schechter RS (1977) Alkyl benzene sulfonates for producing low interfacial tensions between hydrocarbons and water. J Colloid Interface Sci 59:525–531
28. Doe P, El-Emary M, Wade WH, Schechter RS (1977) Surfactants for producing low interfacial tension II. Linear alkylbenzene sulfonates with additional alkyl groups. J Am Oil Chem Soc 54:570
29. Doe P, El-Emary M, Wade WH, Schechter RS (1978) Surfactants for producing low interfacial tension III. Di- and Tri-alkylbenzene sulfonates. J Am Oil Chem Soc 55:513

30. Graciaa A, Barakat Y, El-Emary M, Fortney L, Schechter RS, Yiv S, Wade WH (1982) HLB, CMC and phase behavior as related to hydrophobe branching. J Colloid Interface Sci 89:209–216
31. Johansson I (2004) Does Hydrophobe Branching make a Surfactant more or less Hydrophilic? Spec Chem Mag 11:38–40
32. Tropsch J, Baur R (2004) How does branching influence surfactant properties? Isotridecanols as surfactant base alcohols. CD Proc 6th World Surfactant Congress CESIO, Berlin, Germany, June 21–23
33. Salager JL, Bourrel M, Schechter RS, Wade WH (1979) Mixing rules for optimum phase behavior formulations of surfactant–oil–water systems. Soc Petrol Eng J 19:271–278
34. Wade WH, Morgan JC, Jacobson JK, Schechter RS (1977) Low interfacial tension involving mixtures of surfactants. Soc Petrol Eng J 17:122
35. Hayes M, Bourrel M, El-Emary M, Schechter RS, Wade WH (1979) Interfacial tension and behavior of nonionic surfactants. Soc Petrol Eng J 19:349–356
36. Antón RE, Salager JL (1985) An Improved Graphic Method to Characterize a Surfactant. J Dispers Sci Technol 6:245–253
37. Salager JL, Antón RE (1983) Physico-Chemical Characterization of a Surfactant – a quick and precise method. J Dispers Sci Technol 4:253–273
38. Koukounis C, Wade WH, Schecheter RS (1983) Phase Partitioning of anionic and nonionic Surfactant Mixtures. Soc Petrol Eng J 23:301–310
39. Salager JL (1977) Physico-chemical properties of surfactant–oil–water mixture: phase behavior, microemulsion formation and interfacial tension. PhD Dissertation, University of Texas at Austin
40. Martinez G (1986) Existencia de Comportamiento trifásico en sistemas surfactante–agua–aceite. MSc Thesis, Universidad de Los Andes, Mérida Venezuela
41. Salager JL, Antón RE, Sabatini DA, Harwell JH, Acosta E, Tolosa L (2005) Enhancing Solubilization in Microemulsions – State of the Art and Current Trends. J Surfact Deterg 8(1):3–21
42. Barton AF (1983) Handbook of Solubility Parameters and other Parameters. CRC Press, Boca Raton
43. Hansen CM (1967) The three-dimensional solubility parameter – Key to paint component affinities I. Solvents, polymers, and resins. J Paint Technol 39:104
44. Graciaa A, Lachaise J, Sayous JG, Grenier P, Yiv S, Schechter RS, Wade WH (1983) The partitioning of complex surfactant mixtures between oil–water–microemulsion phases at high surfactant concentration. J Colloid Interface Sci 93:474–486
45. Graciaa A, Lachaise J, Bourrel M, Osborne-Lee I, Schechter RS, Wade WH (1987) Partitioning of nonionic and anionic surfactant mixtures between oil/microemulsion/water phases. SPE Reserv Eng 2:305–331
46. Graciaa A, Andérez JM, Bracho C, Lachaise J, Salager JL, Tolosa L, Ysambertt F (2006) The Selective Partitioning of the Oligomers of Polyethoxylated Surfactant Mixtures between Interface and Oil and Water bulk Phases. Adv Colloid Interface Sci 123–126:63–73
47. Becher P (1977) Emulsions: Theory and Practice, reprint 2nd ed. Robert Krieger Publishing Co. Huntington NY
48. Wade WH, Morgan J, Schechter RS, Jacobson JK, Salager JL (1978) Interfacial tension and phase behavior of surfactant systems. Soc Petrol Eng J 18:242
49. Bourrel M, Chambu C, Schechter RS, Wade WH (1982) The Topology of Phase Boundaries for oil/brine/surfactant Systems and its Relationship to Oil Recovery. Soc Petrol Eng J 22:28–36

50. Bourrel M, Chambu C (1983) The Rules for Achieving High Solubilization of Brine and Oil by Amphiphilic Molecules. Soc Petrol Eng J 23:327–338
51. Kunieda H, Shinoda K (1985) Evaluation of the hydrophile–lipophile balance (HLB) of nonionic surfactants I. Multisurfactant systems. J Colloid Interface Sci 107:107–121
52. Kahlweit M, Strey R, Firman P (1986) Search for tricritical points in ternary systems: Water-oil-nonionic amphiphile. J Phys Chem 90:671
53. Kahlweit M, Strey R, Firman P, Hasse D, Jen J, Schomacker R (1988) General patterns of the phase behavior of mixtures of H2O, non polar solvents, amphiphiles and electrolytes. Langmuir 4:499
54. Andérez JM, Bracho CL, Sereno S, Salager JL (1993) Effect of surfactant concentration on the properties of anionic–nonionic mixed surfactant–oil–brine systems. Colloid Surface A 76:249–256
55. Shinoda K, Saito H (1968) The effect of temperature on the phase equilibria and the types of dispersions of the ternary system composed of water, cyclohexane, and nonionic surfatant. J Colloid Interface Sci 26:70–74
56. Salager JL, Miñana-Perez M, Perez-Sanchez M, Ramirez-Gouveia M, Rojas CI (1983) Surfactant–oil–water systems near the affinity inversion. Part III: The two kinds of emulsion inversion. J Dispers Sci Technol 4:313
57. Antón RE, Castillo P, Salager JL (1986) Surfactant–oil–water systems near the affinity inversion. Part IV: Emulsion inversion temperature. J Dispers Sci Technol 7:319
58. Miñana-Perez M, Jarry P, Perez-Sanchez M, Ramirez-Gouveia M, Salager JL (1986) Surfactant–oil–water systems near the affinity inversion. Part V: Properties of emulsions. J Dispers Sci Technol 7:331
59. Shinoda K, Arai H (1967) The effect of phase volume on the phase inversion temperature of emulsions stabilized with nonionic surfacatnts. J Colloid Interface Sci 25:429
60. Salager JL, Márquez N, Antón RE, Graciaa A, Lachaise J (1995) Retrograde Transition in the Phase Behavior of Surfactant–oil–water systems produced by an alcohol scan. Langmuir 11:37–41
61. Ysambertt F, Antón RE, Salager JL (1997) Retrograde Transition in the phase behavior of surfactant–oil–water systems produced by an oil EACN scan. Colloid Surf A 125:131–136
62. Cash L, Cayias JL, Fournier G, MacAllister D, Shares T, Schechter RS, Wade WH (1977) The application of low interfacial tension scaling rules to binary hydrocarbon mixtures. J Colloid Interface Sci 59:39–44
63. Queste S, Salager JL, Strey R, Aubry JM (2007) The EACN scale for oil classification revisited thanks to fish diagrams. J Colloid Interface Sci 312:98–107
64. Cayias JL, Schechter RS, Wade WH (1976) Modeling Crude Oils for Low Interfacial tension. Soc Petrol Eng J 16:351–357
65. Graciaa A, Lachaise J, Cucuphat C, Bourrel M, Salager JL (1993) Interfacial segregation of ethyl oleate/hexadecane oil mixture in microemulsion systems. Langmuir 9:1473–1478
66. Graciaa A, Lachaise J, Cucuphat C, Bourrel M, Salager JL (1993) Improving Solubilization in microemulsion with additives – Part 1: The Lipophilic Linker role. Langmuir 9:3371–3374
67. Graciaa A, Lachaise J, Cucuphat C, Bourrel M, Salager JL (1993) Improving Solubilization in microemulsion with additives – Part 2: Long chain alcohols as lipophilic linkers. Langmuir 9:669–672

68. Acosta E, Uchiyama H, Sabatini D, Harwell JH (2002) The Role of Hydrophilic Linker. J Surfact Deterg 5:151–157
69. Acosta E, Mai PD, Harwell JH, Sabatini DA (2003) Linker-modified microemulsions for a variety of oils and surfactants. J Surfact Deterg 6:353–363
70. Binks BP, Fletcher PDI, Taylor DJF (1998) Microemulsions Stabilized by Ionic/Nonionic Surfactant Mixtures. Effect of Partitioning of the Nonionic Surfactant into the Oil. Langmuir 14:5324–5326
71. Antón RE, Salager JL, Graciaa A, Lachaise J (1992) Surfactant–oil–water systems near the affinity inversion – Part VIII: Optimum Formulation and phase behavior of mixed anionic-nonionic systems versus temperature. J Dispers Sci Technol 13:565
72. Antón RE, Mosquera F, Oduber M (1995) Anionic–nonionic surfactant mixture to attain emulsion insensitivity to temperature. Prog Colloid Polym Sci 98:85
73. Antón RE, Rivas H, Salager JL (1996) Surfactant–oil–water systems near the affinity inversion – Part X: Emulsions made with anionic–nonioic surfactant mixtures. J Dispers Sci Technol 17:553
74. Kunieda H, Solans C (1997) How to prepare microemulsions: Temperature-insensitive microemulsions. In: Kunieda H, Solans C (eds) Industrial Applications of Microemulsions. Marcel Dekker, New York
75. Cratin PD (1969) A quantitative characterization of pH dependent systems. Ind Eng Chem 61:35–45
76. Antón RE, Graciaa A, Lachaise J, Salager JL (1996) Phase behavior of pH-dependent systems containing oil-water and fatty acid, fatty amine or both. 4th World Surfactants Congress, Barcelona, Spain June 3–7, 1996. Proceedings Vol 2, 244–256, Edited for A.E.P.S.A.T. by Roger de Llúria, Barcelona, Spain
77. Mendez Z, Antón RE, Salager JL (1999) Surfactant–oil–water systems near the affinity inversion. Part XI: pH sensitive emulsions containing carboxylic acids. J Dispers Sci Technol 20:883–892
78. Bravo B, Marquez N, Ysambertt F, Chavez G, Caceres A, Bauza R, Graciaa A, Lachaise J, Salager JL (2006) Phase behavior of fatty acid/oil/water systems: Effect of the acid chain length. J Surfact Deterg 9:141–146
79. Rivas H, Gutierrez X, Ziritt JL, Antón RE, Salager JL (1997) Microemulsion and optimum formulation occurrence in pH-dependent systems as found in alkaline enhanced oil recovery. In: Solans C, Kunieda H (eds) Industrial Applications of Microemulsions, Chap 15. Marcel Dekker, New York, pp 305–329
80. Deng S, Yu G, Jiang Z, Zhang R, Ting YP (2005) Destabilization of droplets in produced water from ASP flooding. Colloids Surf A 252:113–119
81. Acevedo S, Ranaudo MA, Gutierrez LB, Escobar G (1996) Adsorption of high and low molecular weight natural surfactants at the crude water–oil interface and their influence on γ-pH and γ-time behavior. In: Chattopadhay AK, Mittal KL (eds) Surfactants in Solutions. Marcel Dekker, New York, pp 221–231
82. Acevedo S, Gutierrez X, Rivas H (2001) Bitumen-in-water emulsion stabilized with natural surfactants. J Colloid Interface Sci 242:230–238
83. Liu Q, Dong M, Ma S, Tu Y (2007) Surfactant enhanced alkaline flooding for western canadan heavy oil recovery. Colloids Surf A 293:63–71
84. Bourrel M, Bernard D, Graciaa A (1984) Properties of binary mixtures of anionic and cationic surfactants: Micellization and Microemulsions. Tenside Deterg 21:311–318
85. Antón RE, Gómez D, Graciaa A, Lachaise J, Salager JL (1993) Surfactant–oil–water systems near the affinity inversion, Part IX. Optimum Formulation and Phase behavior of mixed anionic–cationic systems. J Dispers Sci Technol 14:401–416

86. Upadhyaya A, Acosta EJ, Scamehorn JF, Sabatini DA (2006) Microemulsion phase behavior of anionic–cationic surfactant mixtures: Effet of tail branching. J Surfact Deterg 9:169–179
87. Antón RE (1992) Contribution to the study of the phase behavior of systems containing a surfactant mixture, oil and water (in French). Doctoral dissertation, University of Pau, France

Part II

New Paradigms for Spreading of Colloidal Fluids on Solid Surfaces

Anoop V. Chengara · Alex D. Nikolov · Darsh T. Wasan (✉)

Department of Chemical Engineering, Illinois Institute of Technology, Chicago, Il 60616, USA
wasan@iit.edu

1	Introduction	117
2	Spreading of Aqueous Trisiloxane Solution on a Hydrophobic Surface	119
2.1	Spreading Rate Behavior	122
2.2	Concentration Dependence of Spreading	124
2.3	Vertical Film Climbing Experiments	126
3	Displacement of Oil from Solid Surface by Micellar Nanofluid	128
3.1	Disjoining Pressure Isotherms	131
3.2	Statics of Contact Line Position	132
3.3	Effect of Micelle Volume Fraction on Meniscus Profile	133
3.4	Effect of Particle Size on Meniscus Profile	135
3.5	Effect of Capillary Pressure on Meniscus Profile	136
4	Conclusions	138
	References	138

Abstract Colloidal fluids are used in a variety of technological contexts. For example, their spreading and adhesion behavior on solid surfaces can yield materials with desirable structural and optical properties. The well-established concepts of spreading and adhesion behavior of simple liquids do not apply to colloidal fluids containing nanometer-sized particles, surfactant micelles, proteins, polymers, vesicles, microemulsions, and solvents. This paper reviews recent progress in the spreading of colloidal/nano-fluids over solid surfaces with emphasis on two applications: the spreading of aqueous trisiloxane surfactant solutions (i.e., superspreaders) on hydrophobic solid surfaces driven by the surface tension gradient, and the spreading of thin colloidal films containing nanoparticles on hydrophilic surfaces driven by the structural disjoining pressure gradient (i.e., film tension gradient). These two mechanistic paradigms of dynamic spreading of colloidal fluids on solids are elucidated with experimental observations and mathematical modeling.

1
Introduction

In practical applications, the fast, uniform and complete spreading of a liquid helps ensure efficient transfer of a solute onto a solid or modifies the solid's surface characteristics through the deposition of the solvent. The spreading rate and extent depends on the mutual affinity between the solid surface and the liquid that is in contact with it. According to classical thermodynamics [17], complete spreading will occur when the spreading coefficient (S) is zero (achieved under true equilibrium conditions) or positive (achieved under experimentally limited non-equilibrium conditions) when defined as

$$S = \sigma_{SV} - \sigma_{SL} - \sigma_{LV}, \qquad (1)$$

where σ refers to interfacial tension and the subscripts S, L and V refer to the solid, liquid and vapor phases, respectively. Complete spreading is characterized by a zero three-phase contact angle. When the spreading coefficient is negative, partial spreading occurs and the final three-phase contact angle is finite. However, the thermodynamic condition on the value of S does not provide information about the dynamics of spreading. Clearly, the rate of spreading of a liquid on a solid is an important consideration in practice – for example, if a surface is to be cooled by spreading a thin film of liquid on it, the rate of heat removal will be governed by the rate at which the liquid film increases the contact area of heat transfer. It is therefore important to study the dynamics of spreading as well as the final configuration of the liquid on the solid.

Surfactants play an important role in enhancing a liquid's ability to wet and spread on solid surfaces. Surfactants are used as emulsifiers and dispersants in pesticide formulations to assist with the delivery of agrochemicals to a spray mixture. A nonionic surfactant is often added to a tank mix containing a herbicide for the purpose of improving the spreading and wetting of the spray deposit on leaf surfaces [2]. Similarly, surfactants are used in detergent formulations to remove oily stains from fabrics and glass or metal surfaces.

In this study, we focus on the spreading behavior of a colloidal/nano-fluid on a solid surface. The colloidal fluid referred to here is an aqueous solution containing surfactant at a concentration greater than the critical micelle concentration (CMC). At concentrations above the CMC, surfactant monomers aggregate into spherical micelles that have a diameter of about 5–10 nm [34]. In addition, the air–liquid interface of the surfactant solution also contains adsorbed surfactant molecules. Thus, the spreading behavior of such a colloidal solution on a solid surface can be due to surface forces (surface tension gradient resulting from the nonuniform distribution of surfactant molecules on the surface) as well as film forces (osmotic pressure gradient resulting from the nonuniform distribution of surfactant micelles in the liquid film). It is to be expected that the surface force (surface tension gradient) will be strongly determined by the nature of the surfactant layer adsorbed on the air–liquid interface. The extent to which surface and film forces in the spreading liquid manifest themselves is also determined by the hydrophobicity of the solid surface – a strongly hydrophobic surface does not allow for the formation of a thin liquid film, making the film forces relatively unimportant to the spreading process, though they may dominate the spreading on hydrophilic surfaces.

The above mentioned spreading mechanisms are illustrated with specific examples of practical interest. The spreading of concentrated aqueous trisiloxane surfactants on a strongly hydrophobic surface like polystyrene is examined in light of a surface tension gradient driven mechanism (Marangoni effect). This mechanism is used to explain some observed experimental features that include: (i) an extent of spreading in excess of predictions based on purely capillary forces (ii) the time dependence of the spreading radius and (iii) maximumization of the rate and extent of spreading as a function of surfactant concentration.

We also describe the spreading of a thin surfactant laden aqueous film on a hydrophilic solid, i.e., one in which the dynamic contact angle is small. In such a case, the osmotic pressure gradient generated by the nonuniform distribution of surfactant micelles in the liquid film can drive the spreading process. The motivation for this study comes from the need to understand the detergent action involved in the removal of an oily soil from a soiled surface. This paper presents an overview of our recent work.

2
Spreading of Aqueous Trisiloxane Solution on a Hydrophobic Surface

As mentioned earlier, surfactants have the ability to reduce the surface tension of water, thereby enabling spray solutions to more effectively wet waxy leaf surfaces and increase the amount of spray retained on the leaf. As a result, surfactants can make spray applications more effective by improving the delivery of agrochemicals.

One relatively new class of surfactant, the trisiloxane alkoxylate superspreaders, has the unique ability to spread spray solutions across difficult-to-wet leaf surfaces to a greater degree than conventional surfactants. Among the known surfactants, the trisiloxane ethoxylate with 8 ethylene oxide groups (TS8EO) has been found to be one of the best water-spreading agents for hydrophobic surfaces in general [1, 59]. In the agricultural industry, this fact has been used to advantage in delivering water-based pesticide more effectively to the waxy surface of leaves by co-formulating them with such surfactants [28]. The overall spread area achieved by an aqueous droplet containing a trisiloxane surfactant can be as much as 50 times greater than water and 25 times more effective than a conventional surfactant on a leaf surface [42]. This increase in spreadability can provide an improvement in spray coverage on target plants and even allow growers to use less water to achieve the same or better degree of coverage as compared to conventional treatments [45].

There are numerous studies on the spreading of trisiloxane surfactants on hydrophobic surfaces (for two comprehensive reviews, see [21, 22]), but the mechanism of spreading is still not fully understood. A number of explanations have been proposed for this superspreading phenomenon, mostly related to the configuration of the trisiloxane molecules at the interfaces [1, 20]. Zhu et al. [59] postulated that the spreading on hydrophobic surfaces is facilitated by the formation of a precursor film of water ahead of the contact line. Extensive experiments were carried out by [44] on solid substrates of different hydrophobicity by using trisiloxane surfactants of different degrees of ethoxylation in the hydrophilic tail. These experiments led them to conclude that spreading behavior was only weakly dependent on the microstructure of the micellar dispersion and more dependent on the length of the hydrophilic chain. It has been suggested that the formation of vesicle type aggregates is a requirement for trisiloxane-mediated superspreading [21]. It has also been suggested that surfactant vesicles and/or other aggregates disintegrate and provide an efficient delivery of surfactant molecules to the contact surfaces (or the spreading front) and enhance spreading on a liquid hydrophobic surface [43]. Svitova et al. [46] postulate that a wetting transition occurs on the solid surface at the surfactant concentration at which superspreading occurs, through an analogy with the classical wetting transition observed for pure fluids with surfactant concentration playing the role of temperature as the variable. An explanation of the thermodynamics of superspreading based on the competition between the van der Waals' forces and the spontaneous curvature of the adsorbed surfactant molecule has been advanced [24] to account for the observed spreading behavior as a function of the hydrophilic chain length [55]. Churaev et al. [11] believe that the spreading behavior of aqueous trisiloxane dispersions is a result of a disjoining pressure gradient created by the surfactant aggregates. Rosen and Wu [41] and Wu and Rosen [58] correlated the spreading pressure and spreading behavior of a mixture of trisiloxane-N-alkyl pyrrolidinone surfactants on polyethylene at concentrations below CMC. Kumar et al. [31] concluded from dynamic surface tension measurements that the kinetics of adsorption of un-aggregated molecules at the air–liquid interface is similar for trisiloxane molecules that spread very quickly (i.e., those with 8 ethoxy groups) and for those that do not (i.e., 12 ethoxy groups), suggesting that a different type of aggregate adsorption must supply the high surface concentration of surfactant at the expanding interface. Dong et al. [14] showed that the structure of surfactant aggregates at the solid–liquid interface are similar for trisiloxanes as well as for surfactants of the ethoxylated surfactants with hydrocarbon tail groups of the form $C_m E_n$, suggesting that the superspreading properties of trisiloxanes are not caused by the formation of a different type of aggregates. Our work [6, 7, 35] suggests that a spontaneously generated surface tension gradient at the expanding air–liquid interface aids the spreading process.

It has long been known that the spatial variation in surface tension at a liquid/vapor surface results in added tangential stresses at the surface; this results in a surface traction that acts on the adjoining fluid, giving rise to the fluid motion in the underlying bulk liquid [32, 48]. If the liquid/air surface tension near the expanding radial edge of the spreading drop is higher than near the center of the drop, then the difference in surface tension 6 establishes a gradient and will create a Marangoni flow.

An estimate of the dynamic contact angle during the very early stages of spreading provides a clue to the underlying mechanism [35]. Figure 1(a) shows photographs of the profile of the spreading drop at different times. We observe that the contact angle is high (about 30°) during the high rate of spreading and remains approximately constant in the short time scale considered. A high dynamic contact angle implies that the air–liquid interfacial tension at the periphery of the spreading drop is not very low. This picture is different from that of classical wetting, which requires the dynamic contact angle to be very low (close to zero) as a result of a low air–liquid surface tension near the contact line. Thus, at least in the initial stages of spreading, the probable mechanism is that of a Marangoni flow. When the spreading starts, the drop's surface area at the three-phase contact region suddenly becomes increased (Fig. 1(b)). Consequently, the local surfactant surface concentration decreases, resulting in a higher surface tension at the leading edge rather than in the middle (which gets stretched less). The resulting radial surface tension gradient causes the drop to spread radially outward. We now examine

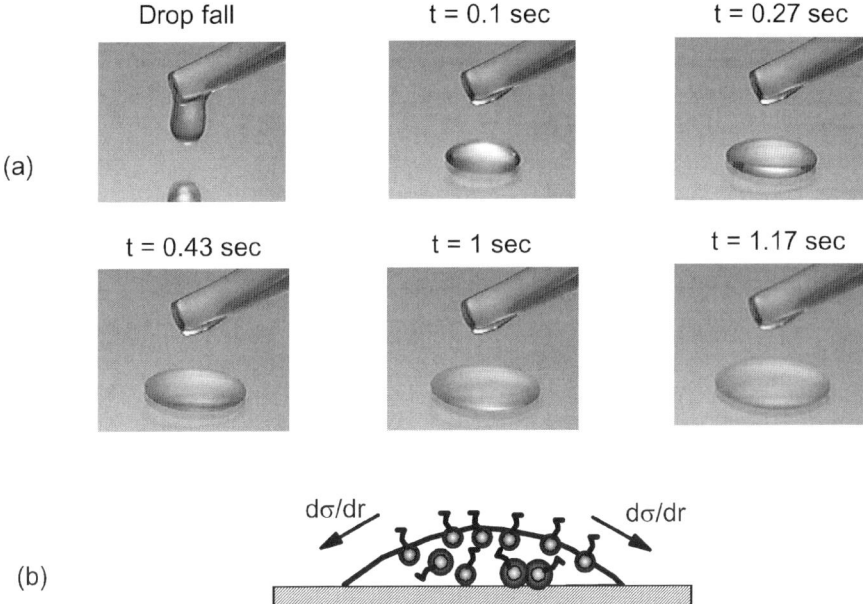

Fig. 1 a Photographs showing dynamic contact angle of 0.1 wt. % Silwet L77 drop and **b** schematic of evolution of surface tension gradient

whether this mechanism is consistent with observed experimental data on the spreading rate and behavior and concentration effect.

2.1
Spreading Rate Behavior

An interesting aspect of the spreading behavior of an aqueous trisiloxane on a solid surface like polystyrene is the time dependence of the drop's radius, which can be described by a $R(t) \propto t^{0.5}$ relationship, implying that the spreading area increases linearly with time [35, 59]. This is much stronger than predicted by models based on a constant surface tension of the drop, which follows a $R(t) \propto t^{0.1}$ dependence [25]. Additional "de-oiling" experiments [6] were performed in order to alter the force balance at the contact line and to check if such a change influences the spreading rate. When a drop of 0.1 wt. % Silwet L-77 (an example of TS8EO) was placed on a thin layer (30 microns) of decalin, a circular spot was formed in less than a minute of drop deposition, whose radius then increased with time. This was caused by the rupture of the thin, unstable film of decalin between the surfactant solution and the solid (polystyrene) surface bringing the Silwet drop into contact with the solid surface and resulting in its spreading. The difference in this experiment over the standard spreading experiment is that in this case, the spreading aqueous drop displaces decalin, while in the latter case it displaces air. The drop radius as a function of time is plotted in Fig. 2 for the two types of spreading experiments. Despite the fact that the balance of forces at the contact line in these two experiments are different, the spreading rate and its time dependence remains the same. This suggests that the driving force is localized along the air–aqueous interface which is the same in the two cases. The contact line is more stable in the case of the "de-oiling" experiment, enabling the spreading to proceed longer.

To explain this spreading rate behavior, Nikolov et al. [35] postulated that the excess driving force (assuming that capillary and hydrostatic forces balance each other) is a radial surface tension gradient, which can be approximated as

$$\frac{\partial \sigma}{\partial r} \approx \frac{\Delta \sigma(t)}{R(t)} \approx \frac{kt}{R(t)}, \tag{2}$$

where $\Delta\sigma(t)$ is the difference between surface tension near the contact line and at the apex of the drop. The last equality in Eq. 2 is rationalized as follows. During spreading, the expansion of interfacial area is higher near the contact line than near the apex as a consequence of the radial 8 flow profile. This causes a greater depletion of surfactant from the interfacial region near the edge of the drop than from the region near the apex, and the difference in surfactant concentration between edge and apex increases with time (i.e., as spreading proceeds). Thus, for the early time period (before diffusion from

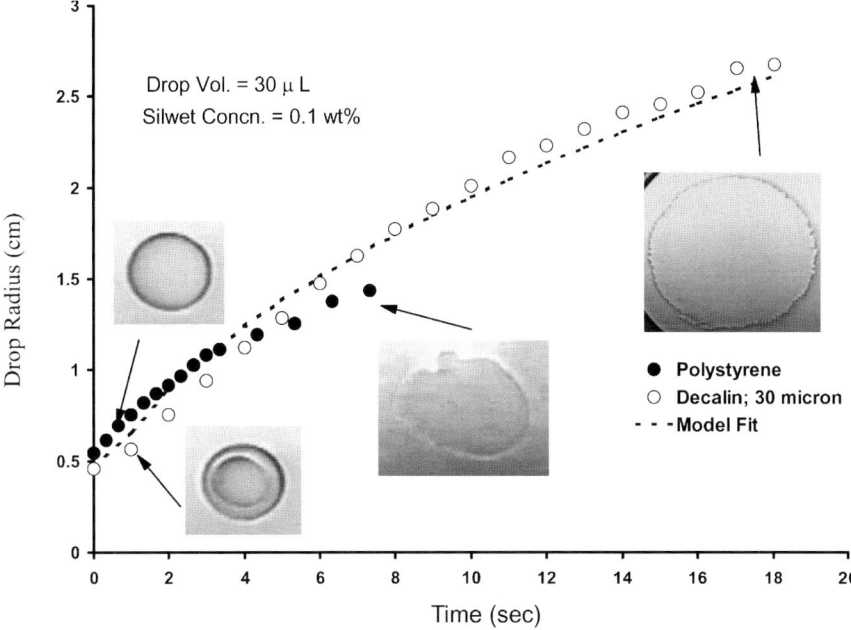

Fig. 2 Comparison of model prediction with experimental data on radius of Silwet L-77 drop in the spreading and de-oiling experiments. The *dotted line* represents Eq. 3 with $\alpha = 0.6$ cm/sec$^{1/2}$

the bulk has restored the surfactant concentration to its equilibrium value), the difference in surface tension $\Delta\sigma(t)$ is expected to increase with time. We assume the simplest form, which is linear, i.e., $\Delta\sigma(t) = kt$, where k is a constant. By approximating the geometry of the spreading drop as a flat pancake, the radius is predicted to follow:

$$R(t) \approx \left(\frac{kV}{\mu\pi}\right)^{1/4} t^{0.5} \approx \alpha t^{0.5}, \tag{3}$$

where V is the volume and μ is the viscosity of the drop. To fit the maximum spreading rate of an aqueous drop containing trisiloxane surfactant at a concentration of 0.1 wt. % (Fig. 2), the value of k is only 0.15 dyn/cm–sec. This in turn gives a value of 0.6 cm/sec$^{1/2}$ for the fitted parameter α in Eq. 3 for this value of surfactant concentration (0.1 wt. %). Over the time scale of the experiment (20 s), the maximum difference in surface tension is only $kt = 3$ dyn/cm, which translates to an average surface tension gradient of 1 dyn/cm^2 over the drop's surface during that time period. This is a small surface tension gradient, making it difficult to verify using combined hydrodynamic and surfactant transport equations; the predictions are very sensitive to the type of surface equation of state used [7].

Rafai et al. [38] also obtained the $R(t) \sim t^{1/2}$ dependence employing the same Marangoni force mechanism but did so by considering the length along which the surface tension gradient varies to be the height instead of the radius of the droplet. However, it must be noted that the $R(t) \sim t^{1/2}$ relationship by itself does not constitute evidence of a Marangoni stress driven spreading. Such a time dependence has also been derived by Von Bahr et al. [54] without assuming a surface tension gradient but rather by assuming a form of time dependency for the contact angle variation. Thus, it appears that the form of the rate behavior can be rationalized by different mechanisms, but the surface tension gradient mechanism is consistent with the other observations on trisiloxane solutions, such as the maxima in spreading rate and area as function of surfactant concentration.

2.2
Concentration Dependence of Spreading

Figure 3 plots the spreading rate (determined as the slope of the spreading area $A(t)$ curve) as a function of surfactant concentration. There is a maximum in the spreading rate at 0.1 wt. %, although the decrease in rate at higher concentration is not sharply pronounced. This maximum in the spreading rate with surfactant concentration had also been previously observed [44, 59]. A stronger maximum is observed [6] when the final area of the drop is plotted as a function of concentration (Fig. 4). The concentration at which the maximum spreading occurs is the same irrespective of whether the surfactant drop displaces air or decalin. This finding also suggests that the driving force is localized along the air–aqueous interface instead of at the contact line. A qualitative explanation for the maximum in spreading rate and final area are given below.

At low surfactant concentration (below 0.05 wt. %), the expanding air–aqueous interface is uniformly depleted of surfactant due to adsorption limitation from the bulk to the interface, resulting in a high dynamic surface tension of the expanding interface. The driving force, being the difference of surface tensions near the contact line and the apex of the drop, is small, giving rise to a low rate and degree of expansion. As the bulk surfactant concentration increases, it is easier to increase the surface concentration of molecules by adsorption at the interface near the apex region (which undergoes less expansion) than at the interface near the contact line (which undergoes maximum stretching). This leads to an increase in the difference of surface tension between the contact line region and the apex, providing a higher driving force and higher rate and extent of spreading (e.g., 0.1 wt. %). At a still higher bulk concentration (e.g., 0.8 wt. %), the surface concentration of surfactant molecules cannot be changed appreciably from its equilibrium value at any location on the interface even under dynamic conditions of surface expansion, since the adsorption from bulk to the surface is fast. Thus, the

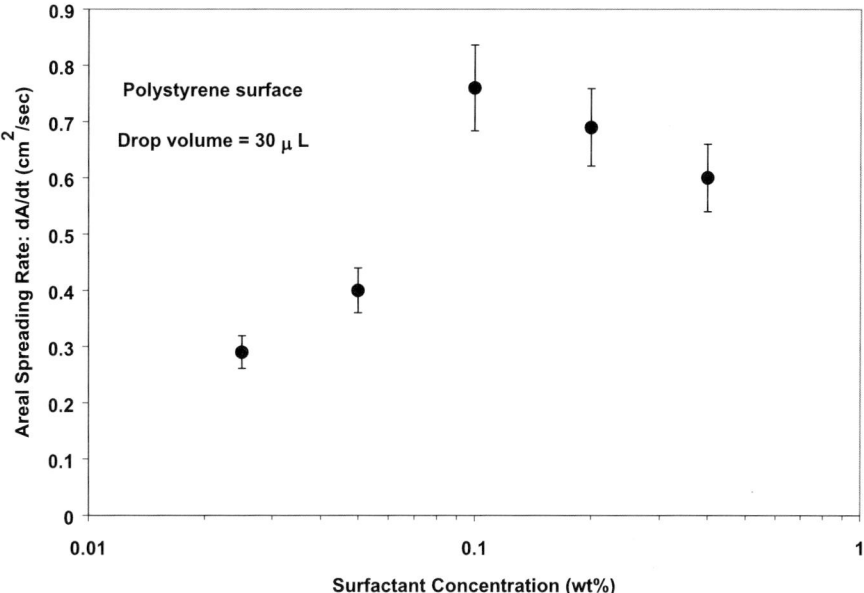

Fig. 3 Spreading rate (on area basis) of Silwet drop on polystyrene as function of surfactant concentration

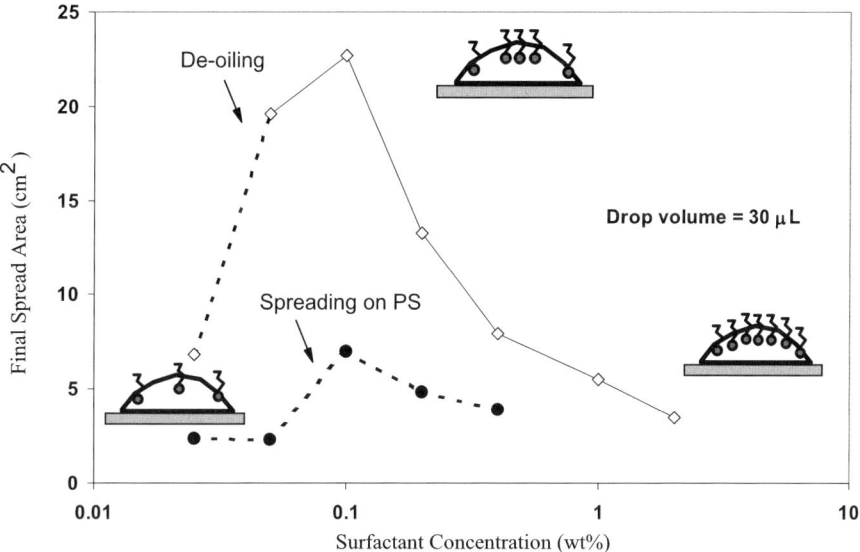

Fig. 4 Plot of total area of drop as a function of Silwet concentration in spreading and de-oiling experiments on polystyrene surface

driving force for spreading decreases at the higher concentration and lasts for a shorter duration of time, resulting in a lower total spread area.

2.3
Vertical Film Climbing Experiments

The spreading of Silwet L-77 as a thin film on polystyrene in a vertical geometry [7], as opposed to a drop on a horizontal surface, offers additional insight into the spreading mechanism. In this scheme, the gravitational force opposes spreading, unlike the drop geometry where it aids spreading. A plot of the length of spreading measured upwards from the free surface for a 0.1 wt. % aqueous dispersion shows a $L(t) \approx 0.4 t^{0.5}$ dependence, which is very close to that observed in the case of a radially spreading drop. Figure 5 plots the climbing rate expressed as the (constant) slope of $L^2(t)$ vs. time for various surfactant concentrations. The climbing rate shows a maximum as a function of concentration, with the peak value occurring at 0.1 wt. % just as in the case of the drop geometry. However, the maximum is much more pronounced because unlike the drop geometry, where gravity aids the spreading and contributes to the rate, here it opposes it; thus, the fact that the intrinsic spreading rate of Silwet L-77 on a strongly hydrophobic surface like polystyrene is much lower at concentrations above 0.1 wt. % is clearly visible.

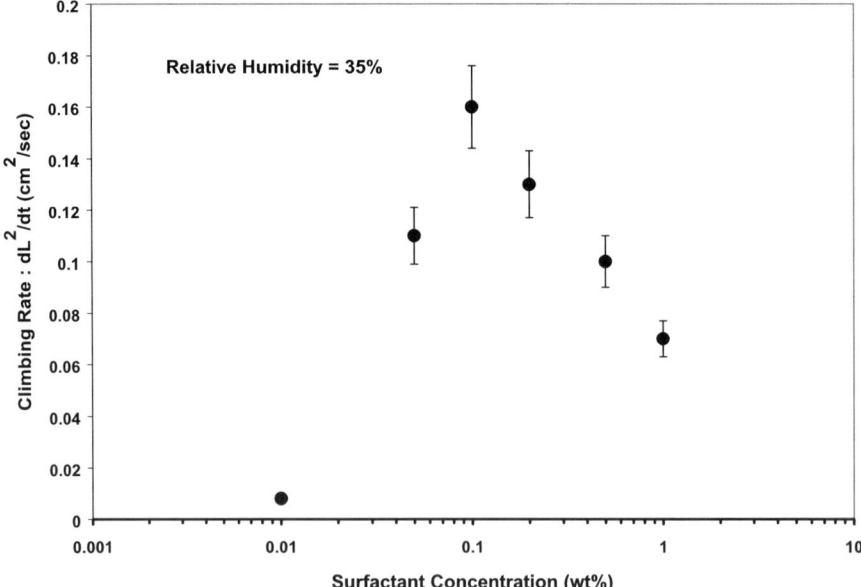

Fig. 5 Plot of climbing rate of film as function of Silwet L-77 concentration

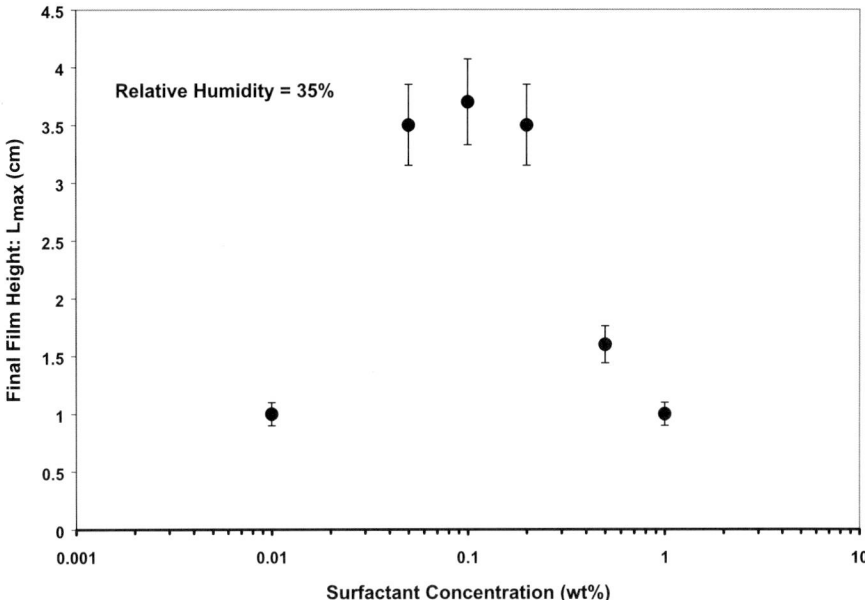

Fig. 6 Plot of final height of climbing film as function of Silwet L-77 concentration

A similar result is observed when the maximum height achieved (L_{max}) is plotted as a function of surfactant concentration (Fig. 6). However, the more interesting observation is that for all concentrations above CMC (for which $\sigma = 22$ dyn/cm), the maximum height predicted for perfect wetting conditions (i.e $\theta = 0$), is

$$L_{max} = \sqrt{\frac{2\sigma}{\rho g}}, \qquad (4)$$

which is only 0.2 cm. As seen from Fig. 6, there is an excess force that causes the liquid film to rise above the meniscus limit determined by purely capillary action. This is a significant result, because it is difficult to draw conclusions with the same degree of accuracy about the equilibrium radius of a drop perfectly wetting a solid surface ($\theta = 0$) from our previous work. That inaccuracy was a consequence of the fact that the solution of the Laplace equation requires large values of apex curvature that accompanies a perfectly wetting drop; this results in a large variation in the calculated radius of the drop for very small changes in the contact angle [18]. Thus, while it was not possible to determine if there was a force in excess of the gravitational and capillary force from radius measurement of the aqueous Silwet L-77 solution drop, the method of the climbing film clearly demonstrates the presence of an excess force (i.e., surface tension gradient).

Another explanation has been offered for the observed maximum in spreading area with concentration, based on the concept of autophobicity, i.e., the existence of an adsorbed layer of surfactant at the liquid–solid interface that is not wetted by its own kind [44]. However, the short time scale of the spreading (~ 2 seconds in the high concentration regime) requires that this autophobic layer is laid down in this short period, which is unlikely considering the kinetic limitations on reorientation of molecules [50].

In summary, the high spreading rate observed with trisiloxane surfactants, in addition to the maximum of spreading area as function of concentration, suggests the presence of a surface tension gradient (Marangoni stress) at the air–liquid surface. As spreading proceeds and surfactant is adsorbed at the air–solution interface, the importance of Marangoni stress diminishes, and classical wetting forces may determine the spreading behavior at later stages.

3
Displacement of Oil from Solid Surface by Micellar Nanofluid

The mechanisms involved in the removal of an oil droplet from a solid using a surfactant solution has received considerable attention in the literature. It is generally accepted [49] that two stages are involved in detergency – (i) the "roll-up" of the oily pollutant into the shape of a droplet occurs as a result of the imbalance of forces at the contact line due to the presence of surfactant in the aqueous cleansing solution and (ii) the "emulsification" of the oil droplet in the aqueous solution after detachment from the solid surface. These two mechanisms often occur concurrently in a practical washing application. Experimental studies using $C_{12}E_4$ and $C_{12}E_5$ surfactants to remove hexadecane/squalane mixtures [39] or triolene/hexadecane mixtures [33] from polyester/cotton fabrics have shown that detergent efficiency (defined as percentage of oil mass removed from a substrate) is maximum when phase inversion temperature of the O/W/surfactant system is reached. The phase inversion temperature (PIT) generally corresponds to a minimum in the O/W interfacial tension and is a consequence of the fact that entropy of micelle formation is equal to the entropy of a surface molecule formation of the surfactant [3]. However, experiments using a mixture of decane and triolene [49] show that the above rule relating maximum detergency and PIT is not general; rather the maximum in detergency may be related to whether a microemulsion or a liquid crystal phase is formed. Thompson's experiments resolved two distinct maxima in detergency, each corresponding to the emulsification mechanism (occurring in the PIT region) and the roll-up mechanism (occurring at temperatures below PIT), respectively.

The detachment of a pollutant has also been shown to depend on hydrodynamics. The effect of the speed of movement of the air–water interface on the removal of small solid particles (micron-sized metallic particles) from silicon wafer was studied by [36]. By balancing the viscous forces in the fluid against the forces of adhesion of the metallic particle on the solid as well as the capillary forces at the air–liquid–particle interface, they showed that there exists a critical velocity above which no particles are removed from the solid surface. This is due to the fact that at higher velocities, the time over which the capillary forces counteract the viscous and adhesive forces are too small to deliver the impulse of forces needed to separate the particle from the solid surface. Our practical situation is further complicated by the fact that oily pollutants are largely liquid and therefore, deformable. Mahe et al. (1986) showed that detachment in the laminar flow regime occurred when a critical shear rate (that varies linearly with the contact radius, difference between receding and advancing contact angles, interfacial tension, and inversely with the aqueous phase viscosity and square of the oil drop's radius) is achieved. This suggests that the higher the velocity, the better the probability of detachment, in contradiction of [36] result. In addition, the removal of a mobile interface from a solid surface is related to the instabilities that develop when the interface is deformed. Chatterjee [5] used drop shape analysis to arrive at a critical Eotvos number for the partial detachment of drops resulting from buoyancy-induced instability. Kolev et al. (2003) propose a generalized Young's equation including a line drag force (that is proportional to the contact line velocity) to be valid under the dynamic conditions. They hypothesize that a line drag coefficient is a fundamental property of the system needed to completely specify the problem and is regressed from experimental observations of contact line velocity and dynamic contact angle measurements. In a recent study, Kralchevsky et al. [30] identify the driving force for oil detachment from glass with the diffusion of water molecules onto the glass surface to form a gel-like layer and the retarding viscous force to the line drag coefficient. However, their proposed model of the detachment of oil drops from a glass plate based on the concept of lateral diffusion of water molecules into the glass surface layer and the surface pressure gradient generated at the oil-aqueous solution interface requires a more systematic study including the role of the surfactant micelle concentration on the three-phase contact angle dynamics.

All the above explanations of the detergency mechanism focus on the tangential (interfacial tension) force at the liquid interface and ignore the effect of the surfactant micelles' size, concentration and its confinement in the oil-water–solid wedge on the removal of the oil. In our study, we examined the role of a force normal to the interface (disjoining pressure) created by the ordering of micelles on the removal of an oil droplet from a solid surface [8]. The background for this study is a set of experiments conducted by [27] and its recent explanation by [57]. Using interference microscopy technique, Kao et al. studied the profile of an crude oil–water interface when the oil drop is being removed from a glass surface by a 1 wt. % aqueous solution of C_{16}

alpha-olefin sulfonate surfactant. They found that there were two distinct contact lines – an outer (between the oil droplet, solid and water film) and an inner (between the oil droplet, solid and mixed oil–water film). The existence of a mixed oil–water film was thought to be a result of "diffusion" of the surfactant carrying aqueous solution into the space between the oil droplet and the solid surface. Wasan and Nikolov [57] threw light on the nature of forces that drive the "diffusion" of the aqueous film into the wedge formed by the oil droplet and the solid surface. They showed that in the wedge shaped geometry, micelles present in the aqueous medium could be expected to arrange themselves in well-ordered layers, the extent of ordering between maximum near the vertex of the wedge (where the oil droplet meets the solid surface) and gradually giving way to thermal disorder near the outer end of the wedge. They provided proof of such particle ordering showing both cubic and hexagonal packing patterns from photographs of 1 μm latex particles in a liquid wedge formed between a glass bubble pressed against a glass surface. It had been shown earlier both experimentally [34] and theoretically [51] that nanoparticles/micelles suspended in a fluid could arrange themselves in layers when confined between plane parallel walls or in wedge geometry [52, 53] and that such an arrangement would result in an excess pressure in the film (compared to the bulk liquid) called structural disjoining pressure. Wasan and Nikolov [57] applied this concept of structural disjoining pressure to the case of spreading to show that the integrated effect of the disjoining pressure is to increase the value of spreading coefficient for aqueous film in the wedge. This suggests that the driving force for the "diffusion" of the aqueous film between the oil droplet and solid reported by [27] is actually the excess film energy given by the structural disjoining pressure. It may be noted that the ingress of the aqueous film into the wedge driven by structural disjoining pressure is equivalent to the removal of the oil droplet (pollutant) from the solid surface – in other words, marking a new understanding of the mechanism of detergency.

Chengara et al. [8] extended the understanding of Nikolov and Wasan [57] of spreading (or pollutant removal) to the case of a wedge geometry. In this analysis, the pollutant is idealized as an oil droplet and the soiled surface by a planar smooth solid surface. The oil droplet is assumed to be attached to the solid surface at a finite contact angle. This configuration is then brought into contact with an aqueous surfactant solution. The imbalance of surface tension forces at the three phase contact line results in the reduction of the radius of the oil droplet (the "roll-up" mechanism alluded to earlier), and, consequently, the contact angle measured from the oil side becomes large, i.e., approaches 180°. In other words, the contact angle measured from the aqueous side becomes very small ($\sim 1°$). This results in a thin wedge film confining the surfactant micelles in the aqueous film between the solid surface and the oil droplet. The movement of the contact line is aided by the disjoining pressure gradient that is set up in the wedge film.

3.1
Disjoining Pressure Isotherms

The disjoining pressure in a thin liquid film is defined as the excess pressure in the film relative to that in the bulk solution. The origin of the disjoining pressure is due to the confinement of the micelles in the film region as opposed to their greater freedom of location in the bulk. The confinement effect is easy to visualize by extending the heuristic model of Kaplan et al. [26] for the entropy of hard spheres confined by a hard wall. There is more space accessible to other particles in the system when a reference particle is aligned at the wall, and, consequently, there is a net increase in the entropy of the system as a whole. A similar logic can be extended to the case of two flat surfaces to show that particles tend to align themselves close to the film surfaces. When the film thickness is large enough to accommodate more than one layer of particles, it has been found that particles tend to arrange themselves in regular layers between the film surfaces. The layering arrangement of micelles in the film has been shown experimentally using stepwise thinning probed by light interferometry [56], by Monte Carlo simulations [9] as well as statistical mechanics [19]. The layering arrangement of micelles also gives rise to an excess pressure in the film (called the disjoining pressure) that has been shown to have an oscillatory decay profile with film thickness [29]. An analytical expression for disjoining pressure (P) was given by [51] based on a solution of the Ornstein–Zernike equation:

$$\Pi(h) = \Pi_0 \cos(\omega h + \phi_2)e^{-xh} + \Pi_1 e^{-\sigma(h-d)} \quad \text{for } h \geq d \quad (5)$$

$$\Pi(h) = -P \quad \text{for } 0 < h < d, \quad (6)$$

where h is film thickness, d is the diameter of micelle and all other terms in Eq. 5 are fitted as cubic polynomials in terms of micelle volume fraction (ϕ). In Eq. 6, P refers to the bulk osmotic pressure from micelles. Of more direct use is the film energy (W), which is related to the disjoining pressure in the following manner:

$$W(h) = \int_h^\infty \Pi(h') \, dh'. \quad (7)$$

The expression for film energy given by [51] is

$$W(h) = -P(d-h) - 2\sigma \quad \text{for } 0 < h < d, \quad (8)$$

where σ is the fluid–wall interfacial tension and other quantities are also expressed as a cubic polynomial in micelle volume fraction ϕ. The amplitude of the film energy increases with micelle volume fraction in the bulk due to the corresponding increase in micelle concentration in the film (Fig. 7).

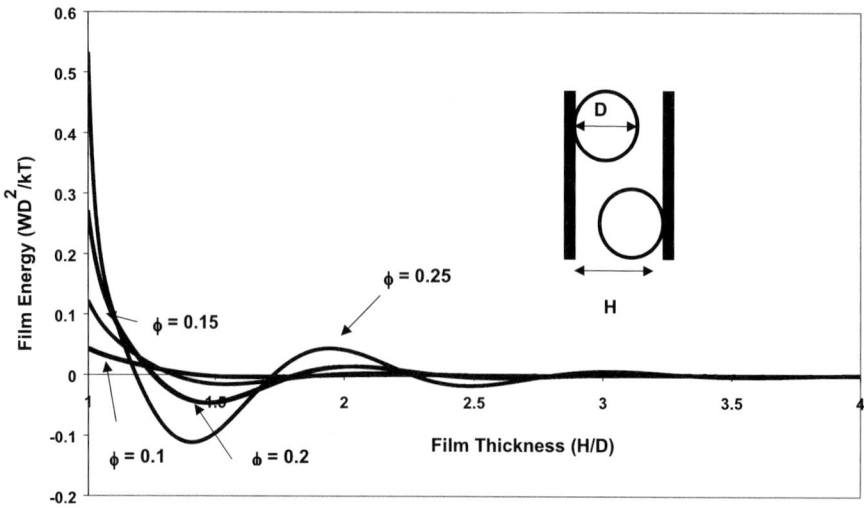

Fig. 7 Film energy profile for a monodisperse suspension as a function of particle volume fraction

3.2
Statics of Contact Line Position

Consider a one dimensional oil droplet pressed against a solid surface and surrounded by an aqueous solution of surfactant (Fig. 8). The shape of the oil–water interface far away from the solid surface is governed by the Laplace equation [18] and close to the solid surface is augmented by an additional term relating to the film energy given by the structural disjoining pressure [12]:

$$\frac{dh}{dx} = \left(\left\{ \frac{\sigma}{C_1 - A(h)} \right\}^2 - 1 \right)^{1/2}, \qquad (9)$$

where C_1 is a constant of integration and

$$A(h) = (P_w^0 - P_0^0)(h_\infty - h) - \frac{\Delta \rho g}{2}(h_\infty^2 - h^2) + W(h), \qquad (10)$$

where h_∞ is the height of the meniscus far away from the wedge, i.e., towards the bulk. This typically corresponds to about 5 micelle diameters, at which film thickness, the structural disjoining pressure contribution, is zero. We use Eq. 8 to describe the film energy term $W(h)$ in Eq. 10. Details of the computational procedure are given in [8]. The conditions for simulation correspond approximately to experimental conditions reported in [57]. For example, we have chosen a micelle volume fraction range between 0.15 and 0.3 while the experimental condition refers to a volume fraction of 0.4. Similarly, the diameter of the nanoparticle is chosen to be 20 nm, while the experimental value

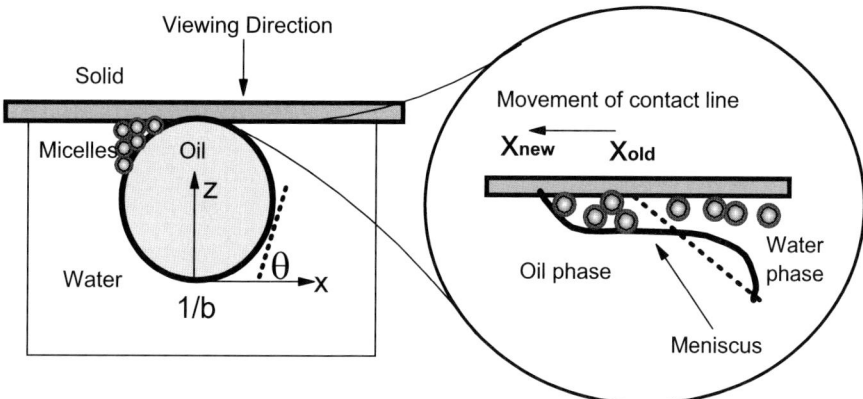

Fig. 8 Configuration of oil drop attached to solid surface in the presence of micellar/nanoparticle solution

for the micelle is 10 nm. The simplified model for film energy that ignores the presence of the solvent molecules, limits the calculation of the meniscus to those cases where the negative (attractive) component of the disjoining pressure is low enough to be balanced by the capillary pressure.

3.3
Effect of Micelle Volume Fraction on Meniscus Profile

The shape of the meniscus profile in the wedge region is shown in Fig. 9, both in the presence and absence of nanoparticles. The latter profile which serves as the base case has been computed by setting the film energy to zero in Eq. 10. The slope of the meniscus is reduced at those radial locations that correspond to film thicknesses characterized by the energy well depletion. Close to the nominal contact line (defined as the location where the height of the film is equal to the diameter of the micelle), the positive contribution of the structural disjoining pressure to the film energy causes the local slope of the meniscus to increase (Fig. 10), as previously reported by [23] for other forms of disjoining pressure isotherms. This deformation causes the convexity of the profile when viewed from the aqueous side.

The effect of the nanoparticle volume fraction on the displacement of the contact line becomes pronounced only at higher volume fractions. For example, the displacement of the contact line is 10 times the nanoparticle diameter or approximately 0.2 μm for a nanoparticle volume fraction of 0.25, while there is no appreciable change in the contact line position when the volume fraction is 0.2. This non-linear dependence of contact line position on nanoparticle volume fraction is consistent with the form of Eq. 10, where the film energy contribution due to structural disjoining pressure is subtracted from the surface energy contribution. The extent of displacement of the con-

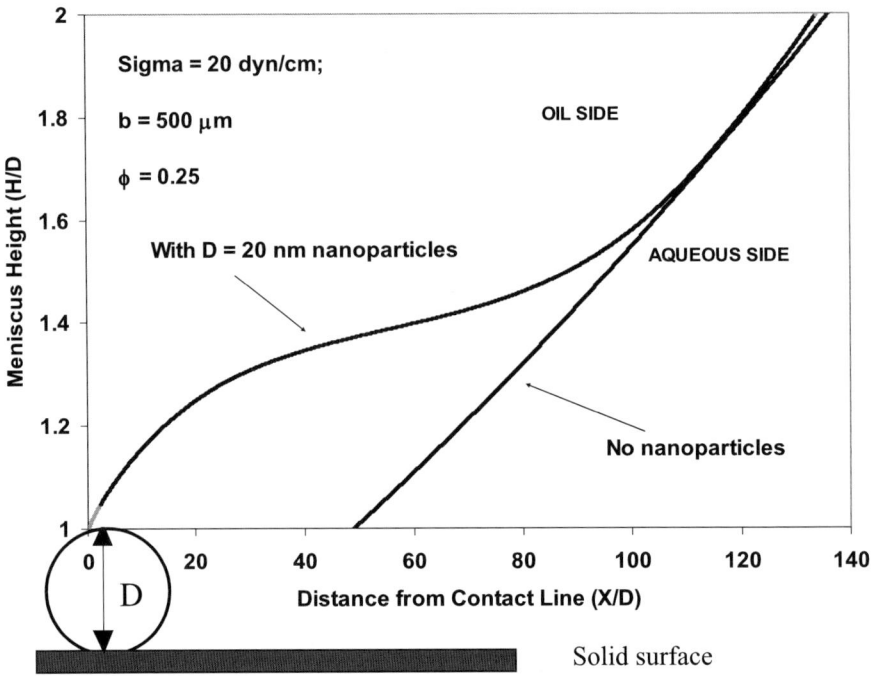

Fig. 9 Effect of nanoparticle on contact line displacement

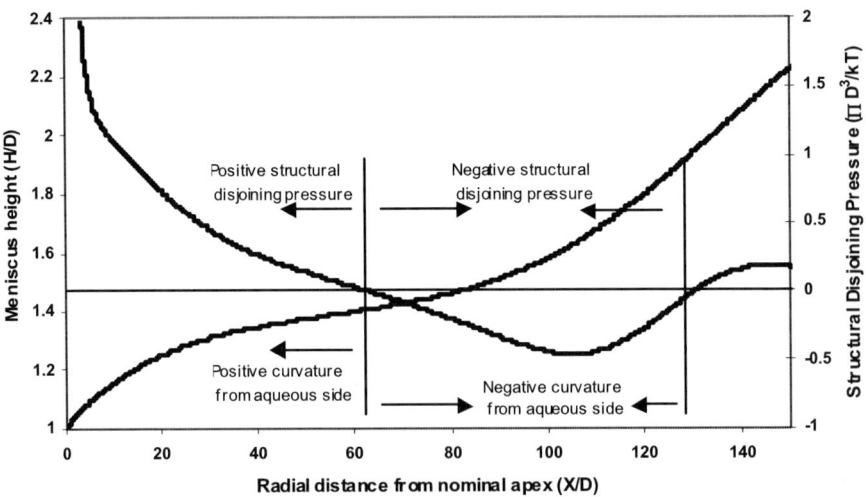

Fig. 10 Correspondence between structural disjoining pressure and meniscus curvature. conditions correspond to those in Fig. 9

tact line by disjoining pressure is significant considering the fact that the entire wedge region is only about 50 times the nanoparticle diameter, or 1 μm.

An illustration of the effect of micelle/nanoparticle volume fraction on contact line motion is found in [57]. They used 0.1 M NaCl solution to reduce the electrical double layer thickness surrounding the NaDS micelle. At a given number concentration of micelles, decreasing the size of each micelle decreases the volume fraction greatly, since the volume of each spherical micelle varies as the third power of the radius. Thus, the addition of electrolyte effectively reduced the micellar volume fraction in the aqueous medium. The authors found that the oil droplet that would otherwise become completely detached from the solid surface, came back to reattach itself to the solid when electrolyte was present. They rationalized this finding as being caused by the inability of the weakened structural disjoining forces to counteract the attraction of the oil drop to the solid surface.

3.4
Effect of Particle Size on Meniscus Profile

It was shown by [51] that the film energy is a function of the particle diameter for a fixed value of volume fraction. The amplitude of oscillation in the

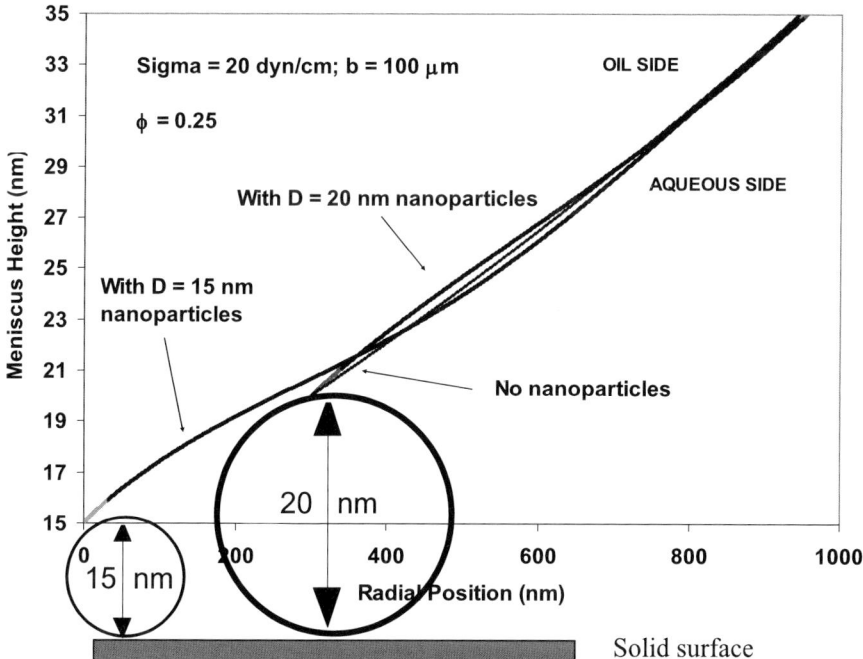

Fig. 11 Effect of nanoparticle diameter on contact line displacement

film energy increases as the particle diameter decreases. This is to be expected since the number of particles in the system increases as the diameter of each particle decreases, for a fixed value of the volume fraction; consequently, more particles are pumped into the film by the entropic forces created by the confinement effect of the film. The effect of particle diameter on the meniscus profile is shown in Fig. 11. When the particle diameter is 20 nm, the meniscus profile coincides with that obtained for the base case (no particles). This is because the capillary pressure is large due to the choice of a small radius of curvature of the oil drop ($b = 100$ µm). However, when the particle size is decreased to 15 nm while maintaining the same volume fraction, the profile of the meniscus shifts towards the oil side by about 0.1 µm.

3.5
Effect of Capillary Pressure on Meniscus Profile

It was noted earlier that the capillary pressure inside the oil drop tends to prevent the spreading of the drop. Thus, for the same spreading force (structural disjoining pressure), we expect lesser displacement of the contact line when the capillary pressure is higher. The capillary pressure of the oil drop can be increased in two ways – by increasing the interfacial tension or by decreasing the radius of curvature. The effect of varying only the interfacial tension on the displacement of the contact line is seen by comparing Fig. 12 with Fig. 9, since both correspond to the same radius of curvature at the apex. The contact line is shifted by 50 nanoparticle diameters or 1 µm in the 20 dyn/cm case (Fig. 9) as compared to 0.1 µm for the 30 dyn/cm case (Fig. 12) when the apex radius of curvature in both cases is held constant at 500 µm.

The effect of increasing only the radius of curvature of the oil drop on the displacement of the contact line while keeping the interfacial tension constant at 20 dyn/cm, is illustrated in Figs. 9 and 11. Figure 11 shows that for a radius of a curvature of 100 µm, there is virtually no movement of the contact line from the base case due to the presence of nanoparticles/micelles even at volume fraction 0.25. However, when the radius of curvature is increased to 500 µm (recall Fig. 9), thereby decreasing the capillary pressure, the presence of nanoparticles at the same concentration moves the contact line by 1 µm.

In summary, we have examined the role of structural disjoining pressure in the movement of a three phase contact line. The movement of the contact line is an integral process in the displacement of one fluid by another. Practical applications include the spreading of a fluid on a solid surface or the removal of a pollutant drop from a solid surface by the action of a surfactant solution.

The origin of the structural disjoining pressure is osmotic in nature. When nanoparticles/surfactant micelles are confined in a thin film, they tend to ar-

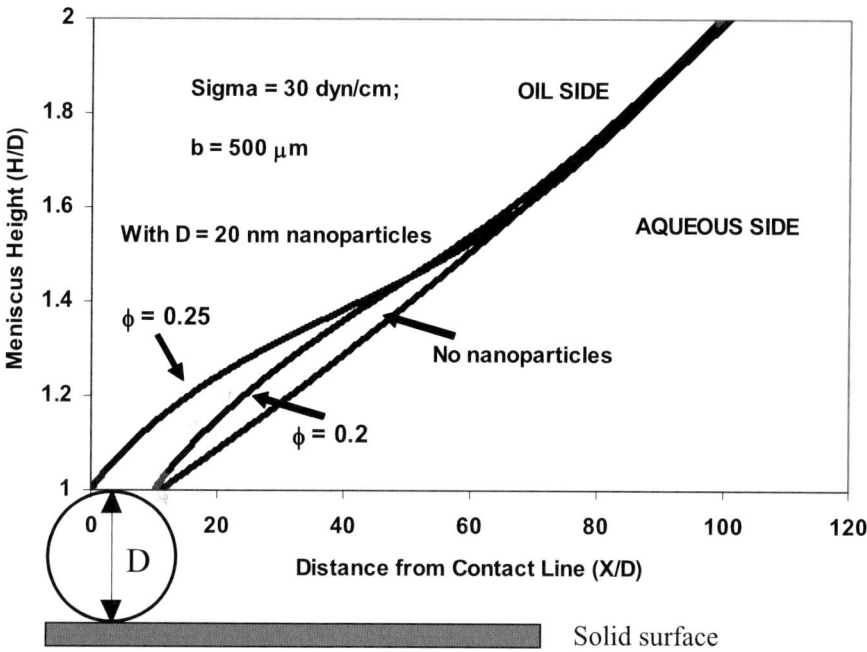

Fig. 12 Effect of nanoparticle volume fraction on contact line displacement at high interfacial tension

range themselves in well-ordered layers. This ordering is a consequence of the fact that this arrangement increases the entropy of the overall suspension by permitting greater freedom for the nanoparticles in the bulk liquid. The result of this arrangement is an excess pressure in the film relative to the bulk; the pressure (and energy) profile show an oscillatory decay with film thickness, the period of oscillations being equal to the effective diameter of the nanoparticle. When the film thickness varies with distance as in a wedge geometry, the disjoining pressure (and film energy) also show a variation with respect to distance from the wedge corner. This gradient of disjoining pressure drives fluid motion into the wedge and results in a movement of the wedge corner (or the nominal contact line).

We have considered the case of a fluid wedge that can deform under the action of the disjoining pressure. Our simulations show that the extent of deformation of the meniscus (or fluid interface) increases with increase in the volume fraction of nanoparticles/micelles, when a decrease in the diameter of micelles and with a decrease in the capillary pressure resisting the deformation is smaller. The resulting deformation of the meniscus causes the contact line to move so that it displaces the fluid that does not contain the micelles (oil) in favor of the fluid that contains it (aqueous surfactant solution).

4
Conclusions

This brief review has described two cases where the spreading behavior of micellar solutions are driven by different forces. In the case of trisiloxane surfactants spreading on hydrophobic surfaces, there is reason to believe that a spontaneously evolving surface tension gradient along the air–liquid interface aids the spreading process. This mechanism has been advanced to qualitatively explain the observed spreading rate behavior as well as the concentration dependence of spreading rate and total covered area. The origin of this surface tension gradient in connection with the surfactant's properties at the air–liquid interface is not yet known and is a subject for further study.

In the case of an anionic surfactant micellar solution displacing an oil drop from a hydrophilic surface, the film energy contributed by the structural disjoining pressure is an important facet that determines the shape of the meniscus and the contact line position. The structural disjoining pressure, which arises from the layering of surfactant micelles/nanoparticles in the thin liquid film near the contact line, causes the fluid interface to deform in such a way that it displaces the oil drop and causes the aqueous solution to spread. This understanding is expected to aid in the synthesis of better detergents by focusing attention on the properties (size, volume fraction) of the micelles in order to maximize the structural disjoining pressure.

Acknowledgements This work was supported in part by the National Science Foundation under Grant No. CTS 01-00854 and by the Office of Science (BER) of the Department of Energy under Grant No. DE-FG0205ER64004.

References

1. Ananthapadmanabhan KP, Goddard ED, Chandar P (1990) A study of the solution, interfacial and wetting properties of silicone surfactants. Colloid Surf 44:281–297
2. Anderson NH, Hall DJ, Wastern NH (1983) The role of dynamic surface tension in spray droplet retention. Proc 10th Int Congr Plant Protect 2:576–581
3. Aveyard R, Binks BP, Clark S, Mead J (1986) Interfacial tension minima in oil–water–surfactant systems. Behavior of alkane–aqueous sodium chloride systems containing AOT. J Chem Soc Faraday Trans 82:125–142
4. Bayer DE (1982) Adjuvants for Herbicides. Weed Science Society of America, Champaign, Illinois
5. Chatterjee J (2002) Critical Eotvos numbers for buoyancy-induced oil drop detachment based on shape analysis. Adv Colloid Interface Sci 98:265–283
6. Chengara A, Nikolov A, Wasan D (2002) Surface tension gradient driven spreading of trisiloxane solution on hydrophobic solid. Colloid Surf A 206:31–39
7. Chengara A (2003) Spreading of colloidal fluids on solid surfaces. PhD thesis, Illinois Institute of Technology

8. Chengara A, Nikolov AD, Wasan D, Trokhymchuk A, Henderson D (2004) Spreading of nanofluids driven by the structural disjoining pressure gradient. J Colloid Interface Sci 280:192–201
9. Chu X L, Nikolov AD, Wasan DT (1995) Thin liquid film structure and stability: The role of depletion and surface-induced structural forces. J Chem Phys 103:6653–6661
10. Chu XL, Nikolov AD, Wasan DT (1996) Errata. J Chem Phys 105:4892
11. Churaev NV, Esipova NE, Hill RM, Sobolev VD, Starov VM, Zorin ZM (2001) The superspreading effect of trisiloxane surfactant solutions. Langmuir 17:1338–1348
12. Davis HT (1996) Statistical mechanics of phases, interfaces and thin films. VCH, New York
13. Derjaguin BV, Churaev NV (1974) Structural component of disjoining pressure. J Colloid Interface Sci 49:249–255
14. Dong J, Mao G, Hill R (2004) Nanoscale aggregate structures of trisiloxane surfactants at the solidliquid interface. Langmuir 20:2695–2700
15. Fox HW, Hare EF, Zisman WA (1955) Wetting properties of organic liquids on high energy surfaces. J Phys Chem 59:1097–1106
16. Gau CH, Zografi G (1990) Relationships between adsorption and wetting of surfactant solutions. J Colloid Interface Sci 140:1–9
17. Harkins WD, Feldman AJ (1922) Films – spreading of liquids and the spreading coefficient. J Am Chem Soc 44:2665–2685
18. Hartland S, Hartley RW (1976) Axisymmetric liquid–liquid interfaces. Elsevier, Amsterdam
19. Henderson D, Sokolowski S, Wasan DT (1997) Second order Percus–Yevick theory for a confined hard sphere fluid. J Stat Phys 89:233–247
20. Hill RM, He M, Davis HT, Scriven LE (1994) Comparison of liquid crystal phase behavior of four trisiloxane superwetter surfactants. Langmuir 10:1724–1734
21. Hill RM (1998) Superspreading. Curr Opin Colloid Interface Sci 3:247–254
22. Hill RM (2002) Silicone surfactants – new developments. Curr Opin Colloid Interface Sci 7:255–261
23. Hirasaki GJ (1991) Wettability: fundamentals and surface forces. SPE Format Evaluat 6:217–226
24. Kabalnov A (2000) Monolayer frustration contributions to surface and interfacial tensions: explanation of surfactant superspreading. Langmuir 16:2595–2603
25. Kalinin VV, Starov VM (1986) Viscous spreading of drops on a wetting surface (English translation). Colloid J USSR 48:907–912
26. Kaplan PD, Rouke JL, Yodh AG, Pine DJ (1994) Entropically driven surface phase separation in binary colloidal mixture. Phys Rev Lett 72:582–585
27. Kao RL, Wasan DT, Nikolov AD, Edwards DA (1988) Mechanisms of oil removal form a solid surface in the presence of anionic micellar solutions. Colloid Surf 34:389–398
28. Knoche M, Tamura H, Bukovac MJ (1991) Performance and stability of the organosilicon surfactant L-77; effect of pH, concentration and temperature. J Agric Food Chem 39:202–206
29. Kralchevsky P, Denkov ND (1995) Analytical expression for the oscillatory structural surface force. Chem Phys Lett 240:385–392
30. Kralchevsky P, Danov KD, Kolev VL, Gurkov VD, Temelska ML, Brenn G (2005) Detachment of oil drops from solid surfaces in surfactant solutions: molecular mechanisms at a moving contact line. Indust Eng Chem Res 44:1309–1321
31. Kumar N, Couzis A, Maldarelli C (2003) Measurement of kinetic rate constants for the adsorption of superspreading trisiloxanes to an air/aqueous interface and the rele-

vance of these measurements to the mechanism of superspreading. J Colloid Interface Sci 267:272–285
32. Marangoni C (1872) Monografia delle bolle liquide (in Italian). Ill Nuovo Cimento 2:239
33. Mori F, Lim JC, Raney OG, Elsik CM, Miller CA (1989) Phase behavior, dynamic contact angle and detergency in systems containing triolene and nonionic surfactants. Colloid Surf 40:323–345
34. Nikolov AD, Wasan DT, Denkov ND, Kralchevsky PA, Ivanov IB (1990) Drainage of foam films in the presence of nonionic micelles. Prog Colloid Polym Sci 82:87–98
35. Nikolov AD, Wasan DT, Chengara A, Koczo K, Policello GA, Kolossvary I (2002) Superspreading driven by Marangoni flow. Adv Colloid Interface Sci 96:325–338
36. O'Brien SBG, Van den Brule BHAA (1991) A mathematical model for the cleansing of silicon substrates by fluid immersion. J Colloid Interface Sci 144:210–221
37. Parker JL, Richetti P, Kekicheff P, Sarman S (1992) Direct measurement of structural forces in a supermolecular fluid. Phys Rev Lett 68:1955–1958
38. Rafai S, Sarker D, Bergeron V, Meunier J, Bonn D (2002) Superspreading: aqueous surfactant drops spreading on hydrophobic surfaces. Langmuir 18:10486–10488
39. Raney K, Benton W, Miller CA (1987) Optimum detergency conditions with nonionic surfactants I. Ternary water–surfactant–hydrocarbon system. J Colloid Interface Sci 117:282–290
40. Raney K, Benton W, Miller CA (1987) Optimum detergency conditions with nonionic surfactants II. Effect of hydrophobic additives. J Colloid Interface Sci 119:539–549
41. Rosen MJ, Wu Y (2001) Superspreading of trisiloxane surfactant mixtures on hydrophobic surfaces 1. Interfacial adsorption of aqueous trisiloxane surfactant – n-alkyl pyrrolidinone mixtures on polyethylene. Langmuir 17:7296–7305
42. Stevens PJG, Kimberely MO, Murphy DS, Policello GA (1993) Adhesion of spray droplets to foliage – the role of dynamic surface tension and advantages of organosilicone surfactants. Pesticide Sci 38:237–245
43. Stoebe T, Lin Z, Hill RM, Ward MD, Davis HT (1997) Superspreading of aqueous films containing trisiloxane surfactant on mineral oil. Langmuir 13:7282–7286
44. Stoebe T, Hill RM, Ward MD, Scriven LE, Davis HT (1996) Surfactant-enhanced spreading. Langmuir 12:337–344
45. Sun JS, Foy CL (1996) Structurally related organosilicone surfactants, their physicochemical properties and effects on uptake and efficacy of primisulfuron in velvetleaf (*Abutilon theophrasti* Medicus). FRI Bulletin 193:225–230 (Proceedings of the fourth international symposium on adjuvants for agrochemicals, 1995)
46. Svitova T, Hill RM, Smirnova Y, Stuermer A, Yakubov G (1998) Wetting and interfacial transitions in dilute solutions of trisiloxane surfactants. Langmuir 14:5023–5031
47. Svitova TF, Hill RM, Radke CJ (2001) Spreading of aqueous trisiloxane surfactant solutions over liquid hydrophobic substrates. Langmuir 17:335–348
48. Thompson J (1855) On certain curious motions observable at the surfaces of wine and other alcoholic liquors. Phil Mag Ser 4:330 (summarized in Maxwell JC (1878) Capillary action. Encyclopedia Brittanica, V, 9 ed, Samuel L Hall, New York)
49. Thompson L (1994) The role of oil detachment mechanisms in determining optimum detergency conditions. J Colloid Interface Sci 163:61–73
50. Tiberg F, Cazabat AM (1994) Spreading of thin films of ordered nonionic surfactants: origin of the stepped shape of the spreading precursor. Langmuir 10:2301–2306
51. Trokhymchuk A, Henderson D, Nikolov A, Wasan DT (2001) A simple calculation of structural and depletion forces for fluids/suspensions confined in a film. Langmuir 17:4940–4947

52. Trokhymchuk A, Henderson D, Nikolov A, Wasan D (2004) Interaction between Macrosphere and Flat Wall Mediated by a Hard-Sphere Colloidal Suspension. Langmuir 20:7036–7044
53. Trokhymchuk A, Henderson D, Nikolov A, Wasan D (2005) In-layer structuring of like-charged macroions in a thin film. Ind Eng Chem Res 44:1175–1180
54. Von Bahr M, Tiberg F, Zhmud BV (1999) Spreading dynamics of surfactant solutions. Langmuir 15:7069–7075
55. Wagner R, Wu Y, Czichocki G, Berlepsch HV, Rexin F, Perepelittchenko L (1999) Silicon-modified surfactants and wetting: II. Temperature-dependent spreading behavior of oligoethylene glycol derivatives of heptamethytrisiloxane. Appl Organomet Chem 13:201–208
56. Wasan DT, Nikolov A (1999) Structural transitions in colloidal suspensions in confined films. In: Manne S, Warr GG (eds) Supramolecular Structure in Confined Geometries. ACS Symp Ser 736. American Chemical Society, Washington, pp 40–53
57. Wasan DT, Nikolov AD (2003) Spreading of nanofluids on solids. Nature 423:156–159
58. Wu Y, Rosen MJ (2002) Superspreading of trisiloxane surfactant mixtures on hydrophobic surfaces 2. Interaction and spreading of aqueous trisiloxane surfactant – n-alkyl pyrrolidinone mixtures in contact with polyethylene. Langmuir 18:2205–2215
59. Zhu S, Miller WG, Scriven LE, Davis HT (1994) Superspreading of water-silicone surfactant on hydrophobic surfaces. Colloids Surf A 90:63–78

Fluorescence Probing of the Surfactant Assemblies in Solutions and at Solid–Liquid Interfaces

Pramila K. Misra[1] (✉) · P. Somasundaran[2]

[1] Center of Studies in Surface Science and Technology, Department of Chemistry, Sambalpur University, 768019 Jyoti Vihar, India
pramila_61@yahoo.co.in

[2] Langmuir Center for Colloids and Interfaces, Henry Krumb School of Mines, Columbia University, New York, 10027, USA

1	Introduction	144
2	Types of Surfactant Assemblies	144
2.1	Micelle	144
2.2	Reverse Micelles	145
2.3	Microemulsions	145
2.4	Vesicles and Liposomes	146
2.5	Solloids or Hemimicelles	147
3	Fluorescence Spectroscopy	147
3.1	Fluorescence Spectroscopic Principles	148
3.2	Characteristics of a Fluorescent Probe	150
3.3	Some Common Fluorescent Probes	150
4	Fluorescence Probing of Surfactant Assemblies	151
4.1	Organization Studies (Concept of CMC, CAC and CSC)	152
4.1.1	Organization in Solution (Concept of CMC and CAC)	154
4.1.2	Aggregation at Solid–Liquid Interfaces	164
4.2	Micropolarity and Solubilization Site of the Probe	176
4.3	Aggregation Number (Hemimicellar Number) and Micellar Size	179
4.4	Micelle Fluidity (Nanoviscosity)	181
5	Conclusions	184
	References	185

Abstract Surfactant assemblies such as micelles, reverse micelles and solloids or hemimicelles, are fluctuating and disordered assemblages of surfactants. These assemblies are formed when the concentration of the surfactants, in polar or nonpolar solvents exceeds a threshold value. Because of their vast applications in both industrial and house-hold affairs, there has been considerable research on the formation and properties of surfactant assemblies in solution and at solid–liquid interfaces. Despite many studies of these organized assemblies using X-ray, NMR, light scattering, conductivity and calorimetry, information on their nanostructures, particularly at solid–liquid interfaces is still very limited. In this article, we review the application of fluorescence techniques to investigate aggregate characteristics such as size, micropolarity, nanoviscosity and aggregation number.

Keywords Fluorescence probing · Hemimicelle · Micellar fluidity · Micelle · Organized assemblies in solution and interfaces · Polarity parameter · Pyrene

1
Introduction

Organizational characteristics of surface-active molecules have been studied by several researchers due to their applications in many areas such as personal care, polymerization, catalysis, drug delivery, separation and purification, enhanced oil recovery and lubrication. The structure of supramolecular organized assemblies formed in different solvents, when a critical concentration is exceeded, determines their properties such as solubilization [1–3], catalysis [1, 4–6], adsorption [7–11] and flocculation [12, 13]. As such, many techniques have been used to determine their structural properties. In this paper, the results obtained using fluorescence probing for properties of assemblies in solution and at solid–liquid interfaces are discussed in detail after a brief review of relevant assemblies formed by them.

2
Types of Surfactant Assemblies

Because of the limited solubility in different solvents surfactants form different types of surfactant assemblies in solutions and on solids. These organized assemblies are formed when different proportions of surfactants, oils, co-surfactants and water are mixed together. The types of surfactant aggregate formed depends on its chemical structure and the nature of the medium.

2.1
Micelle

Above a critical micellar concentration, surfactants readily aggregate to form molecular clusters called micelles with the hydrophobic tails of the surfac-

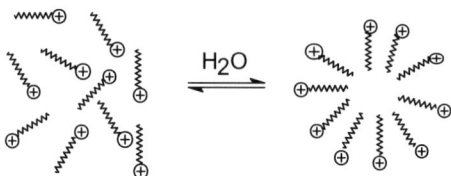

Fig. 1 Micelle

tants directed inwards forming a nonpolar core, and the polar groups in contact with the external aqueous phase (Fig. 1). The central core is disordered and has properties similar to those of the liquid hydrocarbon from which the surfactant is formed. Extensive work [14–18] has been done on the structure and properties of micelles. In some cases, inverted micelles are found to form in pure nonaqueous solvent [19–21].

2.2
Reverse Micelles

Surfactants can aggregate in nonpolar solvents in the presence of small amounts of water with the tails oriented towards the bulk nonpolar solution and head groups interacting with water in the center (Fig. 2). The water pool formed in reverse micelles has been used as a medium to study chemical and biological reactions [22].

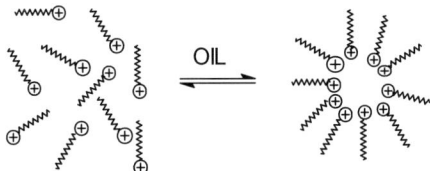

Fig. 2 Reversed micelles

2.3
Microemulsions

Microemulsions are thermodynamically stable, homogeneous, optically isotropic solutions comprised of a mixture of water, hydrocarbons and amphiphilic compounds. The microemulsions are usually four- or three-component systems consisting of surfactant and cosurfactant (termed as emulsifier), oil and water. The cosurfactants are either lower alkanols (like butanol, propanol and hexanol) or amines (like butylamine, hexylamine). Microemulsions are often called swollen micelles (Fig. 3) and swollen re-

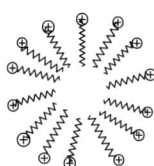

Fig. 3 Swollen micelles

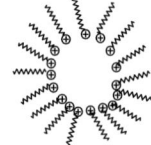

Fig. 4 Swollen reverse micelles

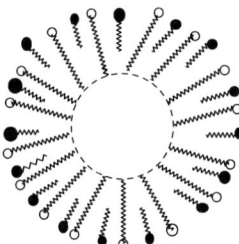

Fig. 5 O/W micelles

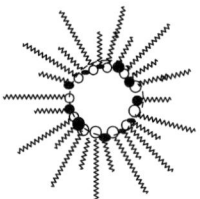

Fig. 6 W/O microemulsion

verse micelles (Fig. 4). The oil-in-water type of microemulsions (Fig. 5) can be thought of as swollen micelles containing a large volume of oil. Similarly, the water-in-oil microemulsion (Fig. 6) can be thought of as swollen reverse micelles containing a large volume of water.

2.4
Vesicles and Liposomes

Vesicles (Fig. 7) can be considered as spherical containers with diameters of the order of 10^{-6} cm and a thickness of 50 Å containing 80 000 to 100 000 surfactant molecules. The surfactants having two alkyl chains with polar groups are able to form these closed bilayers. Once formed, vesicles, unlike micelles do not easily break down into individual surfactants. Depending on their chemical composition, vesicles remain stable for days to weeks.

Liposomes (Fig. 8) are completely closed vesicular bilayer structures formed by exposing phosphoglycerides in water suspension to sonic oscil-

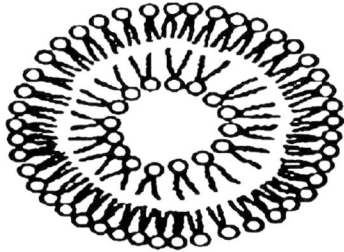

Fig. 7 Vesicles

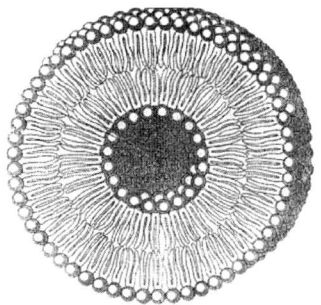

Fig. 8 Liposomes

lations. The liposomes have been reported to form in systems composed of phosphatidylcholine, phosphatidylethanolamine, phosphatidylinositol, and phosphatidic acid.

2.5
Solloids or Hemimicelles

Some surfactants aggregate at the solid–liquid interface to form micelle-like structures, which are popularly known as hemimicelles or in general solloids (surface colloids) [23–26]. There is evidence in favor of the formation of these two-dimensional surfactant aggregates of ionic surfactants at the alumina–water surface and that of nonionic surfactants at the silica–water interface [23–26].

3
Fluorescence Spectroscopy

Numerous techniques measuring properties such as conductance [27–29], surface tension [30–35] and solubilization of a sparingly soluble compound [36–40], and NMR [41–43] have been employed along with calorimetric

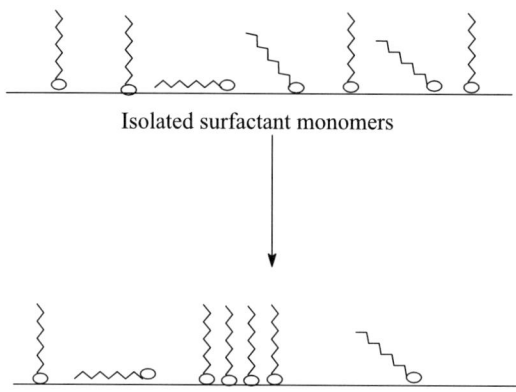

Fig. 9 Hemimicelles

measurements [44–46] with changes in surfactant concentration to study micellization and various spectroscopic properties [37–40]. Recently [47–51], fluorescence spectroscopy has become an important tool because of its ability to explore the structure of organized surfactant systems at the molecular level.

3.1
Fluorescence Spectroscopic Principles

The electronic excitation of a ground-state molecule is achieved by absorption of photons from incident light and this absorption usually produces an excited singlet, S^* in an upper vibrational level. This molecule degenerates to the lowest vibrational state of the first excited state within $10^{-12} - 10^{-10}$ s through thermal relaxation. The excited molecule is different from its ground state due to its large energy content and weak attractive forces. Two types of excited electronic states are produced by energy absorption: states with paired spins (S) are referred to as singlet states, while those with unpaired spins (T) are referred to as triplet states. S_1 and T_1 denote the lowest energy level of the corresponding excited states. The average time that a molecule spends in the excited state before returning to the ground state with or without the radiation of a photon is called the natural lifetime of the molecule. The lifetime of S is about $10^{-8} - 10^{-10}$ s while that of T is from minutes to 10^{-6} s. The excess energy is relinquished through different photophysical pathways within the natural lifetime of the molecule. The pathways are either radiative or nonradiative as illustrated in Fig. 10. An electronically excited molecule can lose its energy by emission of radiation (luminescence) in the following ways [52]:
(i) Fluorescence: a radiative emission process occurring from a thermally equilibrated excited state S_1 to the ground state S_0. The fluorescence

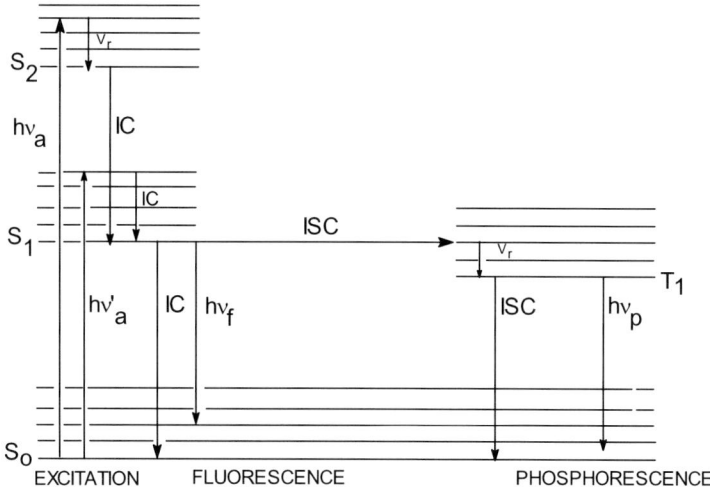

Fig. 10 Energy state diagram showing important photophysical processes following excitation of the luminescent probe

spectrum is located at higher wavelength than those of the absorption because of the energy loss in the excited state due to the vibrational relaxation.
(ii) Phosphorescence: a radiative emission process occurring from T_1 to S_0.
(iii) Excimer fluorescence: a radiative emission process occurring as a result of the decomposition of an excimer. An excimer is a molecular dimer aggregate formed between an excited molecule in the S_1 state and a molecule in the ground state S_0. Excimers are stable in the excited state but are unstable in the ground state. Excimer fluorescence is observed as a broad structureless band that is red shifted with respect to the "normal" fluorescence.
(iv) Delayed fluorescence: a radiative emission process occurring from the decomposition of an excimer formed between excited triplet states.
(v) In some cases a transient complex is formed between the excited species with the ground state of some other species. The decomposition of this complex may lead to the emission of some characteristic radiation.

An electronically excited molecule can also lose its energy through the following radiationless pathways:
(i) Internal conversion: a radiationless process that occurs between states of the same multiplicity (spin states), i.e., $S_n \rightarrow S_0$ or $T_n \rightarrow T_0$. In this process, the molecules lose their energy to the surroundings in the form of heat.
(ii) Intersystem crossing: a radiationless process that occurs with spin inversion, e.g. $S_1 \rightarrow T_1 +$ heat.

(iii) Nonradiative energy transfer: a process whereby an excited molecule (donor) in the singlet or triplet state transfers its energy to a molecule in the ground state (acceptor) thereby bringing the latter into its excited state.

3.2
Characteristics of a Fluorescent Probe

Fluorescence is usually observed in those organic molecules that have a rigid framework and do not have many coupled vibrational energy levels through which the excitation energy can be purged out. Complex organic molecules that are commonly used as luminescence probes typically absorb energy in the spectral range of 250–650 nm which corresponds to transitions with energy changes of 2–5 eV [53]. The utility of the fluorescence technique is due to the fact that numerous probes exist that have a strong and selective affinity for surfactant assemblies. The fluorescence of nonpolar arenes, which are typically used in this studies, are highly sensitive so that a minimal amount of probe is required and thus the system has very little or almost negligible distortion. The suitability of these molecules as fluorescent probes is due to the following:

(i) Owing to their hydrophobic nature, these probes are incorporated exclusively within the surfactant assemblies during their excited lifetimes, i.e., they are short-lived compared to the time they take to exit from a surfactant assembly.
(ii) The micelle may be viewed as being intact during the excited lifetimes of the probes.
(iii) They have a high extinction coefficient and high quantum yield for fluorescence efficiency.
(iv) The placement of the probe in a given matrix does not perturb the matrix structure.
(v) The probes are photo-stable.
(vi) The probes have rigid chemical structure.

3.3
Some Common Fluorescent Probes

A suitable fluorescent probe is an organic molecule, which must change its characteristic parameters with changes in its microenvironment and the parameter must be measurable when the probe is added to the system [54]. The fluorescent probes are categorized as either extrinsic, intrinsic, or covalently bound probes. The intrinsic probes allow a system to be observed without any chemical perturbation. This occurs when the system to be characterized has an in-built fluorescent chromophore unit like tryptophan, tyrosine and phenyl alanine in protein. In some cases the fluorophore is covalently attached to the system like pyrene-labeled poly(acrylic acid). Noncovalently bound fluorophores are called extrinsic probes. Depending on the nature of information they provide, they are either called hydrophobic, polar, excited state proton transfer (ESPT) or anisotropic fluorescent probes.

Several organic molecules, such as pyrene [55–58], Nile red [59, 60], 1,6-diphenyl-1,3,5-hexatriene (DPH) [61, 62], 6-propionyl-2-(dimethylamino) naphthalene (prodan) [63–65], xanthene dyes [66], 7-azatryptophan [67], 2,2,6,6,-tetramethyl-4-hydroxypiperidinium ester of 4-(1-pyrene)butyric acid [68], 8-anilino-1-naphthalenesulfonic acid (ANS) [69–71], octadecylrhodamine B [72, 73], and auramine O [74], Coumarin 153 [75, 76], 2-methylanthracene [77], 1-pyrenesulfonic acid [78], 9-chloromethylanthracene [79], 5-(dimethylamino)1-naphthalene sulfonamide [80], 1-decylpyrene [81], 1,10-bis(1-pyrene) decane [81], tryptophan [82], n-(anthroyloxy) stearic acid [83], dinaphthylpropane (DNP) [84, 85], acridine orange [86], N-phenyl-1-naphthylamine (NPN) [87], pyrene-3-carboxaldehyde [88, 89], 1,4-diphenylbutadiene [90], perylene [91], benzyl-1-pyrenoate [92], 1,3-dipyrenylpropane [93], Fluorazaphore-L [94], 6-dodecanoyl-2-(dimethylamino)-naphthalene (Laurdan) [95], tetrakis (sulfonato phenyl)porphyrin (TTPS) [96], pyrene labeled poly(acrylic acid) [97], Na 11-(3-hexyl-1-indolyl)undecyl sulfate [98], Nitrobenzoxadiazolyl (NBD) [99], N-lissamine rhodamine B [100], K 2-p-toluidinylnaphthalene-6-sulfonate (TNS) [101], 1,3-bis(1-pyreneyl)propane [102], have been used as extrinsic fluorescent probes.

These fluorescent probes have been successful in reporting the structural parameters of surfactant assemblies such as micelles [103], reverse micelles [104], ternary systems [105], swollen micelles [106], microemulsion [107], vesicles [108], liposomes [109], hemimicelles [110], monolayers [111] and bilayers [111].

4
Fluorescence Probing of Surfactant Assemblies

Fluorescence probing of the surfactant assemblies involves the measurement of changes in the emission properties of the molecule (fluorescent probe) or photochemical-intermediate (excimer) such as their spectral distribution (maximum wave length, maximum intensity), quantum yield, fluorescence lifetimes (the time before it starts radiating, and is equal to the time taken by the excited state to decay 1/e of its initial value, e being the base of the natural logarithm) during the transition from the monomeric stage of the surfactants to the aggregates. These quantities are quantitatively related to the various experimental parameters by the following expressions:

$$I_f = I_0(1 - 10^{-\epsilon cl})\Phi_f, \tag{1}$$

where, I_f is the fluorescence intensity, I_0, the intensity of the light, ϵ molar absorptivity, and c is the molar concentration of the solute, l is the path length, Φ_f is the quantum yield of fluorescence, and

$$I_{f(t)} = I_{f(0)} \exp^{-t/\tau}, \tag{2}$$

where $I_{f(0)}$ is the initial fluorescence intensity at time $t = 0$, $I_{f(t)}$ is the fluorescence intensity at time t and τ is the average lifetime of the excited state singlet states.

These responses are sensitive to changes in the microenvironment of the probe, so that the fluorescent probe in different microenvironments displays distinct luminescence properties. These have been well documented [112]. Such data has been applied to micellar photophysics and photochemistry to study the properties of the micellized state by Turro et al. [113]. We discuss here the use of these changes to investigate the characteristic structural parameters such as micropolarity (solubilization sites of the probe), size of the aggregates (aggregation number), and micellar fluidity, hemimicellar number, mechanism of adsorption, etc. In view of the extensive research in this area and hence availability of a large volume of literature, we limit discussions to our contributions to this area as the working model, highlighting some new works.

4.1
Organization Studies (Concept of CMC, CAC and CSC)

At very low concentrations ($< 10^{-5}$ M) surfactants remain soluble due to the hydrophobic hydration of the hydrocarbon chain and solvation of the ionic head. But at high concentrations, the solubilization of the hydrocarbon chain becomes difficult and a pronounced deviation from ideal behavior in dilute solution occurs. When some physical properties like surface tension, electrical conductance, specific heat, osmotic pressure, and electromotive force are plotted as a function of the surfactant concentration, the plot of the type shown in Fig. 11 is obtained. This pronounced change in physical properties occurs due to the aggregation of amphiphile monomer in the solution or at solid–liquid interfaces. The concentration at which this deviation occurs in solution is referred to as the critical micelle concentration (CMC) [114] and where the aggregation at the solid surface begins, i.e., surface colloids form, is referred to as the critical solloids (CSC) concentration [115]. Sometimes, aggregation of two different species occurs at interfaces above a certain concentration, which is referred to as the critical aggregation concentration (CAC) [116, 117]. This aggregation concentration occurs either below(synergism) [118] or above(antagonism) [119] the CMC of the individual species.

Recently, analysis of spectroscopic parameters of solubilized probes has become quite a useful tool to investigate the formation and organization of

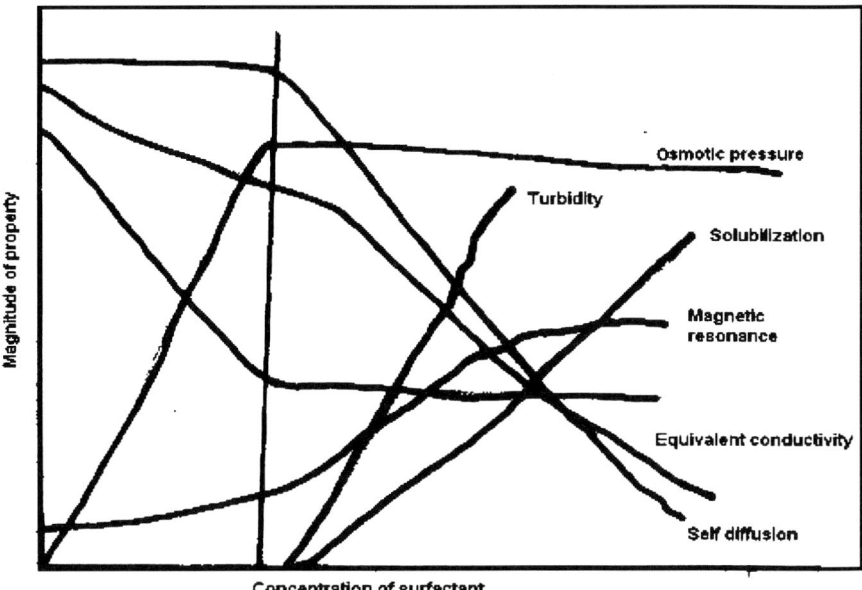

Fig. 11 Schematic representation of the concentration dependence of some physical properties for solutions of micelle-forming amphiphiles

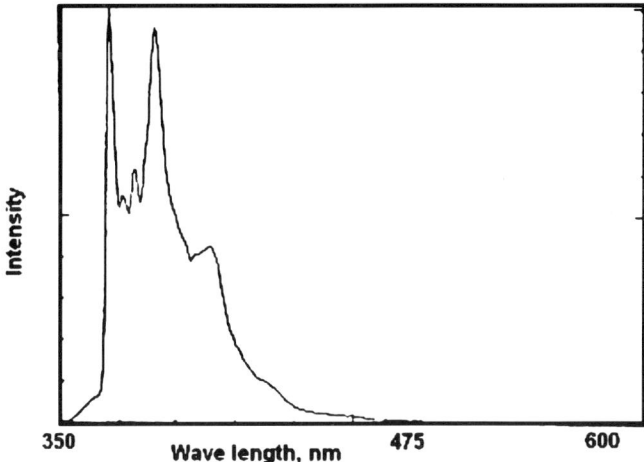

Fig. 12 Typical emission spectrum of pyrene

these assemblies. A literature survey reveals that the symmetrical aromatic molecules such as pyrene or derivatives of pyrene are well studied and frequently used as fluorophores [54–57]. The monomer emission spectrum of pyrene is a structured spectrum with five sharp peaks [120] (Fig. 12). The

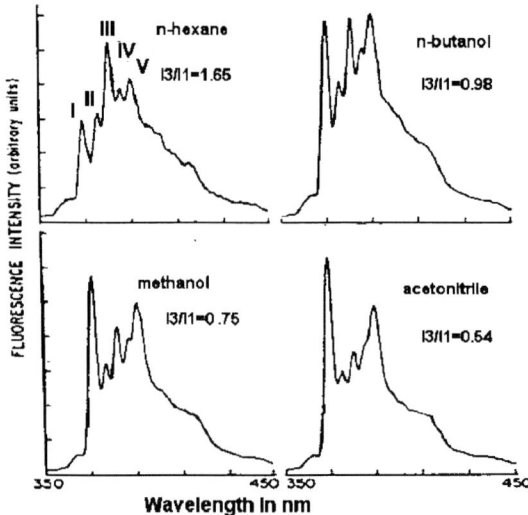

Fig. 13 Emission spectrum of pyrene in different solvents showing the change in peak intensity with the change in the solvent

Table 1 $E_{T(30)}$ values of various solvents and I_3/I_1 of pyrene in corresponding solvents

Solvent	I_3/I_1	$E_{T(30)}$
Water	0.595	63.1
Methanol	0.760	55.5
n-Butanol	0.980	50.2
Dimethyl acetamide	0.984	43.7
n-Hexane	1.650	30.9

change in the intensities of the first and third vibrational emission peak, i.e., the ratio I_3/I_1 (or I_1/I_3), depends on the polarity of the solvent. This ratio is found to decrease with an increase in polarity of the medium (Fig. 13 and Table 1) and in the presence of various micelles [112, 120]. Thus, it is also referred to as the polarity parameter. Usually very low concentrations of the probe, within 10^{-7}–10^{-6} M of pyrene are used [121, 122]. The advantage of a change of the vibrational structure of the pyrene emission spectrum is monitored as a function of surfactant concentration to elucidate the mechanism of aggregation and characterization of surfactant aggregates.

4.1.1
Organization in Solution (Concept of CMC and CAC)

Sahoo et al. [123] have studied the micellar behavior of a series of polyoxyethylated alkyl ethers (a: see below) and some anionic surfactants such as

alkyl olefinic sulfonate (b-i), alkyl benzene sulfonate (b-ii) and polyoxyethylated sodium lauryl sulfonate (b-iii) using pyrene as the fluorescent probe.
(a) Polyoxyethylated alkyl ethers: R–O–(CH$_2$CH$_2$O)$_x$–H, x represents the number of poly(oxyethylene)units
  (i) R = dodecyl, x = 9, 21, 25 (BL-X)
  (ii) R = hexadecyl, x = 15, 20, 25, 30, 40 (BC-X)
  (iii) R = 9-octadecenyl, x = 15, 20, 50 (BO-X)
(b) Anionic surfactants:
  (i) Alkene-1-sulfonate CH$_3$–(CH$_2$)$_{12}$–CH=CH–SO$_3^-$Na$^+$
  (ii) Linear alkylbenzenesulfonate

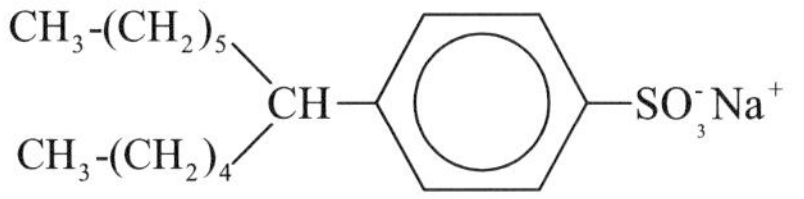

  (iii) Sodium dodecyl [poly(oxyethylene)](2,4)sulfonate
       CH$_3$–(CH$_2$)$_{11}$–O–(CH$_2$–CH$_2$–O)$_{2,4}$–SO$_3^-$Na$^+$

At very low concentrations of the surfactant, the ratio of the intensities of the third peak (380 nm) to that of the first peak (370 nm) is found to be 0.6, which is similar to the characteristics of pyrene monomer in water. The ratio undergoes a gradual increase with an increase in the concentration of the surfactant reaching a value of 0.83–1.05 attaining a plateau at higher concentrations. The plot is of sigmoidal nature (Fig. 14). The constant value above a certain concentration suggests the constant polarity experienced by pyrene, which has been attributed to a consequence of the micelle formation and the concentration at the onset of the plateau is referred to as the CMC. This CMC is found to depend on the number of poly(oxyethylene) units and the hydrocarbon chain length of the hydrocarbon moiety. With an increase in the surfactant concentration, a new peak at around 480 nm appears well below the CMC. The intensity of the peak increases with a subsequent decrease in monomer intensity. After attaining a maximum value much below the CMC, this peak vanishes or diminishes before reaching the plateau (Fig. 14). This peak has been attributed to the formation of pyrene excimers, complexes formed between excited pyrenes with its ground-state species.

The photophysics of formation of pyrene monomer and excimer can be represented as shown in Scheme 1. In sufficiently dilute systems, when there is no micelle

$P \rightarrow P^* \rightarrow P + h\nu$
$P + P^* \rightarrow P_2^* \rightarrow P + P + h\nu$

Scheme 1

in the systems the radiative decay of P^* (excited state pyrene) occurs. Since the probe prefers to stay in the hydrophobic medium, the probe partitions to

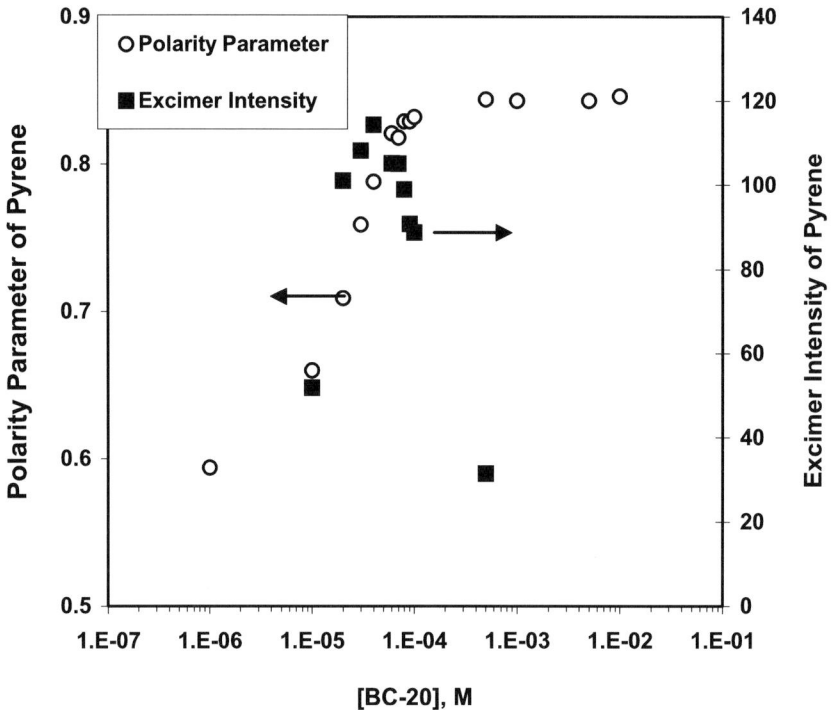

Fig. 14 Plot of polarity parameter (I_3/I_1) and excimer intensity of pyrene at various concentrations of BC-20 (CMC = 8.0×10^{-5} M, $C_{Mex} = 5.0 \times 10^{-5}$ M)

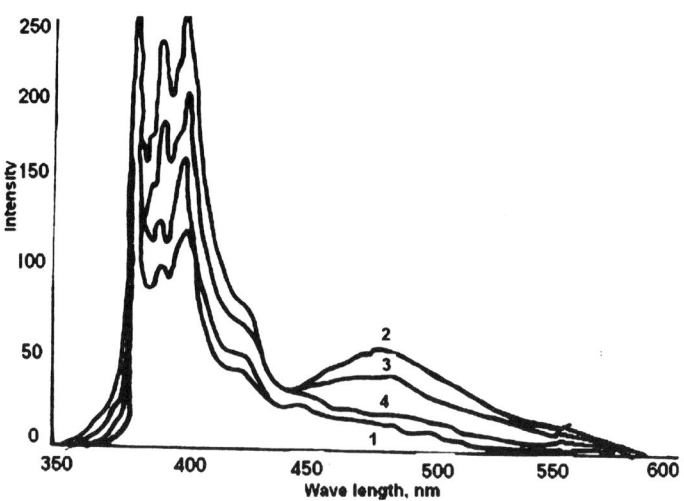

Fig. 15 Emission spectra of pyrene in the presence of various concentrations of BO-20 surfactant (1) 10^{-6} M, (2) 4×10^{-5} M, (3) 2×10^{-4} M, (4) 6×10^{-4} M

the micelle as soon as it is formed, thereby increasing the effective concentration of the probe inside the micelle. With the increase of the concentration of the probe, a P^* can interact with a ground-state P to form P-P^*. In the case of the pyrene the excimer fluorescence can be readily observed as a broad spectrum centered around 480–490 nm that is red-shifted with respect to the monomer fluorescence (360–420 nm, Fig. 15). Pyrene excimers give considerable information on the organization of the surfactants.

The appearance of the excimer peak has been attributed to the formation of a premicelle or nascent micelle aggregate, which because of its small size and hydrophobic nature allows the solubilized excited pyrene to interact with the ground-state pyrene to form an excimer. The peak vanishes in the plateau due to the difficulties for the ground-state pyrene to encounter the excited species in the relatively large microdomain of the micelle. This observation led Sahoo et al. [123] to propose a step-wise formation for the micelles (Scheme 2, Fig. 16)

$M_1 + M_1 \rightarrow M_2$
$M_2 + M_1 \rightarrow M_3$
. . . .
$M_1 + M_{n-1} \rightarrow M_n, \quad n = 2, 3, 4, 5, \dots$
Scheme 2

where M represents surfactant monomer and M_2 to M_n represent the surfactants of aggregates of different size.

Our group [124] has used pyrene and pyrene labeled poly(acrylic acid) as a fluorescent probe to investigate the interaction between poly(acrylic acid) (PAA) and dodecyltrimethylammonium bromide (DTAB) in water. We have measured the I_3/I_1 ratio of pyrene as a function of DTAB in the presence of 1 g/L PAA. A sharp decrease in polarity is found well below the CMC of DTAB (1.3×10^{-2} M, Fig. 17). The onset of this polymer-induced association is referred to as the critical aggregation concentration. The CAC has been measured at various pH and NaCl levels (Fig. 18). It was observed that the

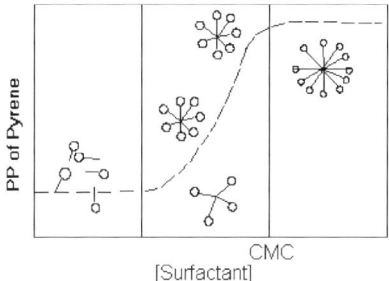

Fig. 16 Schematic representation of the formation of micelles

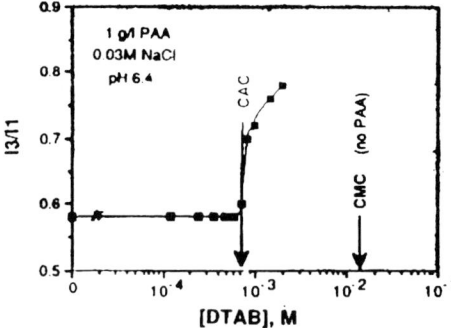

Fig. 17 Ratio of third to first vibrational peak of pyrene emission spectrum as a function of DTAB concentration in the presence of 1 g/L sodium polyacrylate

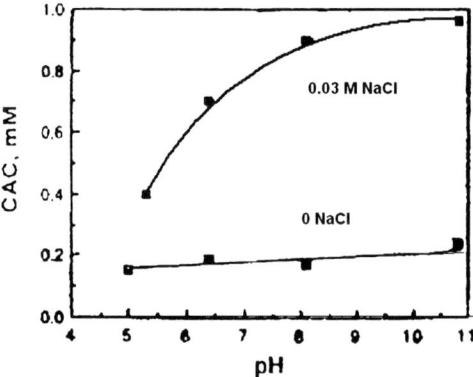

Fig. 18 Critical aggregate concentration, CAC of DTAB in 1 g/L sodium polyacrylate in 0 and 0.03 M added NaCl

addition of salt leads to an increase in the CAC as a result of the electrostatic shielding of the surfactant headgroups by cationic counter ions. The CAC in addition to being a measure of the extent of electrostatic binding also measures the efficiency of association of the polymer with the surfactant depending on the conformational flexibility of the polymer, which in turn depends upon the degree of ionization, α. When the polymer is highly charged (large α), the association with the surfactant becomes difficult due to segment–segment repulsion. But with small values of α, the polymer adopts a coiled conformation thereby allowing an increase in the association with the surfactant. It was suggested that at low NaCl level the electrostatic binding efficiency and segment–segment repulsive contribution offset each other so that the net driving force for the association is the hydrophobic chain interaction, which is therefore independent of pH. But at high salt concentrations, electrostatic binding of the surfactant is suppressed so that the pH-dependent CAC observed is essentially due to the conformational difference of the poly-

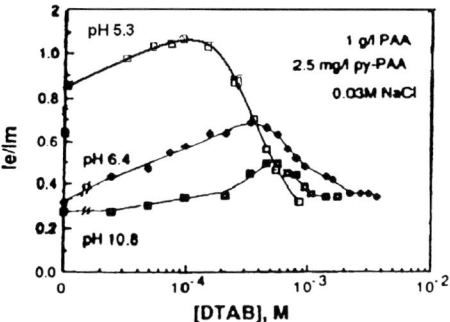

Fig. 19 Excimer to monomer ratio, I_e/I_m of pyrene-labeled poly(acrylic acid), py-PAA, as a function of DTAB concentration in 1 g/L sodium polyacrylate and 0.03 M NaCl

mer. To obtain further information on the conformation of the polymer—due to the surfactant—the ratio of intensity of excimer (I_e) to the monomer (I_m) of the pyrene labeled PAA has been determined as a function of DTAB concentration at three pH values (Fig. 19). The I_e/I_m value is found to increase initially and then decrease. This general trend is observed at all pH values above 5 and at both NaCl levels tested (0 and 0.03 M NaCl). The formation of excimer can take place in two different ways: dynamic excimer formation occurs through the excitation of isolated pyrene groups and subsequent diffusional encounters with ground-state pyrene, or static excimer formation can occur if the pyrene groups on the polymer are in juxtaposition either due to the statistical conformation of the chain or due to the ground-state aggregation of the pyrene groups. On examination of the fluorescence spectra and the excimer decay profiles obtained at the DTAB concentrations before the maximum in I_e/I_m and after the maximum (Figs. 20 and 21), it is seen that the excimer centers around 490 nm in water and at a premaximum concentration of DTAB. Above the maximum the emission wavelength shifts to 486 nm. The premaximum excimer excitation spectrum (emission at 490 nm) is red shifted by about 4 nm compared to the monomer excitation spectrum (emission at 376 nm) as shown in Fig. 20a. In the postmaximum region, the red shift is progressively less with the addition of DTAB (Fig. 20b). The fluorescence decay profiles of the excimer in the premaximum region indicates no initial growth (rise time), whereas in the postmaximum region an initial growth is observed which is characteristic of a diffusion-controlled excimer formation process (nanosecond time scale, Fig. 21A and B). This observation suggests the existence of ground-state pyrene dimers or aggregates at the premaximum DTAB concentration (as well as in water) so that the excimer formation is a static process in this region. With the addition of DTAB above the maximum, the ground-state aggregates dissociate and the absorbing species are the isolated pyrene groups. Thus, at high DTAB concentrations, the excimer formation is dynamic in nature with a characteristic rise time. Since the

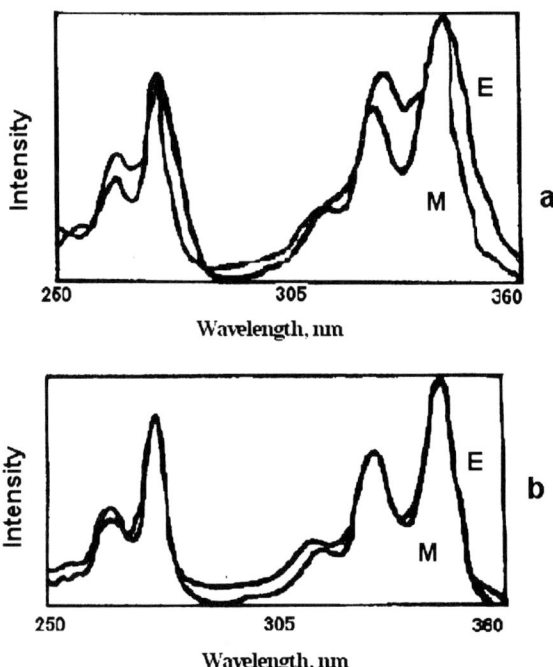

Fig. 20 Excitation spectra of monomer M (376 nm) and excimer E (480 nm), in **a** no DTAB (premaximum) and **b** 0.001 M DTAB (postmaximum): *points* correspond to pH 6.4 in Fig. 19

ground-state pyrene may be due to the hydrophobic interaction the extent of excimer formation depends on both conformational state of the polymer as well as upon the micropolarity of the environment experienced by pyrene. On the basis of these facts Chander et al. [124] proposed that the addition of a low concentration of DTAB leads to the coiling of the polymer, which increases the excimer formation probability and hence increases I_e/I_m. The decrease of I_e/I_m at high concentrations is due to the solubilization of the pyrene moieties in more hydrophobic surfactant–polymer aggregates resulting in a decrease of static excimers. The CAC value obtained from I_e/I_m compares well with the measured micropolarity (I_3/I_1) values (Fig. 22) and proposed interactions of DTAB with poly(acrylic acid) at low and high pH are illustrated in Fig. 23.

Fluorescence probing has also been used by us to investigate the mechanism of vesicle to micelle transformation due to interaction of liposome with sodium dodecyl sulfate micelles. An increase in optical density and hydrodynamic diameter was observed at low SDS concentrations (Fig. 24). The increase was attributed to the incorporation of SDS monomers on liposome vesicles. The point where the hydrodynamic diameter and optical density reached a maximum was proposed to correspond to the saturation of bilayers of the first inflection point. Upon further increase in the SDS concentration

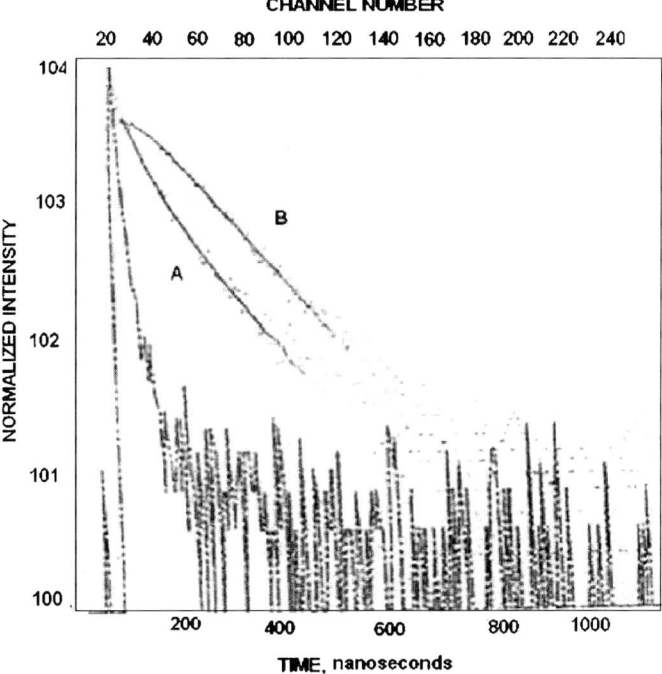

Fig. 21 Fluorescence decay profile of excimer (480 nm) for no DTAB, curve A (premaximum), and 0.001 M DTAB, curve B (postmaximum), corresponding to pH 6.4 in Fig. 19

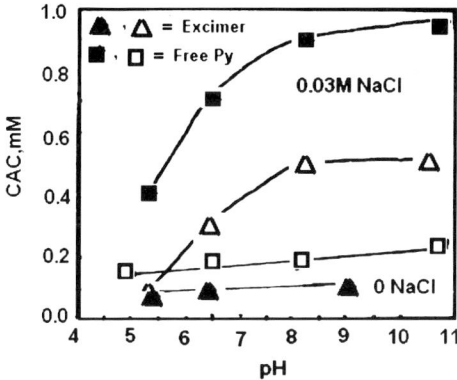

Fig. 22 Comparison of critical aggregate concentration, determined by excimer formation of covalently labeled polymer and micropolarity response of free pyrene

above 1 mM, optical density continuously decreases indicating disruption of the liposome. Light-scattering measurements also suggest that the vesicles and mixed micelles coexist at intermediate concentrations of SDS. However,

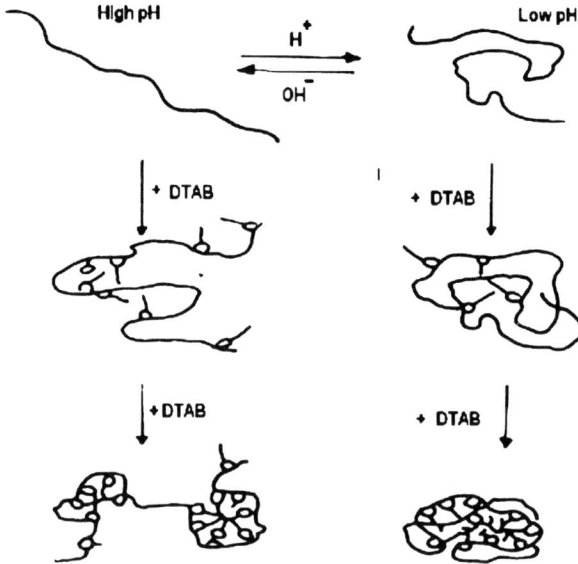

Fig. 23 Schematic representation of the formation of polymer surfactant complexes for the pyrene-labeled (polyacrylic acid)-dodecyltrimethylammonium bromide system for high (high pH) and low (low pH) degrees of ionization

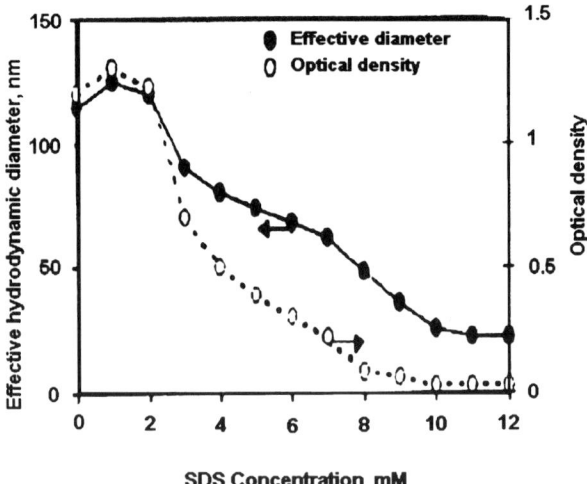

Fig. 24 Change in the optical density and effective hydrodynamic diameter of 1 mM liposome after interaction with SDS for 1 h

as the SDS concentration is increased, liposome concentration was found to gradually decrease which further supports the stepwise-solubilization behavior of liposomes. Results for the surface tension and monomer concentra-

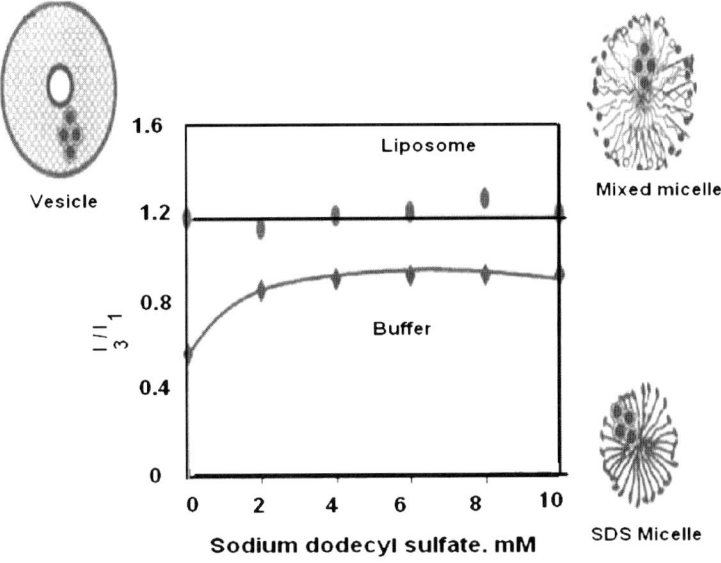

Fig. 25 I_3/I_1 of pyrene in buffered and liposome after interaction with SDS for 1 h

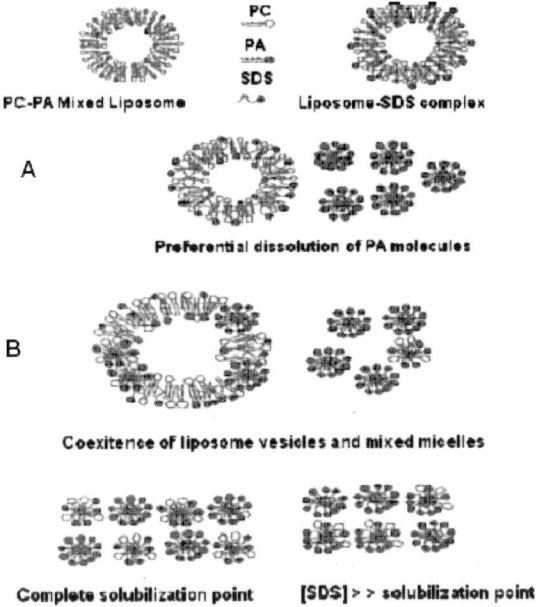

Fig. 26 A Model of the preferential dissolution process of PA at low SDS concentration. **B** Model for the existence of vesicle and mixed micelles at high SDS concentration

tion of SDS suggested that SDS was being consumed by the liposomes for the disintegration processes. Studies [125] using pyrene showed the I_1 peak ($\lambda = 373$ nm) is much higher in the absence of liposome than I_3 ($\lambda = 383$ nm) indicating an aqueous environment for the pyrene molecule. The ratio I_3/I_1 increased and remained constant above 2 mM of surfactant, due to the formation of the SDS micelle (Fig. 25). In the presence of 1 mM of liposome the I_3 peak becomes more intense compared to when it is absent and the ratio does not change even after interaction with the SDS micelles and this is attributed to the presence of pyrene in the mixed micelles. The schematic representation of the solubilization of the liposome is shown in Fig. 26.

4.1.2
Aggregation at Solid–Liquid Interfaces

An insight into the mechanism of surfactant aggregation at solid–liquid interfaces has been obtained in the past by measuring the adsorption isotherm, zeta potential, contact angle and flotation recovery. Information on the nanostructure of the aggregates has been generated more recently by fluorescence probing of the adsorbed layer. As an example, such information obtained using fluorescence is described for the case of adsorption of dodecyl sulfonate on alumina.

A well-studied and often-cited adsorption isotherm of an ionic surfactant is that of sodium dodecylsulfate at the alumina–water interface at pH 6.5 and 0.1 M NaCl. At pH 6.5 and 0.1 M NaCl the S–F isotherm [23]

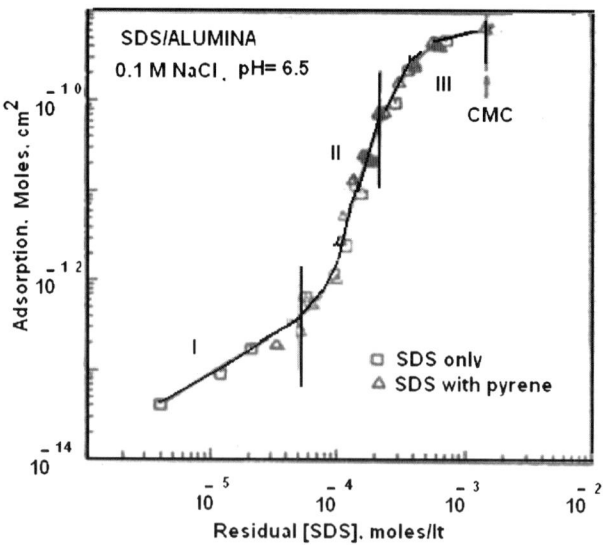

Fig. 27 Adsorption isotherm of SDS on alumina at pH 6.5 in 0.1 kmol/m^3 NaCl

obtained under constant ionic strength conditions for this system is characterized by four distinct regions (Fig. 27), governed by different forces of adsorption:
(i) The first region with a slope unity corresponds to the adsorption of SDS onto the alumina surface through electrostatic interaction between the anionic surfactant and positively charged alumina surface (pzc 9.0) [126].
(ii) The conspicuous increase in the adsorption in the second region is attributed to the formation of surface aggregates through lateral interaction between hydrocarbon chains (solloid or hemimicelle formation).
(iii) The decrease in the slope in region III is due to increasing electrostatic hindrance to the surfactant association following interfacial charge reversal (Fig. 28).
(iv) The plateau adsorption is due to maximum surface coverage as determined by the micelle formation in the bulk or monolayer coverage. Further increase in the surfactant concentration does not increase the adsorption density.

Initially, this mechanism was proposed on the basis of results obtained for zeta potential and flotation (Fig. 29). The formation of the hydrophobic aggregates at the interface was confirmed after the advent of the fluorescence probing technique. The adsorption isotherm is determined in the presence of pyrene as the fluorescent probe and the emission spectra of pyrene in both supernatant and slurries were analyzed after adsorption. The I_3/I_1 of pyrene in solutions of SDS containing 0.1 M NaCl and in the slurry are shown in Figs. 30 and 31. In solution, the ratio remains at around 0.6 till the CMC (as determined by surface tension measurement) is attained. Above CMC, the value becomes ~ 1.0 due to the solubilization of pyrene in micelles. In

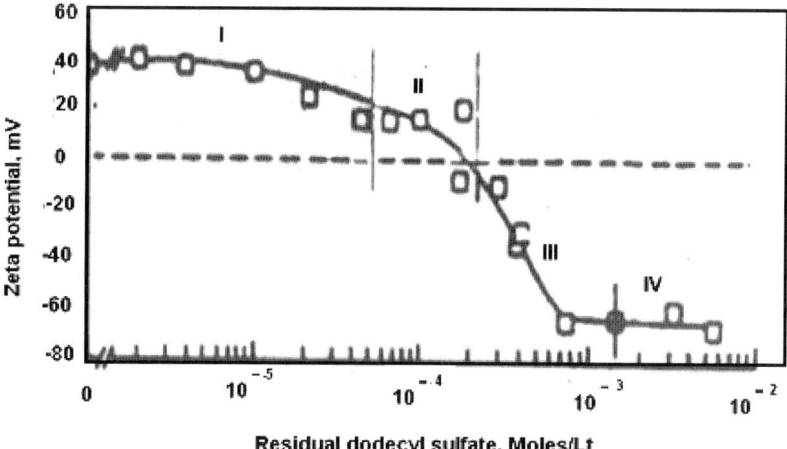

Fig. 28 Zeta potential of alumina as a function of equilibrium concentration of SDS (designation based on the shape of isotherm in Fig. 31)

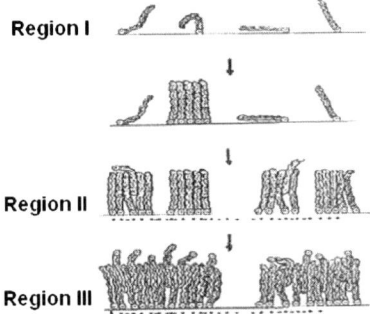

Fig. 29 Schematic representation of the growth of the aggregates for the various regions of the adsorption isotherm of SDS on alumina based on the hypothesis

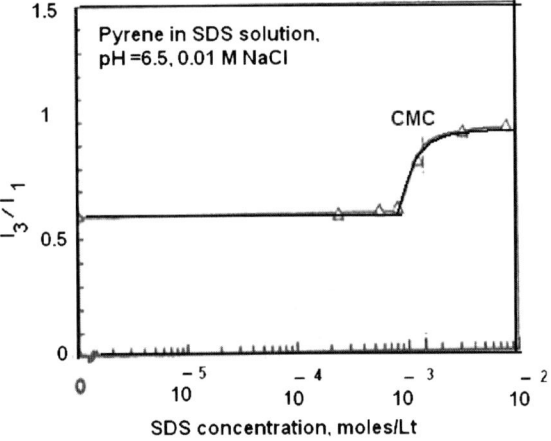

Fig. 30 I_3/I_1 of pyrene in sodium dodecyl sulfate solutions in $0.1\,\mathrm{kmol/m^3}$ NaCl ($\lambda_3 = 383$ nm, $\lambda_1 = 374$ nm)

the case of slurries, the abrupt increase in the ratio of I_3/I_1 well below CMC is indicative of the partitioning of pyrene from the aqueous environment to a relatively nonpolar environment at the alumina interface and the onset of this concentration coincides with the transition from region I to II. The nonpolar environment is a two-dimensional surface aggregate formed by the lateral interaction of the hydrocarbon chain. They are called solloids (surface colloid) or hemimicelles and the concentration at which they start forming is referred to as the solloid or hemimicellar concentration (CSC or CHC). On the basis of the zeta potential and the fluorescence results along with those from electron spin resonance and Raman, the organization of SDS at the alumina–water interface as shown in Fig. 32 has been proposed.

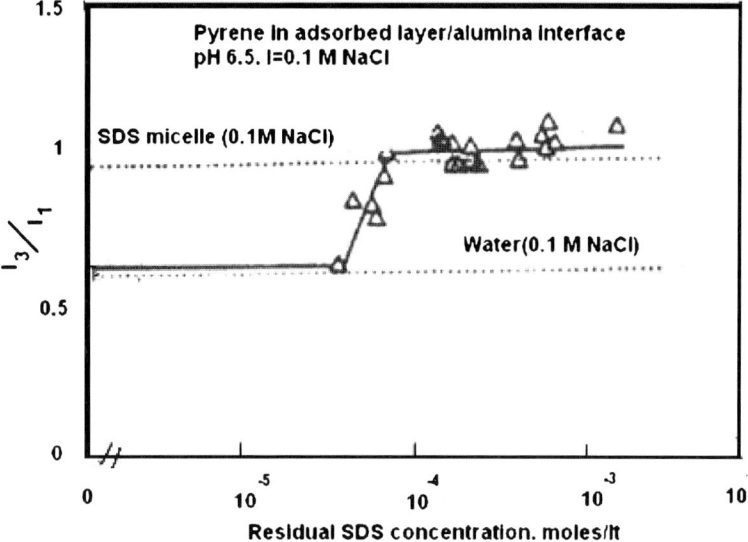

Fig. 31 I_3/I_1 of pyrene in sodium dodecyl sulfate solutions–alumina slurries

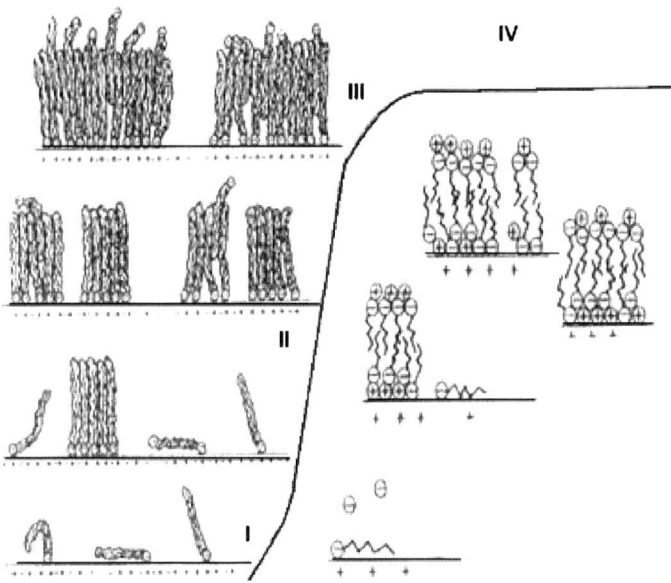

Fig. 32 Schematic representation of the correlation of surface charge and growth of the aggregates for the various regions of the adsorption isotherm of SDS on alumina based on fluorescence data and zeta potential measurement

R—⟨phenyl⟩—O(CH$_2$CH$_2$O)$_{\overline{n}}$—H

OP-X: R= Octyl, n=10 & 30
NP-X : R=Nonyl, n=10,15,20 & 40

Misra et al. [127] have also studied the adsorption behavior of poly(oxyethylated) alkyl phenol (c) on the silica–water interface. The adsorption isotherm is found to depend on the number of poly(oxyethylene) groups present in the surfactant. For surfactants with lower numbers of poly(oxyethylene) groups, a three-region adsorption isotherm is observed, whereas for surfactants with higher numbers a two-region adsorption isotherm is observed (Fig. 33). The initial adsorption is attributed to the hydrogen bonding between oxygen of the oxyethylene units of the nonionic surfactants and –OH groups on the silica surface [128]. The magnitude of this interaction is determined by the number of poly(oxyethylene) units in the chain. The surfactant with more oxyethylene units would therefore adsorb strongly, as observed in the present case. The sharp increase in the adsorp-

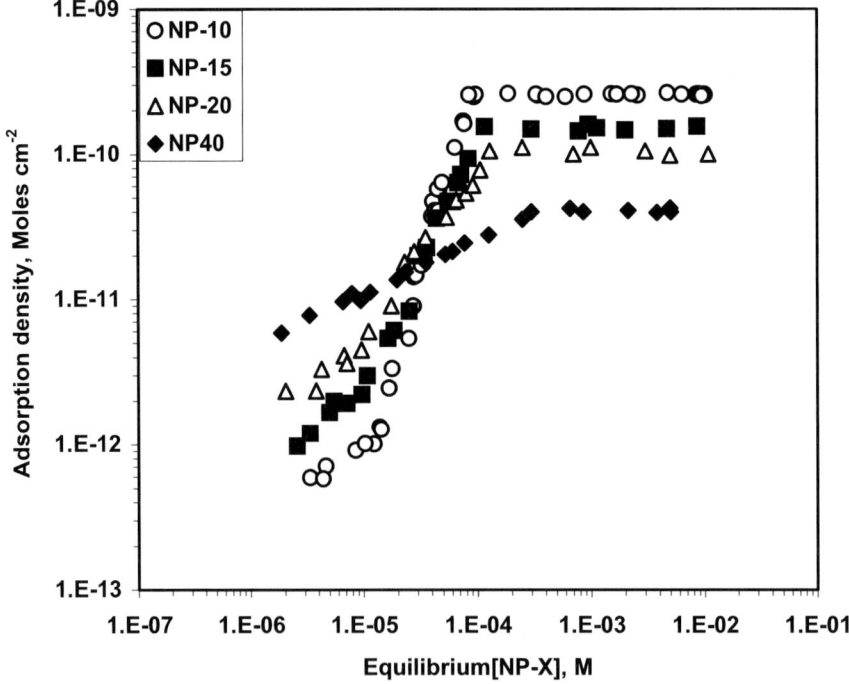

Fig. 33 Comparison of adsorption density with variation of equilibrium concentration for NP-series (Temp. = (23 ± 0.2)°C, I = 0.01 M NaCl, S/L = 0.01, pH = 5.0 ± 0.2)

tion in the second region has been attributed to the formation of solloids (hemimicelles) at the interface. Because of a smaller number of oxyethylene units, these surfactants are more hydrophobic and hence, lateral association of their hydrophobic chains predominates at the interface. The hemimicelle is not formed in the case of higher members of the series due to the decrease in the hydrophobicity. The plateau is due to formation of micelles in solution which limits the monomer concentration, adsorption being determined by the number of monomers only. The adsorption density at the plateau is less for higher members due to constraints in accommodating large numbers of oxyethylene units for the formation of a monolayer. The formation of hemimicelles has been studied by fluorescence probing of interfaces. The adsorption results were obtained in the presence of pyrene and the supernatant and slurries were subjected to analysis of I_3/I_1 of pyrene. The representative plots for both higher and lower members of the series are shown in Figs. 34 and 35. For lower members at low concentrations of surfactant the I_3/I_1 value for the supernatant is around 0.6 indicative of a water-like environment around pyrene. The intensity of pyrene decreases with increased adsorption density and finally becomes zero in the rising arm of the adsorption isotherm in the case of surfactants with a lower number of oxyethylene units. This indicates the absence of pyrene in the bulk solution. The pyrene peak again reappears in

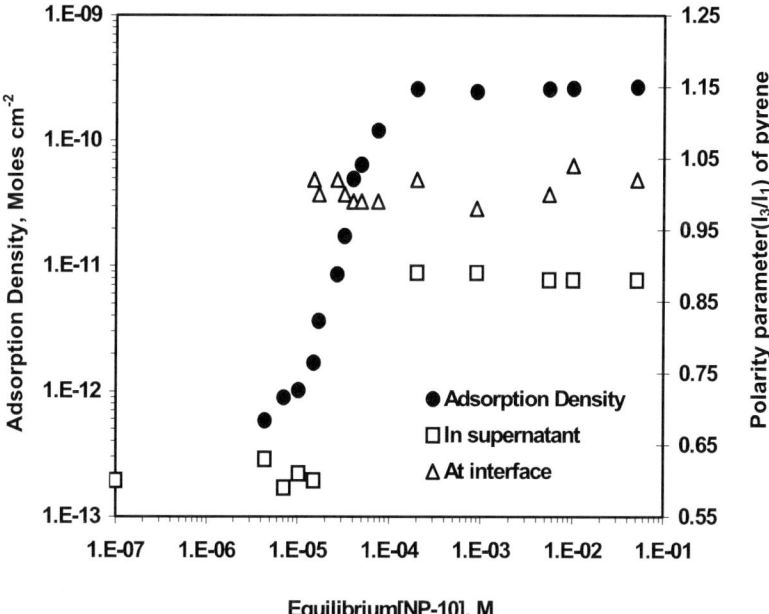

Fig. 34 Plot of equilibrium concentration of NP-10 versus adsorption density, I_3/I_1 of pyrene in bulk and at silica–water interface (pH = 5.0 ± 0.2, $I = 0.01$ M NaCl, Temp = $(23 \pm 0.2\,°C)$, S/L = 0.01)

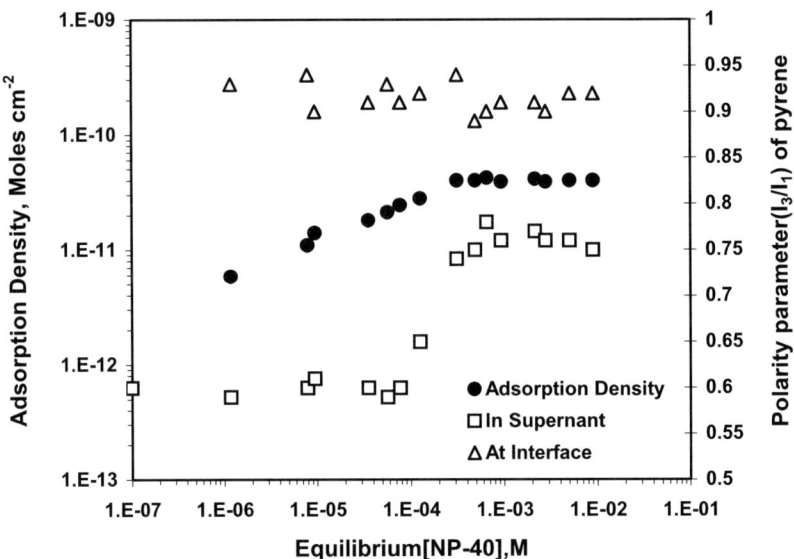

Fig. 35 Plot of equilibrium concentration of NP-40 versus adsorption density, I_3/I_1 of pyrene in bulk and at silica–water interface (pH = 5.0 ± 0.2, I = 0.01 M NaCl, Temp = (23 ± 0.2°C), S/L = 0.01)

the plateau region with a I_3/I_1 value of 0.85. But in contrast to the above for higher members pyrene experiences a water-like environment till the plateau of the adsorption isotherm, with a value of 0.74–0.85 above CMC. Analysis of the fluorescence emission of the slurry reveals that at low concentrations pyrene has a water-like environment and above the sharp rise in the adsorption isotherm, pyrene reports a nonpolar environment. But for the higher members of the series, the value indicates a nonpolar environment throughout the concentration range. The above observations have been attributed to the following: In the case of lower members of the series, at low concentrations, the extent of adsorption is very low and pyrene is therefore, found in bulk water with a I_3/I_1 of 0.6. With the increase in adsorption, pyrene partitions to the silica surface with a I_3/I_1 similar to that of the nonpolar environment due to the formation of hemimicelles. Above CMC, I_3/I_1 = 0.85 is due to the appearance of micelles and surface aggregates. The range of concentration where pyrene is not found in the supernatant is due to the formation of hemimicelles, the mid-point of which is considered to be the critical solloid concentration (CSC). This value increases with an increase in the number of oxyethylene units and is always less than the corresponding CMC (Table 2). In the case of higher member surfactants, I_3/I_1 has a value of 0.6 which indicates that pyrene is present in the water-like environment. The presence of pyrene as indicated by the I_3/I_1 value of 0.6 till CMC is reached suggests low partitioning of pyrene to the interface. But I_3/I_1 of 0.9–1.07 for the slurry

Table 2 CMC, CSC, for NP and OP-series

Surfactants	Number of oxyethylene units	CMC × 10^5, M	CSC × 10^5, M
NP-10	10	8.00	4.35
NP-15	15	10.00	5.33
NP-20	20	12.34	7.11
NP-40	40	30.00	–
OP-10	10	40.00	17.40
OP-30	30	100.00	–

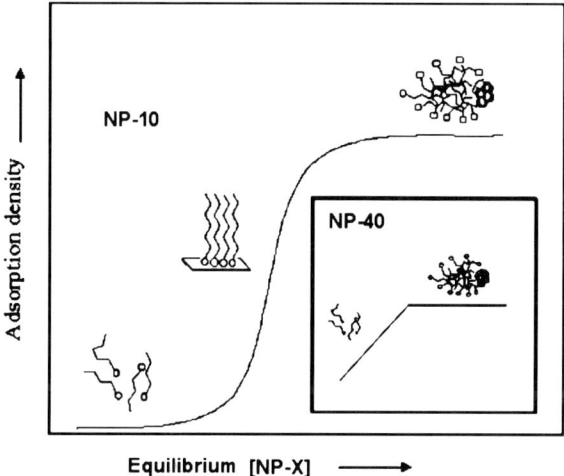

Fig. 36 Schematic diagram of surfactant-adsorption on silica surface as monomer, hemimicelle, and at surfactant concentration above CMC where pyrene is bound to micelle

throughout the adsorption isotherm indicates that pyrene exhibits a nonpolar environment at silica–water interfaces in the presence of higher members of the series. The absence of a solloid region in the presence of them is due to the increased hydrophilicity and bulkiness of the group, which would not favor the formation of hemimicelles at the silica–water interface. This is supported by the hemimicellar number obtained using the Gu and Zhu model [129]. A value of less than one—indicating no hemimicellization—is obtained. Since the adsorption density is quite large at low concentrations and pyrene stays preferably near the oxyethylene chain in nonionic surfactants I_3/I_1 is obtained. On the basis of these facts, the mechanism of adsorption suggested is given in Fig. 36.

A three-stage adsorption isotherm is obtained for tetradecyltrimethylammonium chloride (TTAC) [130] at the alumina interface at pH 10 (Fig. 37). At

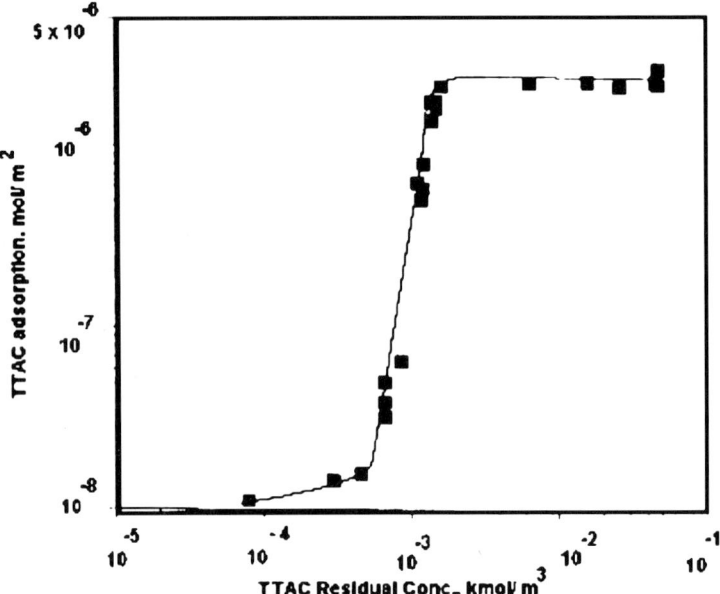

Fig. 37 Adsorption isotherm of TTAC at alumina–water interface at pH 10, ionic strength 0.03 M NaCl

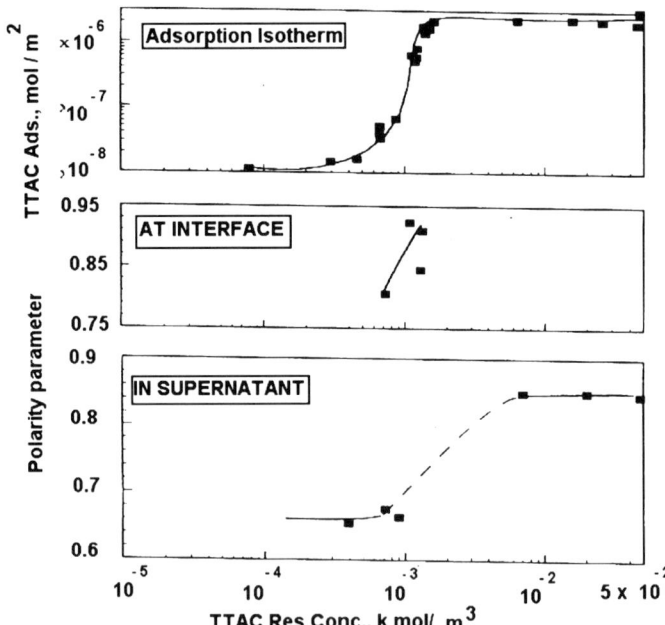

Fig. 38 Adsorption isotherm of TTAC at alumina–water interface at pH 10, ionic strength 0.03 M NaCl and corresponding changes in pyrene monomer fluorescence from the alumina–water interface and the supernatant

pH 10, alumina is negatively charged and hence at low concentrations, the adsorption occurs through electrostatic attraction between the surfactant and the alumina. The sharp increase in adsorption isotherm at around 5×10^{-4} M is attributed to the formation of organized assemblies at the solid–liquid interface. The plateau of the adsorption isotherm is due to the formation of micelles in the solution. To understand the nature of the surface aggregates formed, the adsorption is probed using pyrene fluorescence. The results obtained are shown in Fig. 38. The absence of pyrene from the supernatant and appearance at the interface suggests the formation of surface aggregates at the interface.

Another system where fluorescence was used effectively is that of Aerosol-T (AOT)/graphite in aqueous and nonaqueous media [131]. Fluorescence studies were conducted by coadsorbing the fluorescent probe 1-methyl-8-oxyquinoliumbetaine with surfactant and by measuring the fluorescence response of the probe from the interface as well as in the supernatant after adsorption. The fluorescence spectra from the graphite–cyclohexane interface are found to be extremely low in intensities at all concentrations probably due to the absorption of the emitted light by the graphite surface. However, when the supernatant is analyzed, it is found that there is no residual probe in any of the cases including those where the residual AOT concentration is above the CMC in cyclohexane. This observation clearly shows the phase transfer of the probe to the interface. From the earlier studies, it was shown that the probe can be solubilized only in the polar domains and thus cannot be expected to adsorb at the graphite surface by itself. Thus, the probe must have solubilized in the polar microdomain created by the AOT molecules at the interface in the interparticle aggregate at low concentrations and in the adsorbed micelle at higher concentration. The formation of the aggregate is substantiated by the study of the settling rate of the adsorbed graphite in cyclohexane.

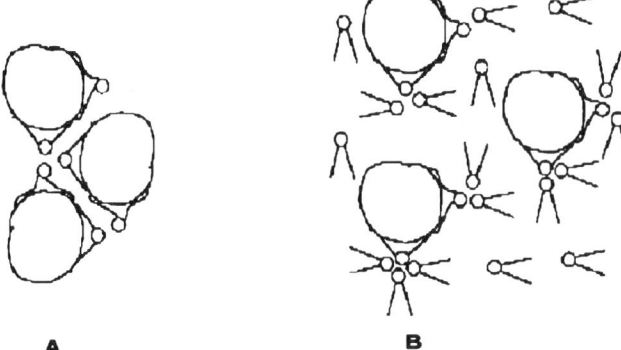

Fig. 39 Schematic diagram showing the formation of **A** interparticle surface aggregates which causes flocculation; **B** aggregates with monomers in solution leading to dispersion

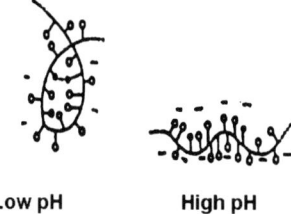

Fig. 40 Schematic representation of the correlation of the extent of excimer formation and intrastrand coiling of pyrene labeled poly(acrylic acid)

The settling rate of the dispersions in cyclohexane initially increases with AOT adsorption but later decreases. The initial increase is attributed to the formation of interparticle surfactant aggregates (Fig. 39A). At higher concentrations, the adsorbed molecules aggregate with excess of surfactant in solution rather than with molecules on the particle, so that flocculation ceases to occur and the dispersion is restabilized. The schematic representation of the surfactant assemblies at the interface is as shown in Fig. 39B.

The conformation of polymers poly(acrylic) acid at the alumina–water interface has also been studied using fluorescence spectroscopy [132]. Since PAA can exist in different conformations depending on the pH, solvent and ionic strength (Fig. 40) [133], its adsorption onto solids can be effected by the above parameters. The pyrene-labeled polymer was used for this investigation. The excimer formation depends on the interaction of the excited pyrene of the polymer pendant group with another pyrene group in the ground state, which in turn is dependent on pH as shown in Fig. 40. At the low pH, since the macromolecule has a coiled conformation, there is better probability for the excimer formation between pyrene groups, but a low probability for the excimer formation at high pH due to electrostatic repulsion between ionized

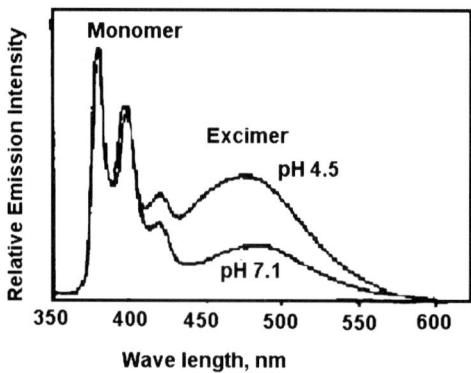

Fig. 41 Fluorescence emission spectra of adsorbed polymer at two pH values

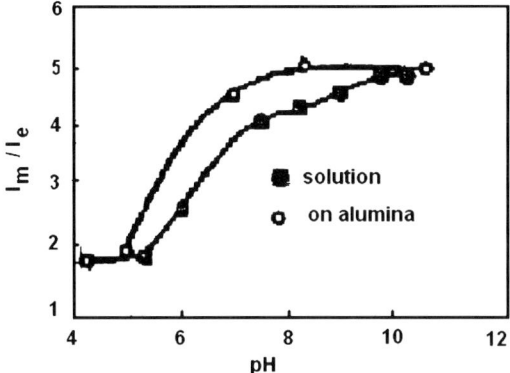

Fig. 42 Monomer to excimer ratio as a function of pH aqueous solution of polymer and adsorbed polymer on alumina

carboxylate groups in the polymer and subsequent stretching of the polymer chain (Fig. 40). This difference is also reflected in the fluorescence spectra of the pyrene labeled poly(acrylic acid) (PPAA) (Fig. 41). It is observed that the adsorption of PPAA from aqueous solution onto positively charged alumina is rapid enough so that the adsorbed polymer adopted a conformation

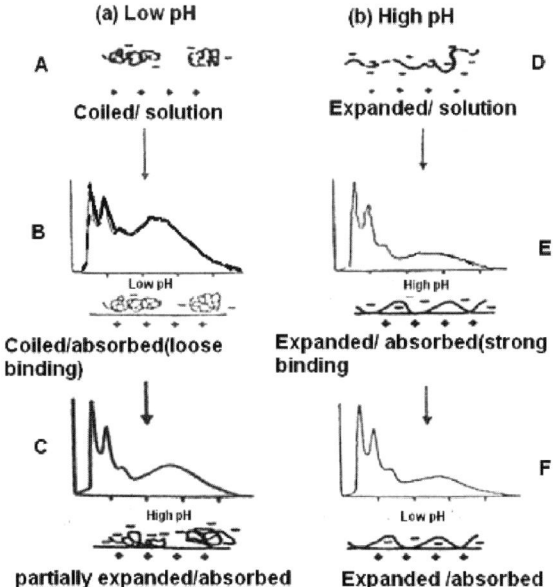

Fig. 43 Schematic representation of the adsorption process of pyrene labeled polyacrylic acid on alumina at different pH

that reflected its state in solution prior to adsorption (Fig. 42). The figure shows a similar monomer/excimer ratio curve as a function of pH for the solution and the adsorbed cases. Further, it is also observed that polymer adsorption at high pH is an irreversible process with respect to the conformation as the efficiency of the excimer conformation remains unaffected by the change in the pH to acidic values. Indeed, polymer adsorbed at low pH could be stretched to some extent by increasing the pH. These processes are represented mechanistically in Fig. 43.

4.2
Micropolarity and Solubilization Site of the Probe

The polarity within a surfactant assembly will be quite different from that of the bulk solution. It is useful to know the micropolarity of these assemblies for such applications where different substrates are compartmentalized inside these surfactants. The micropolarity of the surfactant assemblies can be determined using any fluorescence probe whose emission characteristics change with solvent polarity. The emissions of the probe are measured in solvents of known polarities and the polarity of the surfactant assemblies is determined by comparison.

The micropolarity of a number of poly(oxyethylated) nonionic surfactants and anionic surfactants have been determined by fluorescence probing using pyrene [123]. The I_3/I_1 of pyrene in various organic solvents are determined and these values are plotted against the solvent polarity parameter, $E_{T(30)}$ [134]. The $E_{T(30)}$ value of a solvent is the energy of the intermolecular charge transfer band in kJ mol^{-1} of 1-ethyl-4-carbomethoxypyridinium iodide when dissolved in that solvent. These values are indicative of the polarity of the medium. This is used to determine the calibration curve to predict the $E_{T(30)}$ value of the environment experienced by pyrene (Fig. 44). The solvent polarity parameter, $E_{T(30)}$ values of these surfactant micelles have been evaluated to be 49–56 kJ mole^{-1}, which is intermediate between those of water and hexane. The micropolarities of the micelles as experienced by pyrene are therefore found to be alcohol-like and pyrene is suggested to reside on the protruded fatty patches with substantial amounts of penetrated water around it. On the basis of this observation the structures of those micelles are proposed to be of Menger-type [15] (Fig. 45). From the micropolarity studies it is concluded that pyrene is solubilized in the palisade layer of the micelle but moves deeper in the case of an anionic micelle in order to avoid unfavorable interactions with point charges of the surface.

Zhang et al. [135] have studied the physicochemical behavior of mixtures of n-dodecyl-β-D-maltoside with anionic, cationic and nonionic surfactants in aqueous solutions. To acquire information on the property of mixed micelles, the characteristic change of pyrene with changes in polarity was monitored. The polarity parameter at low concentrations was found to be 0.5–0.6.

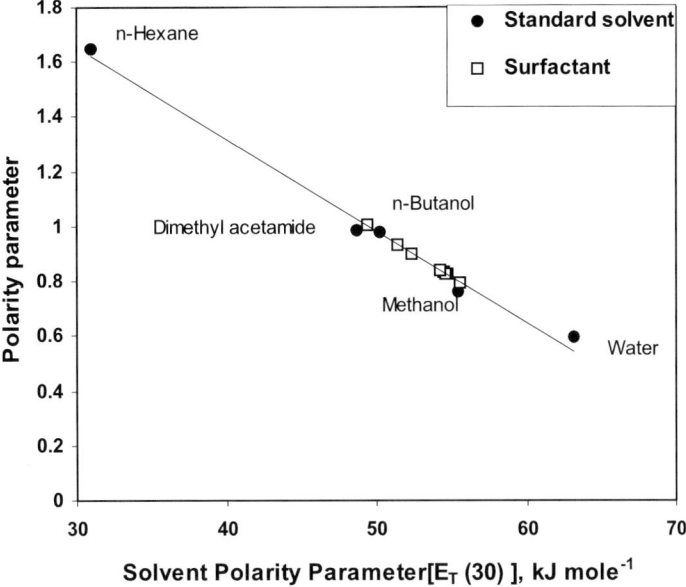

Fig. 44 Plot of (I_3/I_1) of pyrene versus solvent polarity [$E_{T(30)}$] parameter of some of the selected solvents and extrapolated values of the surfactants under study

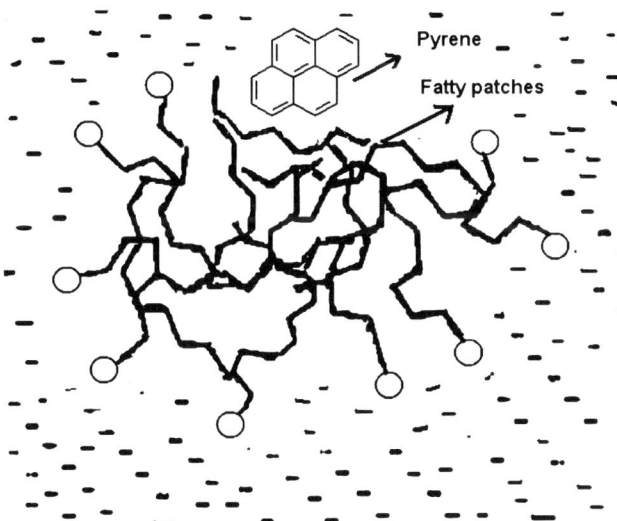

Fig. 45 Schematic representation of solubilization of pyrene in hydrophobic patches of a menger micelle, the exposure of the hydrophobic tail to water results in formation of hydrophobic patches at the micellar surface

The polarity parameter increased around the CMC. The I_3/I_1 ratio for n-dodecyl-β-D-maltoside/sodium dodecylsulfate mixtures above the CMC is similar to that for sodium dodecylsulfate and the n-dodecyl-β-D-maltoside system suggesting similar hydrophobicity for micelles of n-dodecyl-β-D-maltoside, sodiumdodecyl sulfate and their mixtures. The polarity parameter is less than that of hydrocarbons suggesting that the interface, as expected, is a mixture of hydrocarbon chains and small amounts of water. The addition of salt increases the polarity of the solvent as well as that of the interface. For the n-dodecyl-β-D-maltoside/dodecyltrimethylammounim bromide system, the I_3/I_1 ratio for n-dodecyl-β-D-maltoside above the CMC is higher than that for dodecyltrimethylammoinum bromide suggesting the interface of n-dodecyl-β-D-maltoside micelles is more hydrophobic than that of dodecyltrimethylammonium bromide micelles, possibly because of the bulky nature of the trimethylammonium head group. In the case of the mixtures, the predominance of n-dodecyl-β-D-maltoside in mixed micelles increases the hydrophobicity for the mixtures. Thus, the polarity parameter of n-dodecyl-β-D-maltoside/dodecyltrimethylammonium bromide mixed micelles indicates a more hydrophobic interface. The polarity parameter of the n-dodecyl-β-D-maltoside/dodecyltrimethylammonium bromide mixed micelle is only slightly lower than that of n-dodecyl-β-D-maltoside. In the case of n-dodecyl-β-D-maltoside/pentaethyleneglycolmonododecylether systems, I_3/I_1 for both the n-dodecyl-β-D-maltoside and pentaethyleneglycolmonododecylether are similar at concentrations higher than the CMC. Since pentaethyleneglycolmonododecylether is predominant in mixed micelles, the polarity parameter of the mixed micelles is close to its parameter. Thus, in the case of mixtures of sugar-based surfactants with ionic ones, mixed micellar characteristics are concluded to be similar to that of nonionic sugar-based surfactants.

The polarity of the solloids formed by cationic, anionic and nonionic surfactants have been determined using fluorescence spectroscopy [127, 135]. It is seen that pyrene partitions to both micelles in the supernatant and the solloids at the silica–water interface in the case of the nonionic surfactants. Both the aggregates provide a hydrophobic atmosphere for pyrene molecules. In contrast to these observations it has been reported that pyrene is solubilized in bulk micelles and does not partition to the alumina–water interface in the presence of tetradecyltrimethylammonium chloride solloids, whereas pyrene is preferentially solubilized in the solloids of anionic SDS surfactant at the alumina–water interface. This indicates that solloids of anionic surfactants like SDS have high solubilizing power, whereas those of the cationic surfactants like TTAC have less. The nonionic surfactant solloids, on the other hand, have intermediate solubilizing power.

4.3
Aggregation Number (Hemimicellar Number) and Micellar Size

The aggregation number is the number of surfactants constituting a surfactant assembly and this in turn determines the size of the aggregates. This is, therefore, one of the important structural properties in addition to micellization. Determination of aggregation number and micellar size using florescence spectroscopy involves the monitoring of the decrease of fluorescence intensity of the probe by a quencher. The quenching occurs through photochemical reactions between the fluorescent probe and the quencher inside the aggregates. The quencher can be the fluorescent probe itself (through excimer formation) or another species (exciplex) having a suitable excited state equivalent to the emitted energy. Photochemical reactions commonly studied in such studies are diffusion-controlled quenching reactions [136, 137]. There are two methods available for measuring aggregation number and micellar size: steady-state fluorescence quenching [138, 139] and time-resolved fluorescence quenching [140, 141].

Turro and Yekta [142] discussed a simple method for measuring aggregation number by steady-state fluorescence. In this method, the decrease of the fluorescence intensity of a micelle-bound probe is monitored as a function of the quencher concentration and is fitted to the equation [143]

$$\ln I_0/I = N[Q]/C_s - \text{CMC} \,, \qquad (3)$$

where N is the aggregation number, C_s is the total concentration of the surfactant and $[Q]$ is the quencher concentration.

In the time-resolved quenching method, the decay kinetics of the monomer and the excimer emission are monitored in the presence of a micellized medium. If the micellar system is viewed as a group of individual micelles with probe occupancies 0, 1, 2, 3, etc., the probability of micelles with n probes, P_n, may be related to n, the average number of probes per micelle by Poissonian statistics through relation,

$$P_n = n^n \exp(-n)/n! \,. \qquad (4)$$

This model yields the following relation for the time-dependent monomer emission.

$$I_{m(t)} = I_{m(0)} \exp[-k_0 t + n(\exp(-k_e t)) - 1] \,, \qquad (5)$$

where k_0 is the reciprocal life time of excited state pyrene in the absence of excimer formation, k_e is the intramicellar encounter frequency of pyrene in excited and ground states and $I_{m(t)}$ and $I_{m(0)}$ represent the intensity of monomer emission at time t and zero time, respectively. Knowledge of n leads to the aggregation number N for the adsorbed layer as given by the expression

$$n = [P]/[\text{Agg}] = [P]N/([S] - [S_{eq}]) \,, \qquad (6)$$

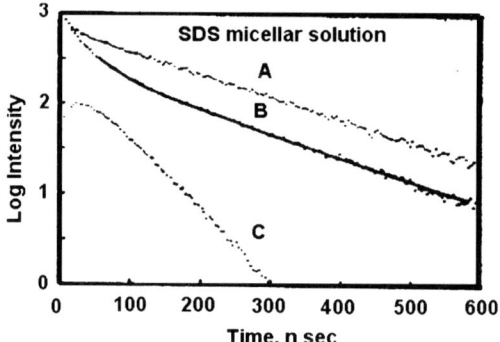

Fig. 46 Pyrene monomer and excimer decay profiles in SDS micellar solutions: [SDS] = 8.2×10^{-2} kmol m^{-3}, [NaCl] = 10^{-1} kmol m^{-3}, CMC = 1.5×10^{-3} kmol m^{-3}, pyrene levels are indicated as the ratio of micellized SDS to added pyrene; emission monitored at 383 nm for monomer and 480 nm for excimer. (A) Monomer emission for SDS/Py = 2160, (B) monomer emission for SDS/Py 108; (C) excimer emission for SDS/Py = 108

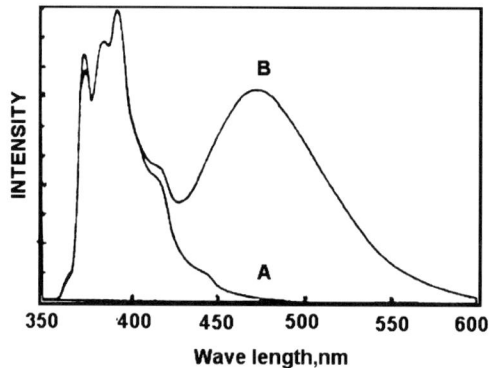

Fig. 47 Pyrene emission spectra in SDS micellar solutions under varying levels of pyrene (A and B correspond to pyrene levels as indicated in Fig. 46)

where [P] is the total concentration of pyrene. [Agg] is the concentration of the aggregates, and [S] – [S$_{eq}$] is the concentration of the adsorbed surfactant.

Chandar et al. [144] carried out a kinetic analysis based on the above relation from the decay profiles of pyrene in the adsorbed state for different regions of the adsorption isotherm of dodecyl sulfate on alumina (Fig. 27). Figure 46 represents the decay curves for monomer (curves A and B) at two pyrene levels and excimer (curve C) corresponding to curve B. The fluorescence spectra of these two cases are given in Fig. 47. The aggregation number was calculated from n and is marked on the adsorption isotherm shown in Fig. 48. The aggregates in region II appear to be of relatively uniform size (N = 121 to 128). But in region III, there is a marked growth in the aggregate size (N = 166 to 356). These results have special significance with respect to

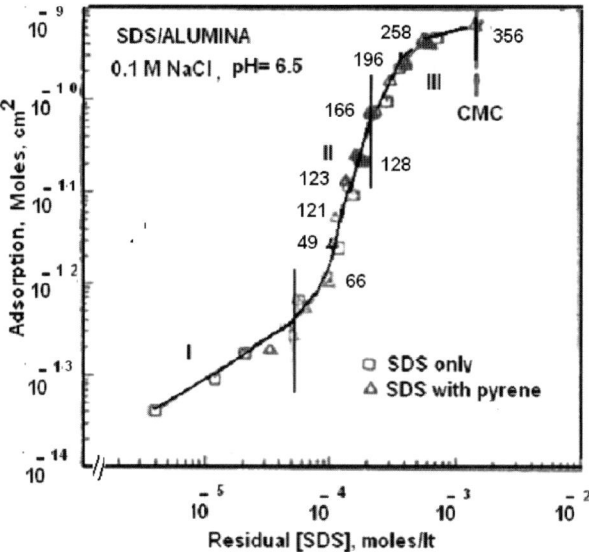

Fig. 48 Surfactant aggregation numbers determined at various adsorption densities (average number at each adsorption density shown along the adsorption isotherm)

the evolution and structure of the adsorbed layer. Region II and above is characterized by the surfactant solloids (hemimicelles of limited size). Here, the surface is still not fully occupied and enough positive sites remain available. Further adsorption occurs mainly by increasing the number of aggregates as revealed by an approximately constant aggregation number. The transition from region II to III corresponds to the isoelectric point of the mineral and the adsorption in region III apparently occurs through the growth of the existing aggregates. This is possible by the hydrophobic interaction between the tails of the already-adsorbed surfactant molecules and the new ones. Once the positive surface sites are neutralized, further adsorption under these conditions takes place with a reverse orientation of the surfactant molecules as shown in Figs. 28 and 32.

4.4
Micelle Fluidity (Nanoviscosity)

The fluidity (nanoviscosity) in an organized surfactant assembly on solids can be substantially different from that in the bulk aqueous phase and hence, the diffusional resistance experienced by the probe in the micelle will be considerably different from that faced in the bulk solution [145]. Measurement of the viscosity or fluidity of the interior of a micelle is based on measurement of fluorescence properties that depend on the mobility of the probe in the interior. A commonly used method for such studies involves the intramolecular

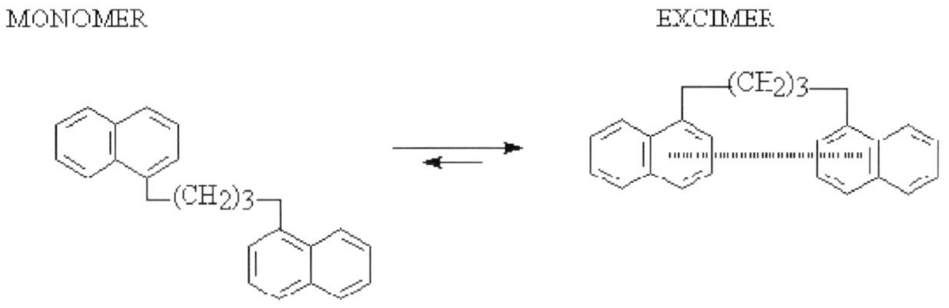

Fig. 49 Schematic representation of intramolecular excimer formation of dianaphthypropane (DNP)

excimer formation of bischromophoric probes [146–150]. The extent of excimer formation is dependent on the diffusional mobility of the probe groups involved in the process. This mobility is decreased in a medium of high rigidity or viscosity so that the extent of excimer formation decreases as viscosity is increased. Chandar et al. [144] have measured the nanofluidity of solloids formed by dodecylsulfate on an alumina surface using the excimer probing properties of 1,3-dinaphthyl propane (DNP). The intramolecular excimer formation of DNP depends on the motion of the arene groups in order to achieve the excimer conformation formation as shown in Fig. 49 and thus is a function sensitive to the viscosity of its neighborhood. Excimer formation at low probe concentrations can be considered to be unimolecular (intramolecular) so that only the medium viscosity determines the extent of excimer formation. Thus, the monomer to excimer intensity ratio I_m/I_e obtained from the steady-state emission spectrum of DNP is used to estimate the effective fluidity of the microenvironment of the probe using values obtained with solvents of known viscosity. Usually, the I_m/I_e value of the test surface is compared with those of ethanol–glycerol mixtures of known viscosity (Fig. 50). The yields of monomer to excimer (I_m/I_e) are determined in solution and in the adsorbed layer for various regions of the adsorption isotherm. The results

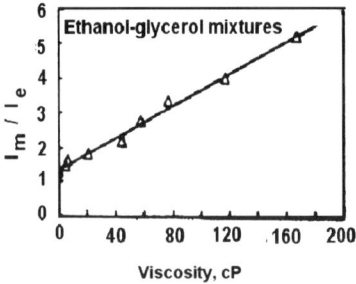

Fig. 50 Monomer to excimer ratio (I_e/I_m) of DNP in ethanol–glycerol mixtures (viscosities of the solvent measured by capillary flow method)

Table 3 Monomer to excimer ratio of DNP and corresponding viscosities in various systems

Test 1: SDS–micelle (0.1 M NaCl)				
[SDS], M	CMC, M	SDS/Py*	I_m/I_e	η, cP
8.2×10^{-2}	1.5×10^{-3}	785	1.7	80
Test 2: SDS–alumina hemimicelle (0.1 M NaCl), pH 6.5 (no DNP in supernatant)				
Adsorption density	Equilibrium concentration	SDS/Py**	I_m/I_e	η, cP
4.5×10^{-10}	6.0×10^{-4}	672	3.3	90
4.5×10^{-10}	6.0×10^{-4}	1680	3.3	90
Test 3: SDS–silica (0.1 M NaCl), pH 6.0 (I_m/I_e in supernatant = 1.7)				
Adsorption density mol/cm^2	Equilibrium concentration M	SDS/Py**	I_m/I_e	
0	8.2×10^{-2}	785	1.8	

* ratio of micellized SDS added to DNP;
** ratio of adsorbed SDS added to DNP;
viscosities based on I_m/I_e in ethanol–glycerol mixtures

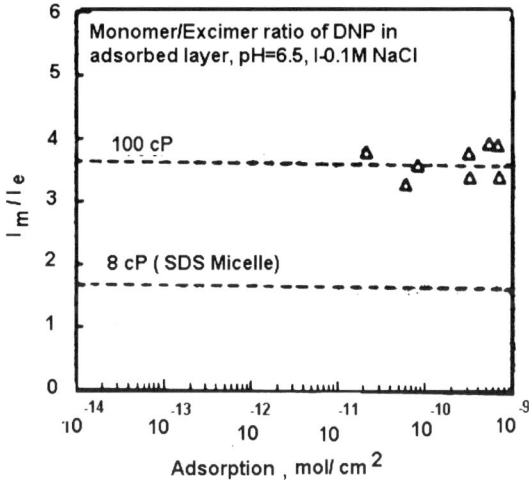

Fig. 51 Monomer to excimer ratio (I_e/I_m) of DNP in SDS-alumina slurries as a function of SDS adsorption density ($\lambda_m = 340$ nm, $\lambda_e = 420$ nm)

are shown in Table 3. The extent of excimer formation in the adsorbed layer is significantly lower than that in the micellar solution. On the basis of the I_m/I_e value, a high viscosity value of 90 to 120 cp is obtained for the adsorbed layer compared to 8 cp for micelles (Fig. 51). It is to be noted that variation of the level of DNP had no effect on the I_m/I_e value suggesting that the excimer formation occurs unimolecularly (intramolecular). Also a SDS-DNP micellar

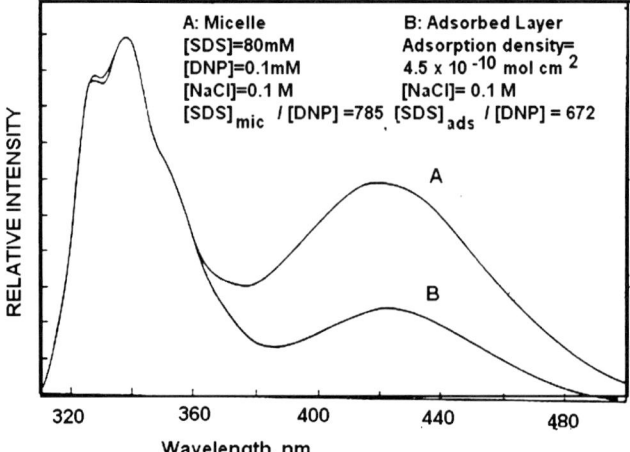

Fig. 52 DNP fluorescence spectra in (A) SDS micellar solution and (B) SDS–alumina slurry: [SDS]/[DNP] refers to the micellized or adsorbed SDS to added DNP

solution added to nonadsorbing silica particles gives almost the same I_m/I_e as that of the supernatant and of SDS micellar solution suggesting that the presence of solid particles does not interfere with the measurement; DNP is apparently present exclusively in micelles (Fig. 52). It is clear that the fluorescence technique is able to reveal nanostructural properties of surfactant assemblies that are valuable for many applications.

5
Conclusions

The success of fluorescence spectroscopy as a convenient tool stems largely from the fact that some chromophore molecules are highly sensitive to their microenvironment and their fluorescence characteristics provide information on the surrounding environment. The use of fluorescence probing to elucidate the mechanism of formation of the surfactant assemblies has been discussed herein alongside methods to determine parameters such as critical micellar concentration, critical aggregation, critical solloid (hemimicelle) concentration, nanopolarity, viscosity, aggregation number, and size of the aggregates. The review is limited to our work on the characterization of surfactant assemblies using fluorescence spectroscopy. The fluorescence technique, along with electron spin resonance and the resonance Raman technique, used by us for the same system, provides information on the fundamentals of surfactant and polymer aggregation and conformation behavior in solutions and at solid–liquid interfaces.

Acknowledgements One of the authors (P.K.M.) is grateful to the Indian National Science Academy for sanctioning a visiting fellowship at Central Leather Research Institute (CLRI), Chennai. Discussions with Prof. A.B. Mandal, Director Grade Scientist and Head, Chemical Laboratory (CLRI), Chennai, India and his constant support in various ways are highly appreciated. We also acknowledge the support of the US Department of Energy and the National Science Foundation.

References

1. Fendler JH, Fendler EJ (1975) Catalysis in Micellar and Macromolecular system. Academic, New York
2. Todorov PD, Kralchevsky PA, Denkov ND, Broze G, Mehreteab A (2002) J Colloid Interface Sci 245:371
3. Pena AA, Miller CA (2001) J Colloid Interface Sci 244:154
4. Thomas JK (1980) Chem Rev 80:283
5. Misra PK, Mishra BK, Behera GB (1992) Int J Chem Kinet 24:533
6. Misra PK, Mishra BK, Behera GB (1991) Int J Chem Kinet 23:639
7. Sahoo L, Misra PK, Somasundaran P (2002) Ind J Chem 41A:1402
8. Rosen MJ (1978) Surfactants and interfacial phenomenon. John Wiley and Sons, New York
9. Boissier C, Loefroth JE, Nyden M (2002) Langmuir 18:7313
10. Misra PK, Panigrahi S, Somasundaran P (2006) Int J Min Processing 80:229
11. Misra PK, Somasundaran P (2004) J Surf Dets 7:373
12. Furusawa K, Sato A, Shirai J, Nashima T (2002) J Colloid Interface Sci 253:273
13. Misra PK, Dash U, Somasundaran P (2008) Int J Min Processing, communicated
14. Hartley GS (1936) Aqueous Solution of Paraffin Chain Salts, A Study of Micelle Formation. Herman and Co, Paris
15. Menger FM (1979) Acc Chem Res 12:111
16. Dill KA, Flory PJ (1981) Proc Natl Acad Sci USA 78:676
17. Gruen DWR (1981) J Colloid Interface Sci 84:281
18. Fromherz P (1981) Ber Bunsenges Phys Chem 85:891
19. Krishnan RSG, Thennarasu S, Manadal AB (2005) J Phys Chem B 108:8806
20. Mandal AB, Jayakumar R, Manoharan PT (1993) J Chem Soc Chem Commun :853
21. Mandal AB, Jayakumar R, Jeevan RG, Manoharan PT (1994) J Chem Soc Faraday Trans 90:2725
22. Pileni MP (ed) (1989) Structure and Reactivity in Reversed Micelles. Elsevier, Amsterdam
23. Somasundaran P, Fuerstenau DW (1966) J Phys Chem 70:90
24. Somasundaran P, Krishnakumar S (1994) Colloids Surfs 93:79
25. Chandar P, Somasundarn P, Turro NJ (1987) J Colloid Interface Sci 117:31
26. Huang L, Somasundaran P (1996) Colloids Surfs 117:235
27. Hamdiyyah MA, Rehman IA (1985) J Phys Chem 89:2377
28. Zana R, Levy H (1995) J Colloid Interface Sci 170:128
29. Sugihara G, Arakawa Y, Tanaka K, Lee S, Moroi Y (1995) J Colloid Interface Sci 170:399
30. Paredes S, Tribout M, Sepulveda L (1984) J Colloid Interface Sci 88:187
31. Tolin VN, Hoeberges H, Leonis J, Paredes S (1982) J Colloid Interface Sci 85:597
32. Sulthana SB, Bhat SGT, Rakshit AK (2000) Bull Chem Soc Japan 73:281

33. Sulthana SB, Rao PVC, Bhat SGT, Nakano TY, Sugihara G, Rakshit AK (2000) Langmuir 16:980
34. Kratzal K, Finkelmann H (1996) Langmuir 12:1765
35. Jun-Fu LI, Min G, Ducker WA (2001) Langmuir 17:4895
36. Fendler JH, Fendler EJ (1975) Catalysis in Micellar and Macromolucular System. Academic, New York
37. Donchi KF, Robert GP, Ternai B, Derrick P (1980) Aust J Chem 33:2199
38. Panda LN, Behera GB (1983) J Ind Chem Soc LX :194
39. Das R, Guha D, Mitra S, Kar S, Lahiri S, Mukherjee S (1997) J Phys Chem A 101:4042
40. Nagarajan R, Wang CC (1996) J Colloid Int Sci 178:471
41. Carlesson I, Edlund H, Persson G, Lindstrom B (1996) J Colloid Interface Sci 180:598
42. Desndo MA, McGarvey B, Reeves LW (1996) J Colloid Interface Sci 181:331
43. Yuan HZ, Zhao S, Cheng GZ, Zhang L, Miao XJ, Mao SZ, Yu JY, Sen LF, Du YR (2001) J Phys Chem B 105:4611
44. Couderc S, Li Y, Bloor DM, Holz-Warth JF, Wyn-Jones E (2001) Langmuir 17:4818
45. Bai G, Wang J, Yan H, Li Z, Thomas RK (2001) J Phys Chem B 105:3105
46. Akio O, Shigeyoshi M , Makoto A (2001) J Phys Chem B 105:2826
47. Kanthimati M, Deepa K, Nair BU, Mandal AB (2000) Bull Chem Soc 73:1769
48. Gehlen MH, De Schryver FC (1993) Chem Rev 93:199
49. Sivkumar A, Somasundaran P (1994) Langmuir 10:131
50. Sanchez FG, Ruiz CC (1996) J Luminescence 69:179
51. Wong JE, Duchscherer TM, Pietraru G, Cramb DT (1999) Langmuir 15:6181
52. Rabek JF (1980) Experimental Methods in Polymer Chemistry, Chap 16, Vol XXV. Wiley, New York
53. Singer LA (1982) Fluorescence Probes of Micellar Systems – An Overview. In: Mittal KL, Fendler EJ (eds) Solution Behaviour of Surfactants, Vol 1. Plenum, New York, p 73
54. Lackowicz JR (1999) Principles of Fluorescence spectroscopy. Plenum, New York
55. Tsubone K, Ghosh S (2004) J Surfactant Detergent 7:47
56. Zhang Y, Lam YM,Tan WS (2005) J Colloid Interface Sci 285:74
57. Burke SE, Rodgers MP, Palepu R (2001) Mol Phys 99:517
58. Xu JP, Chen WD, Shen JC (2005) Macromol Biosci 5:164
59. Gautier S, Boustta M, Vert M (1999) J Controlled Release 60:235
60. Watkins DM, Sayed-Sweet Y, Klimash JW, Turro NJ, Tomalia DA (1997) Langmuir 13:3136
61. Roy S, Mohanty A, Dey J (2005) Chem Phys Lett 414:23
62. Konopásek I, Večeř J, Strzalka K, Amler E (2004) Chem Phys Lipids 130:135
63. Karukstis KK, Frazier AA, Martula DS, Whiles JA (1996) J Phys Chem 100:11133
64. Komorek U, Wilk KA (2004) J Colloid Interface Sci 271:206
65. Karukstis KK, McCormack SA, McQueen TM, Goto KF (2004) Langmuir 20:64
66. Hasegawa M, Sugimura T, Shindo Y, Kitahara A (1996) Colloids Surfs A 109:305
67. Guharay J, Sengupta P (1996) Biochem Biophys Res Commun 219:388
68. Hrdlovic P, Horinova L, Chmela S (1995) Canadian J Chem 73:1948
69. Roy S, Dey J (2003) Langmuir 19:9625
70. Griffiths PC, Roe JA, Bales BL, Pitt AR, Howe AM (2000) Langmuir 16:8248
71. Carnero RC, Molina-Bolivar JA, Aguiar J, MacIsaac G, Moroze S, Palepu R (2003) Colloid Polym Sci 281:531
72. Nakashima K, Fujimoto Y, Anzai T (1995) Photochem Photobio 61:592
73. Carnero RC (1995) Colloid Polym Sci 273:1033
74. Angelescu D, Vasilescu M, Caidararu H, Caragheorgheopol A, Khan A (2003) Ser Chem 12:829

75. Kumbhakar M, Goel T, Mukherjee T, Pal H (2005) J Phys Chem B 109:18528
76. Hara K, Baden N, Kajimoto O (2004) J Phys Condensed Mat 16:S1207
77. Carnero RC, Garcia SF (1994) J Colloid Interface Sci 165:110
78. Caldararu H, Caragheorgheopol A, Vasilescu M, Dragutan I, Lemmetyinen H (1994) J Phys Chem 98:5320
79. Marchi MC, Negri RM, Bilmes SA (2003) J Sol-Gel Sci Tech 26:131
80. Seo T, Take S, Akimoto T, Hamada K, Iijima T (1991) Macromolecules 24:4801
81. Varadaraj R, Valint P, Bock J, Zushma S, Brons N (1991) J Colloid Interface Sci 144:340
82. Shobini J, Mishra AK, Chandra N (2003) J Photochem Photobiol B 70:117
83. Miyagishi S, Asakawa T, Nishida M (1987) J Colloid Interface Sci 115:199
84. Somasundaran P, Turro NJ, Chandar P (1986) Colloids Surfs 20:145
85. Lei X, Zhao G, Liu Y, Turro N (1992) Langmuir 8:475
86. Piasecki DA, Wirth MJ (1993) J Phys Chem 97:7700
87. Brito RM, Vaz WL (1986) Anal Biochem 152:250
88. Rosen MJ, Mathias JH, Davenport L (1999) Langmuir 15:7340
89. Ruiz CC (1999) Colloids Surf A 147:349
90. Ruiz CC, Aguiar J (1999) Mol Phys 97:1095
91. Evertsson H, Nilsson S (1998) Carbohydrate Polym 35:135
92. Medeiros GMM, Costa SMB (1996) Colloids Surf A 119:141
93. Zana R (1999) J Phys Chem B 103:9117
94. Gramlich G, Zhang J, Winterhalter M, Nau WM (2001) Chem Phys Lipids 113:1
95. Lemp E, Zanocco AL, Gunther G (2003) Colloids Surf A Phys Eng Aspects 229:63
96. Basu S, Paul TK, Roy S (1989) Ind J Chem 28A:729
97. Dan FA, Françoise MW, Nicoleta G (1999) Colloids Surf A 149:339
98. Fanghaenel E, Ortman W, Behrmann K, Willscher S, Turro NJ, Gould IR (1987) J Phys Chem 91:3700
99. Raghuraman H, Pradhan SK, Chattopadhyay A (2004) J Phys Chem B 108:2489
100. Baptista ALF, Coutinho PJG, Oliveira MEDDR, Gomes JINR (2000) J Liposome Res 10:419
101. Miyagishi S, Asakawa T, Nishida M (1987) J Colloid Interface Sci 115:199
102. Ruiz CC, Molina-Bolivar JA, Aguiar J, Peula-Garcia JM (2004) Colloids Surf A 249:35
103. Haldar J, Aswal V, Goyal PS, Bhattacharya S (2001) J Phys Chem B 105:12803
104. Dutt GB (2004) J Phys Chem B 108:7944
105. Karukstis KK, McCormack SA, McQueen TM, Goto KF (2004) Langmuir 20:64
106. Liang P, Thomas JK (1988) J Colloid Interface Sci 124:358
107. Rushforth DS, Sanchez-Rubio M, Santos-Vidals LM, Wormuth KR, Kaler EW, Cuevas R, Puig JE (1986) J Phys Chem 90:6668
108. Zhai L, Zhang J, Shi Q, Chen W, Zhao M (2005) J Colloid Interface Sci 284:698
109. Cocera M, Lopez O, Estelrich J, Parra JL, De la M A (2000) Langmuir 16:4068
110. Sivakumar A, Somasundaran P (1994) Langmuir 10:131
111. Jesús S, Félix MG, Alicia A (2005) Biophys Biochem Acta Biomembr 1711:12
112. Carley AF, Davies PR, Roberts MW, Thomas KK (1990) Hydroxylation of molecularly adsorbed water at Ag(111) and Cu(100) surfaces by dioxegen-photoelectron and vibrational spectroscopic studies. Surf Sci 238:L467
113. Goddard ED, Turro NJ, Kuo PL (1985) Langmuir 1:352
114. Lianos P, Zana R (1981) J Colloid Interface Sci 84:100-7
115. Kunjappu JT, Somasundaran P (1989) Colloids Surf 38:305
116. Abuin E, Leon A, Lissi E, Varas JM (1999) Colloids Surf A 147:55
117. Khatua D, Ghosh S, Dey J, Ghosh G, Aswal VK (2008) J Phys Chem 112:5374

118. Bakshi MS, Singh K (2005) J Colloid Interface Sci 287:288
119. Bakshi MS, Singh J, Kaur G (2005) J Colloid Interface Sci 285:403
120. Kalyanasundaram K, Thomas JK (1977) J Chem Soc Faraday Trans 99:1312
121. Ueno M, Kimoto Y, Ikeda Y, Momose H, Zana R (1987) J Colloid Interface Sci 117:179
122. Leaver IH, Jurdana L (1992) J Colloid Interface Sci 153:552
123. Sahoo L, Sarangi J, Misra PK (2002) Bull Chem Japan 75:859
124. Chandar P, Somasundaran P, Turro NJ (1988) Macromolecules 21:950
125. Deo N, Somasundaran P, Itagaki Y (2005) Ind Eng Chem Res 44:1181
126. Kosmul M (2001) J Colloid Interface Sci 238:225
127. Misra PK, Mishra BK, Somasundaran P (2003) J Colloids Int Sci 265:1
128. Misra PK, Mishra BK, Somasundaran P (2005) Colloids Surf 252:169
129. Gu T, Zhu B (1990) Colloid Surf 44:81
130. Huang L, Somasundaran P (1996) Colloids Surf A 117:235
131. Krishnakumar S, Somasundaran P (1996) Colloid Surf 117:227
132. Chander P, Somasundaran P, Turro NJ, Waterman KC (1987) Langmuir 3:296
133. Arora KS, Turro NJ (1987) J Polym Sci B 25:243
134. Koswer EM (1965) Reactions through charge transfer complexes. In: Taft RW (ed) Progress in Physical Organic Chemistry, Vol 3. Wiley Interscience, New York, p 81
135. Zhang R, Zhang L, Somasundaran P (2004) J Colloid Interface Sci 278:453
136. Ray P, Bhattacharya SC, Moulik SP (1998) J Photochem Photobiol A 116:85
137. Medhage B, Almgren M, Alsins J (1993) J Phys Chem 97:7753
138. Hierrezuelo JM, Aguiar J, Ruiz CC (2004) Langmuir 20:10419
139. Zanette D, Frescura VL A. (1999) J Colloid Interface Sci 213:379
140. Lang J, Jada A, Malliaris A (1988) J Phys Chem 92:1946
141. Ranganathan R, Peric M, Bales BL (1998) J Chem Phys B 102:8436
142. Turro NJ, Yekta A (1978) J Am Chem Soc 100:5951
143. Infelta P (1979) Chem Phys Lett 61:88
144. Baden N, Kajimoto O, Hara K (2002) J Phys Chem 106:8621
145. Chander P, Somasundaran P, Turro NJ (1987) J Colloid Interface Sci 117:31
146. Turro NJ, Gratzel M, Braun AM (1980) Angew Chem Int Ed Engl 19:675
147. Turro NJ, Aikawa M, Yekta A (1979) J Am Chem Soc 101:772
148. Emert J, Behrens C, Goldenberg M (1979) J Am Chem Soc 101:771
149. Zacharisse K (1978) Chem Phys Lett 57:429
150. Lewis FD, Zhang Y, Letsinger RL (1997) J Am Chem Soc 119:5451

Surfactant-Mediated Fabrication of Optical Nanoprobes

Parvesh Sharma[1,2] · Scott Brown[1] · Manoj Varshney[1,3] · Brij Moudgil[1] (✉)

[1]Particle Engineering Research Center and Materials Science and Engineering,
University of Florida, Gainesville, 32611, USA
bmoudgil@perc.ufl.edu

[2]Department of Chemistry, St. Stephen's College, Delhi University, India

[3]Department of Anesthesiology, Shands Hospital, University of Florida,
Gainesville, FL 32611, USA

1	Introduction	190
2	Imaging Agents	190
3	Design of Nanoparticle for Bio-Imaging Applications	191
4	Surfactant System-Based Nanoparticle Synthesis	192
4.1	Nanoparticle Formation Process in Microemulsions	193
5	Dye-Doped Silica Nanoparticles	195
5.1	Dye-Doped Silica Nanoparticle Synthesis Using Nonionic Surfactant-Based Microemulsion Systems	196
5.2	Dye-Doped Silica Nanoparticle Synthesis Using Ionic Surfactant-Based Microemulsion Systems	200
6	Gold Nanoparticles	203
6.1	Gold Nanoparticle Synthesis Using Nonionic Surfactant-Based Microemulsion Systems	204
6.2	Gold Nanoparticle Synthesis Using Ionic Surfactant-Based Microemulsion Systems	205
6.3	Gold Nanoparticle Synthesis in Other Surfactant Aggregate-Based Systems	209
7	Quantum Dots	211
7.1	Quantum Dot Synthesis Using Nonionic Surfactant-Based Microemulsions	213
7.2	Quantum Dot Synthesis Using Ionic Surfactant-Based Microemulsions	213
8	Conclusion	220
	References	220

Abstract Modern bio-imaging techniques often employ contrast agents to improve the image quality and also to provide specific information about anatomical structure and/or the function of biological systems. Quantum dots, fluorescent dye-doped silica and gold nanoparticles are important examples of new nanoparticulate-based imaging agents that have overcome many of the limitations of conventional contrast media such as organic dyes. These agents have the ability to provide enhanced photostability and sensitivity in combination with sufficient in vitro and in vivo stability. Surfactant-mediated methods are one of the most versatile strategies for synthesizing nanosized contrast agents. Microemulsion-mediated synthesis, in particular, offers a widely applicable approach to produce a variety of engineered optical nanoprobes presenting good control over nanoparticle size, design and robust surface derivatization. Herein the authors provide a review of surfactant chemistry and strategies, with a particular focus on microemulsions, for generating luminescent nanoprobes, such as quantum dots, fluorescent silica and gold nanoparticles for bioimaging applications.

1
Introduction

A number of noninvasive imaging approaches such as computed tomography (CT), magnetic resonance imaging (MRI), positron emission tomography, single photon emission CT, ultrasound and optical imaging are currently being used for both in vitro and in vivo diagnosis [1]. In particular, fluorescence microscopy and MRI have attracted increased attention from researchers due to their respective abilities to provide highly sensitive detection at the cellular level and three-dimensional imaging of biological structures and processes. The need for contrast agents for selectively highlighting tumors and diseased tissues has fueled research for developing novel bioimaging nanoparticles. The wide variety of methods available for the synthesis of nanoparticles [2–5] can broadly be divided into either lithographic top-down approaches (patterning bulk materials down to nanoscale) or the bottom-up approaches (forming nanostructures starting from molecular precursors). Amongst the latter methodologies, surfactant-mediated synthesis has emerged as a broadly applicable method for producing relatively monodisperse optical nanoprobes for bioimaging applications. Nanoprobes such as dye-doped silica nanoparticles, quantum dots (QDs) and gold nanoparticles have been synthesized using surfactant templates. Herein the authors provide a review of surfactant chemistry and strategies employed for generating luminescent nanoprobes for bioimaging via surfactant-mediated synthesis.

2
Imaging Agents

Optical contrast agents (e.g., organic fluorescent dyes) are frequently used to stain portions of biological samples in order to obtain a greater understanding of the molecular, cellular, and physiological changes at hand. Colorimetric and fluorescent markers are two important classes of these agents. Colorimetric contrast agents modify the light absorption resulting in a contrast of observed color. Fluorescent agents also modify light absorption, but subsequently re-emit a portion of the absorbed light at higher (lower energy) wavelengths. This phenomenon is leveraged to achieve greater sensitivity with the use of fluorescent markers. By not illuminating samples with light at the emission wavelengths of the fluorophore(s) present, an enhanced signal to noise ratio is achieved. In recent years, fluorescent microscopy has become a mainstay in biological sciences; however, many challenges still exist in the application of fluorescent markers in biological systems. For instance, organic dyes are prone to rapid photobleaching limiting their application to long-term bioimaging investigations [6, 7]. Many dyes also have relatively broad emission spectra complicating their integration into multicolor imaging applications. In bioimaging, additional artifacts such as high light scattering via tissue interfaces, autofluorescence, and absorption by hemoglobin (Hb) in a mid-visible wavelength range are observed. The application of near infrared (NIR) range (650 to 900 nm) contrast agents can overcome many of these issues and are preferred for bio-imaging thick tissues. Hence, the development of NIR imaging agents such as NIR quantum dots and dyes (e.g., cyanine dyes) have recently attracted significant attention [8, 9]. Recent years have also seen the emergence of multifunctional contrast agents [10, 11] that attempt to leverage the benefits of multiple imaging modalities with the use of a single contrast agent. For instance, fluorescent and MRI active probes allow for simultaneous detection by MRI and fluorescence microscopy combining the advantages of three-dimensional anatomical resolution at cellular level from MRI with the high sensitivity offered by fluorescence microscopy in a single particle.

3
Design of Nanoparticle for Bio-Imaging Applications

Optical nanoprobes are preferably designed as core/shell nanostructures where the optically active material is located in the core. A general schematic of such a nanoparticle is indicated in Fig. 1 and the typical synthesis steps are as follows:

(i) Synthesis of an optical core: The desired optical component is either synthesized (e.g., the gold nanoparticle or quantum dot) or constrained in a matrix (e.g., fluorescent dyes) using surfactant-based systems.
(ii) Synthesis of shell: A shell that performs multiple functions is coated over the optical core. The shell protects the optical core from the external environment, thus improving its photostability (particularly for organic dyes) and sometimes helps to enhance the optical properties (e.g., by providing a higher bandgap material for QDs). It also provides a substrate for the attachment of various chemical species (e.g., antibodies,

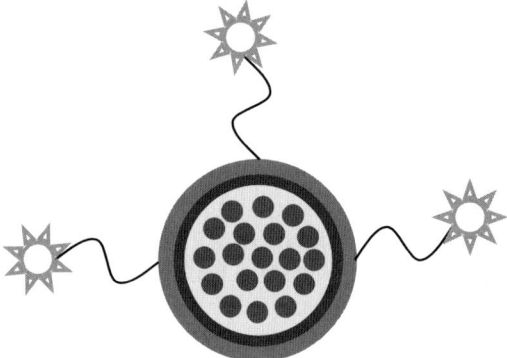

Fig. 1 Schematic design of nanoparticle for bioimaging containing the fluorescent dye ● encapsulated in an inorganic matrix ○ surface modified with another shell ○ and conjugated to biomolecules ✻

peptides, targeting ligands, etc.) for specific biological functions and biocompatibility.

(iii) Surface modification: This step is often used to improve the dispersion of nanoparticles in aqueous environments for better efficacy in biological systems. Strongly hydrated polymers such as polyethylene glycol (PEG), PEG-derivatized phospholipids, etc. [12, 13] are employed to enhance the aqueous dispersion of the QDs as wells as reduce nonspecific adhesion to biological cells.

(iv) Bioconjugation: For targeted delivery of the nanoparticles, it is often necessary to attach suitable molecular agents or biomolecules to the surface of the nanoparticles such as antibodies, peptides, and enzymes [14] in order to provide selective adhesion to the intended target.

4
Surfactant System-Based Nanoparticle Synthesis

Surfactants provide several types of well-organized self-assemblies, which can be used to control the physical parameters of synthesized nanoparticles, such as size, geometry and stability within liquid media. Established surfactant assembles that are commonly employed for nanoparticle fabrication are aqueous micelles, reversed micelles, microemulsions, vesicles [15, 16], polymerized vesicles, monolayers, deposited organized multilayers (Langmuir–Blodgett (LB) films) [17, 18] and bilayer lipid membranes [19](Fig. 2).

Out of these assemblies, microemulsion droplets and swollen micelles have been widely used as nanoreactors for inorganic nanoparticle synthesis [20–23]. These self-assembled nanosized beakers/droplets provide a robust and tunable environment that permits size-controlled encapsulation of

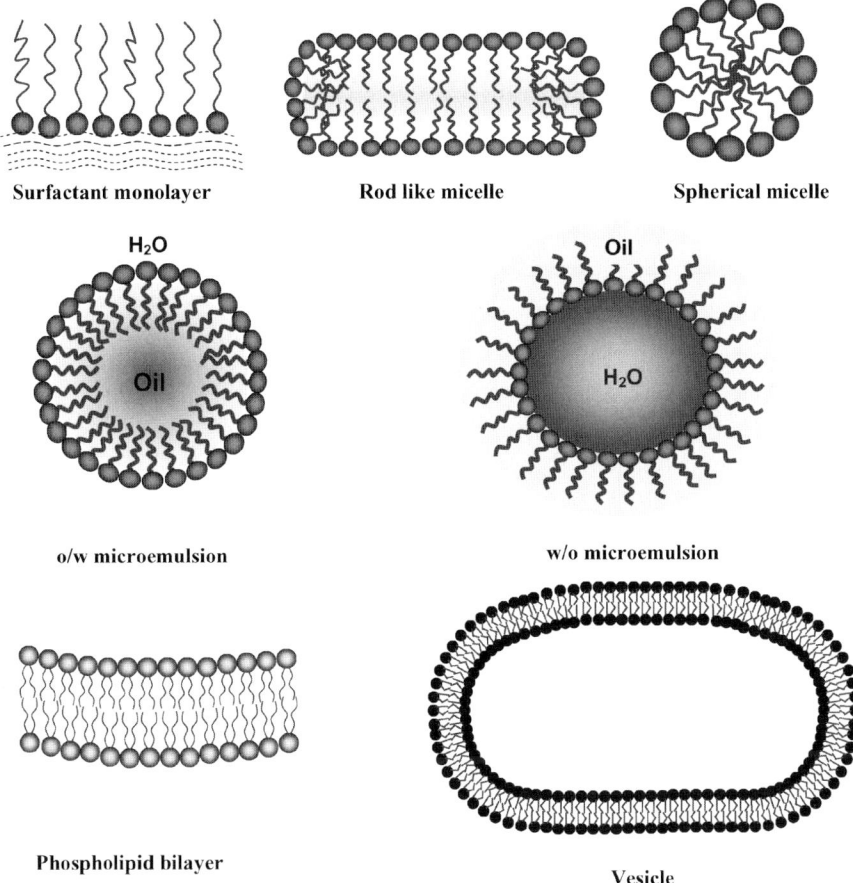

Fig. 2 Schematic representation of some organized structures formed using surfactants

luminophores during nanoparticle synthesis, subsequent surface derivitization (for stability and bioconjugation), and facile integration of multimodal functionalization. Microemulsions have been used extensively to develop different types of core/shell structure, quantum dots and gold nanoparticles as discussed in the following sections.

4.1
Nanoparticle Formation Process in Microemulsions

Microemulsions allow for the synthesis of nanoparticles of controlled particle size with a narrow size distribution, and are amicable to simple functionalization routes for decorating the resulting nanoparticle surfaces. Being a soft chemical method, these methods are well suited for encapsulating or surface anchoring biologically sensitive molecules (e.g., DNA, proteins, peptides). Since hydrophilic particles offer improved biocompatibility and are more amicable for modification with biological agents, "water-in-oil" (w/o) microemulsion systems are preferred for bioimaging nanoparticles synthesis. W/o microemulsions have tremendous scope for controlling the nanoparticle design (e.g., by changing the surfactant, co-surfactant, major or minor phase composition, or temperature). Various microemulsion-mediated methods have been used to prepare nanoparticles of metals, metal oxides and some polymeric materials. However, the pronounced versatility of microemulsion synthesis is somewhat diluted by the difficulties associated with upgrading this synthesis route to the bulk scale economically. A number of books [16, 24, 25] and reviews [4, 26–28] have covered this subject in the past and the readers are advised to refer to those for a complete understanding and application of this route for synthesis of nanoparticles.

In w/o microemulsion, the water molecules reside in the core of the microemulsion as shown in Fig. 3. The radius of such a core is known as the hydrodynamic radius (R_H) and lies in the nanodomain [29]. The water cores are separated from the bulk organic phase by a surfactant layer [20, 30]. These cores act as nanoreactors during nanoprobe synthesis. The size of this core is controlled by varying the mole ratio of water to surfactant (Wo). Such microemulsion-based nanoreactors collide continuously due to thermal Brownian motion; hence a fraction of droplets exist in the form of short-lived dimers [31] that exchange their water content [32]. After a short time these dimers decoalesce resulting in an eventual equilibrium distribution of all the contents [33]. Therefore, when a precipitation reaction is carried out in these w/o microemulsions, as shown schematically in Fig. 3, the product formed acquires nano dimensions. The exact nature of the dynamic process of the chemical reaction in the microemulsions is described by the relative change of two parameters [28]: (i) collision and material exchange, represented by τ_{ex} and (ii) chemical reaction time τ_r. Primarily the reactions in the microemulsion can be classified into those in which the collision-exchange rate is much more than the chemical reaction time, i.e., $\tau_r/\tau_{ex} \gg 1$ (e.g., for the hydrolysis of tetraethylorthosilicate (TEOS) during silica nanoparticle formation) and others in which the chemical reaction rate is nearly equal to collision exchange dynamics, i.e., $\tau_r/\tau_{ex} \approx 1$ (e.g., fast reduction of gold chloride by borohydride). Importantly, both these kinds are affected differently by the flexibility/interfacial rigidity, a characteristic of the surfactant/co-surfactant combination comprising the microemulsion.

The attractive strength between the water cores and the associated kinetic exchange process is also influenced by the length of the alkyl chain of the bulk oil used [34, 35]. For instance, short alkyl chain oils solvate the similar surfactant alkyl chains more as compared to the longer (and branched) alkyl chains thereby reducing the exchange process. For instance, the kinetic exchange rate is accordingly increased by a factor of 10 on substituting cy-

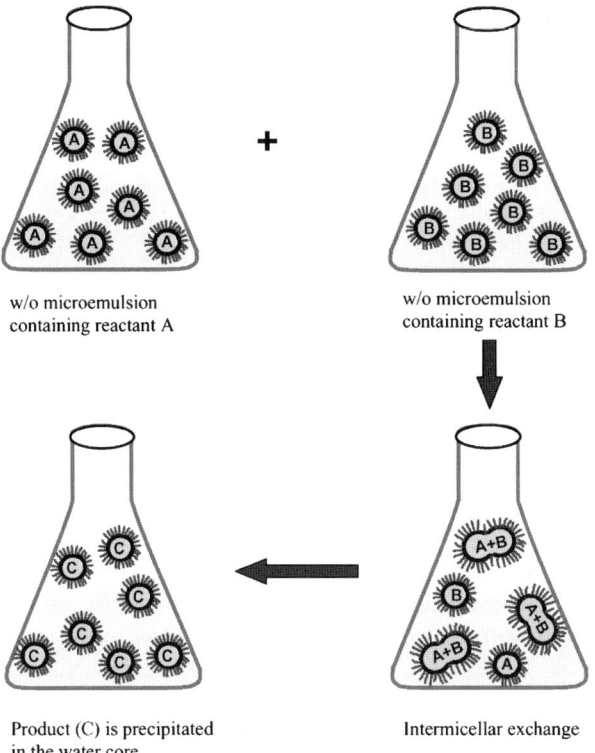

Fig. 3 Preparation of nanoparticles by precipitation reaction from w/o microemulsion

clohexane with isooctane as bulk oil solvent [36]. An additional important feature of using w/o microemulsions is that they inhibit the aggregation process of particles because the surfactant molecules have the ability to adsorb on the particle surface. As a result, the particles obtained in such a medium are generally monodisperse. Largely similar reaction dynamics occur when o/w microemulsion systems are used to prepare particles, except that the starting materials in this case are hydrophobic in nature. Recently, a range of experimental and simulation findings has been summarized by López-Quintela [27]. The synthesis of dye-doped silica, semiconductor quantum dots and gold nanoparticles using the surfactant-mediated approach, primarily the microemulsions, is described in the following sections.

5
Dye-Doped Silica Nanoparticles

Amorphous silica is an optically transparent, biocompatible [37] and non-toxic [38] matrix that permits the encapsulation of the dyes without affecting their spectral properties [11, 39, 40]. In addition, conjugation of biomolecules (e.g., proteins, peptides, antibodies, oligonucleotides [39, 41]) to the silica surface is relatively facile. Silica nanoparticles have been widely used to encapsulate dyes in order to enhance their stability for in vitro and in vivo applications [11, 40–44]. The encapsulation of dyes within the silica core inhibits rapid photobleaching [40, 45], allowing for extended imaging that would not be possible with free dye molecules. Both inorganic (e.g., tris(2,2'-bipyridyl)dichlororuthenium(II)hexahydrate, RuBpy) [11, 41] as well as organic (tetramethyl rhodamine) [46] dyes were incorporated in the silica matrix with suitable modification. These dyes were entrapped either by covalent linkage to the silica [45, 47] (precursor) or simply by encapsulation (physical entrapment) [11, 39, 40]. Our group has recently prepared gadolinium and dye-doped silica nanoparticles, which can be readily bioconjugated [11].

Dye-labeled particles have been used for quantitative real-time studies with confocal fluorescence microscopy, time-resolved phosphorescence anisotropy [48], as photostable biomarkers [40–42], as biosensors [49, 50], DNA hybridization analysis [51], and cancer cell recognition [43] (including fluorescent-linked immunosorbent assay, immunocytochem, immunohistochem [44]). By combining the high-intensity luminescent silica nanoparticles with the specificity of antibody-mediated recognition, ultra-sensitive target detection has been achieved in various fluorescence-labeling techniques, including DNA microarrays and protein microarrays [44].

Traditionally, the sol-gel process has been used for the preparation of silica nanoparticles via the hydrolysis of alkoxides in organic solvents [52, 53]. Similar hydrolysis and condensation carried out in w/o microemulsion offers robust control over the synthesis process. W/o emulsion-mediated sol-gel synthesis is currently used for the fabrication of pure silica, as well as inorganic and organic dye-doped silica nanoparticles. The synthesis of silica and dye-doped nanoparticles is classified in the following sections on the basis of the classification of the head group functionality of the major surfactant used.

5.1
Dye-Doped Silica Nanoparticle Synthesis Using Nonionic Surfactant-Based Microemulsion Systems

Silica particles synthesized in nonionic w/o microemulsions (e.g., polyoxythylene alkyl phenyl ether/alkane/water) typically have a narrow size distribution with the average value between 25 and 75 nm [54, 55]. Both water and surfactant are necessary components for the formation of stable silica suspensions in microemulsions. The amounts of each phase present in the microemulsion system has an influence on the resulting size of the silica nanoparticle. The role of residual water (that is the water that is present in the interface between the silica particle and the surfactant) is considered important in providing stability to the silica nanoparticle in the oil continuum (and in the microemulsion system) since it provides a means of hydrogen bonding between silica particles and surfactants. This facilitates adsorption of the surfactant molecules onto silica particles providing steric stabilization [56]. At a constant surfactant concentration with an increase in the water concentration, the silica particle size generally decreases [54]. For a given water concentration, at low surfactant-to-water molar ratios, increasing the surfactant concentration leads to a reduction of silica particle size, whereas at high molar ratios the opposite trend occurs [54, 55]. Chang [57] et al. conducted a detailed investigation on the effect of type and concentration of surfactant, the concentration of water, and the type of oil on the silica nanoparticle. It was found that silica particles with smaller sizes are synthesized when the variables decrease the condensation of silica reacting species by increasing the number of microemulsion droplets or by increasing the steric hindrance of surfactant films that retard the dynamic exchange of reacting species between droplets. Thus, using the surfactants which are polydisperse mixtures of poly(oxyethylene) alkylphenyl ether molecules with the general chemical structure n-H(CH$_2$)$_n$Ph-(OC$_2$H$_4$)$_m$OH (where for NP-4, $n = 4$, $m = 9$; for NP-5, $n = 5$, $m = 9$; for DP-6, $n = 6$, $m = 12$) it is shown that NP-5 droplets tend to associate and form interdroplet open water channels leading to larger-sized silica particles. However, DP-6 surfactant with longer head and tail groups, provides stronger steric film barriers to the interdroplet exchange of silica reacting species than those of the NP-4 surfactant leading to more nuclei formation and eventually smaller particles than in NP-4 solutions. In a recent report 14 nm silica particles have been reported using the NP-5/cyclohexane/water microemulsion [58].

An increase in the water concentration leads to: an increase in the size of bulk water to bound water, an enhanced number of monomers of the reactants per unit microemulsion droplet, and a decrease in the rigidity of the interfacial film. The effect of changing the water concentration shows a weak dependence on the size of silica particles due to the formation of open water channels in NP-5 solutions. In the case of NP-4 and DP-6 solutions, increasing the water concentration dilutes reacting species within the water pools of droplets, decreasing the rate of condensation between reacting and forming small particles. The effect of water-to-surfactant molar ratio W_0 (and other reaction variables also), on the silica nanoparticle size has been investigated extensively by Arriagada et al. using the polyoxyethylene (5) nonylphenylether (NP-5)/cyclohexane/ammonium hydroxide water-in-oil microemulsion system [59, 60]. It was suggested that intra-micellar nucleation should be increasingly facilitated as W_0 increased and, therefore, the particle size was expected to decrease monotonically as W_0 increased (at low ammonia concentrations). Bagwe et al. [58] similarly observed a decrease in the size from 178 nm ($W_0 = 5$) to 69 nm ($W_0 = 15$) of RuBpy dye-doped silica nanoparticles prepared in Triton X-100(Tx-100)/cyclohexane/water microemulsion.

In another communication using w/o microemulsions containing a nonionic surfactant, it is shown that TEOS hydrolysis and silica-particle growth occur at the same rate, indicating the growth of silica particles is rate-controlled by the hydrolysis of TEOS [54]. The rate of TEOS hydrolysis also depends on the surfactant concentration, which controls the molecular contact between hydroxyl ions and TEOS in solution. Because of the reaction-controlled growth mechanism, the silica-particle size distribution remains virtually same over the growth period.

A large number of reports can be found in the literature for the entrapment of luminophores (e.g., organic and inorganic fluorescent dyes) within the core of the silica nanoparticles. The brightness of dye-doped silica nanoparticle fluorescence can be improved by incorporating high-quantum yield organic dyes having large absorption coefficient values. This is desired because brighter probes can improve image resolution and the detection sensitivity [50, 61]. In a modification of protocol reported by Jain et al. [62], Yang and others [63] developed novel fluorophore-doped silica nanoparticles via the w/o microemulsion method (and sol-gel technique) as efficient and uniform biomarkers. The fluorophores are covalently incorporated in the nanoparticles, and the surface groups on the particles can be used for conjugation with biomolecules. It was shown that providing an inert barrier to the dye (within the silica shell) leads to an enhancement in emission by minimizing the solvent effects responsible for nonradiative decay mechanisms.

In most examples cited above only one type of dye molecule is doped inside silica nanoparticles using the microemulsion method allowing single-target detection. In comparison, multiplexed assays can potentially reduce the time and cost per analysis, allow for simpler assay protocols, decrease the sample volumes required, and, most importantly, make comparison of samples feasible and measurements reproducible and reliable. Tan's group [64] synthesized dual-luminophore-doped silica nanoparticles for multiplexed signaling. Two luminophores, tris(2,2'-bipyridyl)osmium(II)bis(hexafluorophosphate) (OsBpy) and RuBpy, were simultaneously entrapped inside silica nanoparticles prepared in Tx-100 nonionic microemulsion. Having precisely controlled dye ratios, desirable size and surface functionality, endowed the nanoparticles with single-wavelength excitation, dual emission capability thus enabling optical encoding for rapid and high-throughput multiplexed detection.

Using a similar w/o microemulsion system (of Tx-100/cyclohexane) Zhao [46] and co-workers entrapped an organic dye (tetramethylrhodamine) in the silica nanoparticle core. The higher quantum yield of organic dyes makes them useful for entrapment in comparison to inorganic fluorophores. The entrapment of a hydrophobic compound (organic dye) within the cavity of the hydrophilic core of the w/o microemulsion was achieved by conjugating the dye with the hydrophilic dextran molecule. This has the added benefit of making the dye sterically bulky thus preventing leaching from the inorganic core.

Fluorescent europium (III) chelate-doped silica nanoparticles were prepared in a w/o microemulsion consisting of a strongly fluorescent Eu(III) chelate, 4,4'-bis(1'',1'',1'',2'',2'',3'',3''-heptafluoro-4'',6''-hexanedion-6''-yl)-o-terphenyl-Eu(III), in Tx-100, co-surfactant (n-hexanol, n-heptanol or n-octanol)/H_2O (or D_2O)/cyclohexane by controlling the hydrolysis of TEOS [65]. With the change of co-surfactants from n-hexanol to n-heptanol to n-octanol, the diameters of the nanoparticles were observed to decrease gradually from 37 to 29 nm. This was explained on the basis of an increase in the hydrophobicity and the surface activity of the co-surfactant, which corresponds, to an increase of the amount of surfactant, leading to a decrease of the nanoparticle size. Perhaps the lowering of the curvature by the addition of the surfactants also plays a role in decreasing particle size [66, 67].

From our research group Santra et al. [11, 41, 42] reported the development of novel luminescent nanoparticles composed of inorganic luminescent dye RuBpy, doped inside a silica network. These dye-doped silica nanoparticles were synthesized using a w/o microemulsion of Tx-100/cyclohexane/n-hexanol/water in which controlled hydrolysis of the TEOS leads to the formation of monodispersed nanoparticles ranging from 5–400 nm. This research illustrates the efficiency of the microemulsion technique for the synthesis of uniform nanoparticles. These nanoparticles are suitable for biomarker application since they are much smaller than the cellular dimension and they are highly photostable in comparison to most commonly used organic dyes. It was shown that maximum luminescence intensity was achieved when the dye content was around 20%. Moreover, for demonstration

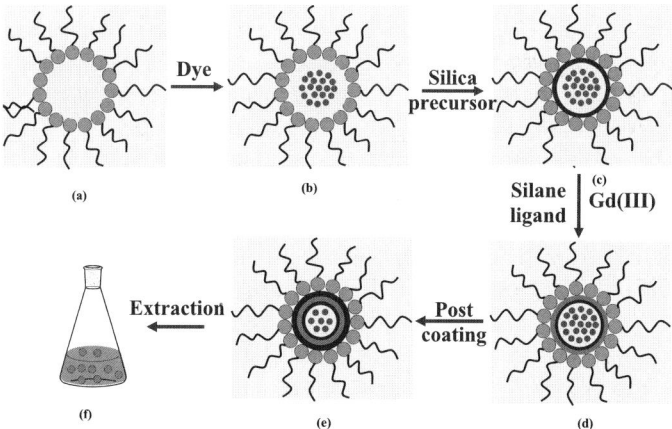

Fig. 4 Schematic illustration of synthesis of multifunctional nanoparticles starting from **a** w/o microemulsion, **b** solubilization of fluorescent dye in the microemulsion core, **c** formation of silica nanoparticle and encapsulation of fluorescent dye, **d** condensation of silane ligand and chelation of Gd(III), **e** post coating with silica, and **f** extraction of nanoparticles

as a biomarker, the dye-doped silica nanoparticle surface was biochemically modified to attach to membrane-bound groups and applied successfully to stain human leukemia cells. In another report Santra et al. [11] engineered multifunctional, single core multiple shell, RuBpy and gadolinium(III)-doped silica (RuBpy: Gd(III)/SiO$_2$) nanoparticles in water-in-oil microemulsion of Tx-100/n-hexanol/cyclohexane. The core of the silica contains the optical component (i.e., RuBpy, an organometallic dye) (Fig. 4). The nanoparticles inner shell was designed to capture the Gd(III) ion by forming a chelate thus preventing loss of Gd(III) due to leaching and possible toxicity. The presence of paramagnetic ion Gd (III) in the nanoparticle makes it detectable by MRI and the Ru(II) and Gd(III), electron dense metal ions impart the radio opaque nature to the core shell nanoparticles. Thus, MRI, CT and diffuse optical tomography can be used to detect these multifunctional probes. In addition these nanoparticles have surface reactive groups that can be modified to attach both ligands and antibodies.

5.2
Dye-Doped Silica Nanoparticle Synthesis Using Ionic Surfactant-Based Microemulsion Systems

Dioctyl sodium sulfosuccinate (AOT)/water/oil is a commonly used anionic surfactant system for dye-doped silica nanoparticle synthesis [68, 69]. A number of factors (e.g., droplet size, droplets concentration, interfacial rigidity, reactant concentration per reverse micelle droplet) are known to affect the particle size. It is reported that droplet size, aggregation number and intermicellar exchange rate are influenced by W_0 [70]. It is also found that no particle formation takes place below $W_0 = 4$. The surfactant molecules at this stage are tightly packed at the interface and the water molecules are strongly bound to the surfactant polar groups and the surrounding counter ions. Up to $W_0 = 10$, water exists as "bound water" (bound to the surfactant) in the spheroidal microemulsion core and free water is available on subsequent increase in water molar ratio. The unavailability of the free water molecules, up to $W_0 = 10$, is confirmed by following the fluorescence spectra of 1,3,6,8-pyrenetetrasulphonic acid. Arriagada et al. [71] have investigated the formation of silica nanoparticles in the AOT/NH$_4$OH/decane w/o microemulsions and compared it to those prepared in the nonionic system. For a nonionic system the increase in W_0 corresponds to availability of more water for TEOS hydrolysis. For a fixed amount of TEOS it results in increased nuclei formation and thus decreased particle size. However, this is not so in the case of anionic surfactant AOT since free water required for the nuclei formation is not available with increasing W_0 (up to $W_0 = 10$), thus a corresponding decrease in the particle size is not observed. Illustrating a further distinction from nonionic systems [55, 72], dispersions in the AOT system were found to be unstable and turned into gel-like materials. Besides AOT, another anionic surfactant that has been commonly used for the preparation of the silica nanoparticles is sodium dodecly sulfate (SDS). SDS readily forms w/o microemulsion in pentanol and the complete phase diagram of this system is known [73]. Frieber et al. [74, 75] have studied the gel formation on hydrolysis of silicon tetraethoxide in inverse micellar solution of water/SDS/pentanol. The effect of molar ratios of water to surfactant, co-surfactant to surfactant, the nature of the surfactant molecule, reactant concentration effect on nucleation rate and dye location inside the nanoparticle have been described recently [58]. Bagwe et al. synthesized dye-doped silica nanoparticles using different microemulsion systems comprised of AOT/heptane/water AOT + NP-5 (polyoxyethylene nonylphenolether)/heptane/water, NP 5/cyclohexane/water; Tx-100/ n-hexanol/cyclohexane/water. The microemulsions used are different in ionic nature as AOT, SDS are anionic surfactants, while NP-5, Tx-100 are nonionic surfactants and n-hexanol acts as a co-surfactant in the Tx-100 microemulsion system. The particles obtained were spherical in shape, and the particle size was smallest in NP-5 (14 nm), followed by those obtained in the AOT system (30 nm), while the largest size was obtained in AOT + NP-5 (130 nm). These variations in particle size are explained on the basis of structure of the microemulsion droplet formed in different surfactant systems as shown in Fig. 5. AOT, a double-tailed anionic surfactant with a head group area smaller than the volume of the hydrocarbon tail forms micelles having spherical core structure [32]. Chang et al. [57] on the other hand have shown that NP-5 microemulsions have a lamellar structure, wherein the water droplets tended to associate together and form interdroplet water channels, which leads to less compartmentalized droplets than those in AOT, enhancing the nucleation rate and thus forming smaller particles.

In addition, the amount of co-surfactant present in the microemulsion system is known to have a strong influence on the particle size. The overall effect of co-surfactant addition is largely dependent on the kind of influence it has on the stability of the microemulsion system and on the interfacial rigidity. In a microemulsion comprising of both AOT and NP-5, the latter acts as a co-surfactant, changing the interfacial rigidity as a result of shielding of the negative charge between two anionic AOT surfactant molecules and also because of an increase in hydrophobic interaction between nonpolar tails of AOT and NP-5. This results in a decrease in microemulsion droplet size [76], and a lowering of intermicellar exchange rate, preventing more nucleation and favoring growth leading to bigger size particles. Interestingly, contrasting results are obtained for MoSx and Ag nanoparticle [77] synthesis using the same microemulsion system [76]. The addition of NP-5, as a nonionic co-surfactant, at small concentrations compared to that of AOT, leads to a substantial decrease of the mean micellar and nanoparticle size. This has been explained on the basis of discontinuity introduced in the interfacial film of the water pool due to a cycloalkane hydrophobic chain in NP-5 that pro-

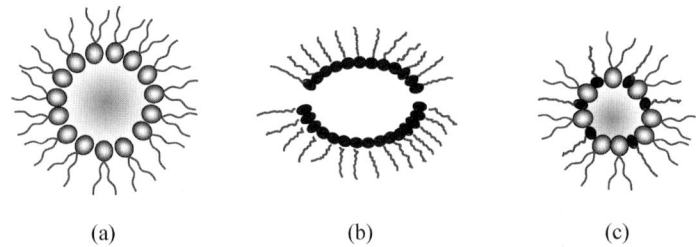

Fig. 5 Simplified illustration of the variation in droplet structure. **a** AOT/heptane/water, **b** NP-5/cyclohexane/water and, **c** (AOT + NP-5)/heptane/water, w/o microemulsions

motes the intermicellar exchange leading to a larger rate of growth during the nucleation stage resulting in smaller-sized particles.

It is well known that the aqueous phase behavior of surfactants is influenced by, for example, the presence of short-chain alcohols [66, 78]. These co-surfactants increase the effective value of the packing parameter [67, 79] due to a decrease in the area per head group and therefore favor the formation of structures with a lower curvature. It was found that organic dyes such as thymol blue, dimidiumbromide and methyl orange that are not soluble in pure supercritical CO_2, could be conveniently solubilized in AOT water-in-CO_2 reverse microemulsions with 2,2,3,3,4,4,5,5-octafluoro-1-pentanol as a co-surfactant [80]. In a recent report [81] the solubilization capacity of water in a Tx-100/cyclohexane/water system was found to be influenced by the compressed gases, which worked as a co-surfactant.

The core of the w/o microemulsion can strongly influence the fluorescence emission of the encapsulated dyes. For instance, an increase in W_0 causes a red shift in fluorescence emission of the RuBpy-doped NP-5/cyclohexane/water microemulsion system [60]. This is explained on the basis of an increase in the polarity of the medium surrounding the dye. It is observed [58] that the emission wavelength of RuBpy dye-doped nanoparticles was more red shifted in AOT and AOT + NP-5 than in NP-5 alone. The dye molecules are in more polar surroundings when they are present in the AOT microemulsion system as compared to the nonionic surfactant, NP-5. Hence, when these dye-doped nanoparticles are dispersed in water, they have less interaction with the surrounding water molecules and show fluorescence emission at a lower wavelength relatively. In another experiment, it was shown that addition of AOT or SDS to particles prepared in the Tx-100/cyclohexane/hexanol/water system led to a red shift in fluorescence emission. Arriagada et al. have found a similar red shift in the fluorescence spectra of RuBpy dye in w/o microemulsion on an increase in the polarity surrounding the probe [60].

Few reports are available in the literature, in which cationic surfactants have been used to prepare the silica nanoparticles [82] and particularly those encapsulating an optical component. In a surfactant template approach Lin et al. [83] prepared Gd(III)-loaded mesoporous silica

(MS) using (hexadecyltrimethylammonium bromide) C_{16}TAB as a template. In order to prevent the precipitation of gadolinium hydroxide under basic conditions, surfactant, $GdCl_3 \cdot 6H_2O$, and TEOS were mixed under acidic conditions. After the hydrolysis of TEOS, the product Gd-MS was precipitated out by raising the pH to 9 with NH_4OH. This is a simple and fast method to make luminescent mesoporous silicates in contrast to earlier methods [84, 85] that involved hydrothermal conditions. Macroscopic forms of mesoporous silica [86] have a variety of applications including catalysis and chromatography. Nanoparticulate forms of mesoporous silicas exhibit remarkable a potential due to improved accessibility of the mesopores and ease in handling the material. Synthesis of these nanomaterials has been carried out by numerous researchers using surfactant micelle templating methods [87–89]. Mesoscopically ordered surfactant/co-surfactant templated metal oxides (silica, titania and zirconia) have been synthesized using C_{16}TAB as a structure-directing agent and 1-octanol as a co-surfactant [90]. Silica nanoparticles having mesopores with a uniform aperture of ca. 2 nm were prepared using a cationic surfactant, cetyltrimethylammonium chloride (CTAC), as a templating agent and a nonionic surfactant triblock copolymer (Pluronic F_{127}; $EO_{106}PO_{60}EO_{106}$) (where EO represents the ethylene oxide block and PO represents the propylene oxide block) as a suppressant of grain growth, respectively [86]. A well-ordered hexagonal arrangement of the mesopores was obtained in nanoparticles by previous hydrolysis of tetraethoxysilane with hydrochloric acid and then subsequent assembly of cationic surfactant micelles and anionic silicate species under basic conditions. The co-existence of a block copolymer (nonionic surfactant) suppressed the growth of the silica particles resulting in a minimum grain size without deformation of the hexagonal mesostructure of ca. 20 nm diameter.

6
Gold Nanoparticles

Gold nanoparticles provide colorimetric contrast through surface plasmon resonance (SPR). SPR results when the incident photon frequency is in resonance with the collective excitation of the conductive electrons of the particle. Depending on particle size, shape, and agglomeration, gold colloids can appear red, violet, or blue as explained by Mie scattering theory [3, 91]. Usually, stable gold colloids with a small particle diameter and no agglomeration are red. Any color change to violet or blue indicates agglomeration and subsequently, in many cases, particle precipitation [92]. The SPR frequency depends on various factors, e.g., particle size [93], shape [94], solvent and ligand [95], dielectric properties [96], aggregate morphology [97], surface functionalization [98] and the refractive index of the surrounding medium [99].

Compared to QDs and organic dyes, gold nanoparticles do not undergo any photodecomposition and show excellent biocompatibility [100, 101] and facile bioconjugation [102]. Various research groups have illustrated that gold-coated silica nanoparticles shift the SPR, in a controlled fashion, to the NIR spectral region, which opens the way for numerous additional bioapplications [103, 104]. For instance, gold nanoshells have been fabricated for near-infrared resonance with antibodies against target antigens conjugated to the nanoshell surface [105]. The antibody–antigen linkages caused the gold nanoshells to aggregate, shifting the resonant wavelength toward the infrared region. Such unique optical (colorimetric) properties of gold nanoparticles are now finding increasing applications for use in the detection of biological systems [106–108].

A number of references including books [3, 92, 109] and reviews [110–112] are available for the synthesis of gold nanoparticles. The book published by Hayat [92] especially deals with the biological and imaging aspects of gold nanoparticles. In comparison to the top-down approach (e.g., metal vapor synthesis, laser ablation etc.) the alternative approach (the bottom-up approach) has flourished in the past decade, based on the principles of self-assembly of building units using colloidal chemistry approaches to the synthesis and rational design of advanced materials.

6.1
Gold Nanoparticle Synthesis Using Nonionic Surfactant-Based Microemulsion Systems

The use of an inorganic phase in w/o microemulsions has received considerable attention for preparing semiconductor and metal particles including the platinum group [113] and the noble metals [114]. As described earlier the main advantage of using w/o microemulsions in the synthesis of gold (or metal) nanoparticles is that the resulting particles have a very narrow size distributions. Furthermore, these particles are prevented from aggregation by the layers of surfactant molecules that cap the surface. The size of the nanoparticles can be easily varied by changing the concentrations of the reagents used, varying the W_0, temperature and time allowed for ripening of particles. Wilcoxon et al. [115] have studied in detail the formation of gold nanoparticles in w/o microemulsions composed of ionic and nonionic surfactants by the reduction of the gold precursor salt with various reducing agents (chemical and photolytic) and compared the results with those obtained in aqueous medium. Chemical reduction was achieved using reducing agents such as hydrazine, sodium borohydride, and metallic sodium (dispersed in oil). Using nonionic surfactants [belonging to the ethoxylated alcohol family, $CH_3\,(CH_2)_{i-1}\,(CH_2CH_2O)_j\,OH$ (commonly referred to as CiEj)], it was observed that by using either too hydrophobic (e.g., $C_{12}E_3$) or too hydrophilic ($C_{12}E_8$) surfactant resulted in poor stability of the Au colloids. By using a more hydrophobic solvent (e.g., with a C_{16} chain) with $C_{12}E_3$, stability and good quality samples were obtained, indicating there is an optimal hydrocarbon number for a given i/j ratio. By varying the concentration of the precursor salt to the surfactant $C_{12}E_4$ (in which the attractive intermicellar interactions are small) and carrying out the reduction with a strong reducing agent like $NaBH_4$, it was shown that the average size was significantly smaller and the cluster size increased significantly with increasing precursor salt concentration. Moreover, on decreasing the surfactant concentration an increase in the colloids size was observed indicating the importance of the precursor salt to surfactant concentration in controlling the particle size.

Additionally, the phase boundaries of hydrophilic nonionic surfactant (e.g., $C_{10}E_8$)/hydrophobic oil (e.g., octane) have also been used to obtain colloidal gold [115]. It was demonstrated that the addition of the gold salt to a single phase inverse micelle of $C_{10}E_8$/iso-octane results in the formation of two phases at room temperature and a single phase at higher temperature. Upon reduction of the metal salt, colloidal gold is formed whose size depends on the reaction temperature relative to the two-phase boundary. Lowering of the temperature results in phase separation of the system into two phases with gold colloids remaining in the upper oil phase and the ionic by-products in the lower surfactant phase. This allows facile separation of the metal colloids from excess surfactant and unwanted by-products. Spirin et al. [116] observed that the reduction of $HAuCl_4$ in water pools of Tx-100 micelles yielded large gold nanoparticles with a wide size distribution consistent with earlier observations [31]. The presence of the sulfite ions in the Tx-100 micelles, however, makes it possible to increase the yield of nanoparticles by several times, to decrease their mean size, and to substantially enhance their stability without addition of any stabilizers. In this case sodium sulfite performs the role of stabilizing the micelles, competing with OH^- nucleophiles and causing surfactant molecules to ionize and acquire the ability to reduce gold.

6.2
Gold Nanoparticle Synthesis Using Ionic Surfactant-Based Microemulsion Systems

Amongst the anionic surfactants, AOT has been widely used for synthesizing gold nanoparticles [116, 117]. It is observed that the gold nanoparticles prepared in the AOT microemulsions [116, 118] aggregate and eventually form a precipitate. This is consistent with earlier observations [115] and explained by the oxidative degradation of gold nanoparticles in micelle water pools. Chiang [119] has shown that the stability of the gold colloids formed in the AOT microemulsion is increased significantly by the addition of a nonionic surfactant sorbitan monooleate (Span 80), which adsorbs strongly on the colloid surface and blocks further growth by causing steric hindrance/coulombic repulsion. Herrera et al. [117] studied the aggregation properties of gold nanoparticles in AOT microemulsions using sulfite ions as the reducing agent. These studies indicated that gold nanoparticles initially formed by intermicellar exchange of reagents to a size of about 8 nm. Subsequently, these nanoparticles appeared to agglomerate, slowing down further intermicellar exchange and thus preventing further growth. Moreover, the microemulsion stability is shown to be influenced strongly by the purity and moisture content of AOT as well as by the presence of dissolved oxygen [115, 116]. For this reason, AOT must be extensively desiccated. Even recrystallized AOT is notorious for absorbing significant amounts of water (thus contributing to change in W_0 and consequent polydispersity) [115]. In the presence of oxygen (in microemulsions based on AOT and Tx-100) gold nano-aggregates oxidize [120] to form $[Au(OH)_4]^-$ ions rather than gold oxides [116]. Furthermore, a change in pH causes spontaneous coagulation and destabilization of the system. Hence, the removal of dissolved oxygen is a necessary condition for w/o microemulsion stability. The micellar stability has been shown to be increased by the addition of thioglycerol [116] (because its molecules interact with the surface of nanoparticles, preventing surface oxidation) or sodium sulfite (which binds the dissolved oxygen, preventing the oxidation in w/o microemulsion, water pools) [121]. Gold nanoparticles have also been prepared in the presence of dopamine hydrochloride in an aqueous micellar solution of SDS under the action of UV radiation [122]. In contrast to the nonionic systems described above the differences in AOT system are: much weaker micellar interactions, less dependence on temperature of reaction, and no complexity with the issue of phase separation.

Gold nanoparticles have also been synthesized by the reduction of tetrachloroauric acid with hydrazine in mixed w/o microemulsions composed of anionic surfactant AOT and nonionic surfactants tetraethylene glycol dodecyl ether ($C_{12}E_4$) [123] and Span 80 [119] in isooctane. It is found that in the presence of co-surfactant $C_{12}E_4$, the Au particle size is larger than that in the absence of $C_{12}E_4$. Moreover, the particle size decreases with increase in the concentration of $C_{12}E_4$ showing that the co-surfactant stabilizes the nanoparticles, and prevents their further growth and precipitation. It was also observed that a decrease in the molar ratio of hydrazine to $HAuCl_4$ resulted in larger Au particles with significantly more polydispersity (which is probably due to the low number of nuclei formation which leads to the larger size of the nanoparticles). In another interesting observation [119] $HAuCl_4$ injection directly into the mixed w/o microemulsions containing the reducing agent, anisotropic gold nanoparticles, such as cylinders, polyhedrons and cubes were obtained at the molar ratio of hydrazine to $HAuCl_4$ of less than 0.5. This is explained on the basis of an alternative structural model, namely, the open water-channel model [124]. According to this the water molecules tend to be localized in channels within the surface of some rod-like micellar aggregates, which provide the site for crystal nucleation leading to different geometries. Alternately, the addition of co-surfactant Span 80 is expected to increase the packing parameter (by increasing the average volume per surfactant molecule) resulting in the formation of cylindrical lamellar intermediates. Interestingly, face-centered cubic gold colloids are obtained by the reduction of $[AuCl_4]^-$ with hydrazine as reducing agent in AOT microemulsions [119, 125]. The soft preparation characteristic of the w/o microemulsion approach is highlighted in the synthesis of hollow gold nanoparticles using AOT microemulsion [125].

Gold nanoparticles from 2.5 to 5 nm sizes have also been prepared by using a biphasic Winsor II [126] (a water-in-oil microemulsion that is in equilibrium with the excess water phase) type microemulsion of diethyl ether/AOT/water. The surfactant, AOT, performs the dual role of forming a microemulsion and the transferring of charged metal ions from the aqueous to organic phase. This provides gold nanoparticles, which are readily dispersed in the nonpolar phase.

Many reports are available where the cationic surfactant CTAB has been used to prepare gold nanoparticles [127–129]. Giustini et al. [130] have characterized the quaternary w/o microemulsion of CTAB/n-pentanol/n-hexane/water. Some salient features of CTAB/co-surfactant/alkane/water system are (1) formation of nearly spherical droplets in the L2 region (a liquid isotropic phase formed by disconnected aqueous domains dispersed in a continuous organic bulk) stabilized by a surfactant/co-surfactant interfacial film. (2) With an increase in water content, L2 is followed up to the water solubilization failure, without any transition to bicontinuous structure, and (3) at low W_0, the droplet radius is smaller than $R°$ (spontaneous radius of curvature of the interfacial film) but when the droplet radius tends to become larger than $R°$ (i.e., increasing W_0), the microemulsion phase separates into a Winsor II system.

Moreover, the reduction of $[AuCl_4]^-$ in cationic surfactants is different from anionic surfactants due to the formation of insoluble ion pairs between $[AuCl_4]^-$ and cationic surfactants [131, 132]. It is shown that this precipitate can be dissolved on increasing the concentration of the surfactant above the CMC [133]. With the aim of estimating the particle-size distribution of nanoparticles formed in the w/o microemulsions, Sato et al. [118] prepared gold nanoparticles using two cationic[cetyltrimethylammonium chloride (CTAC) and dodecyltrimethylammonium chloride (DTAC)] and one anionic (AOT) surfactant. No significant difference was found from using the cationic surfactants and 3–4 nm Au particles were obtained. Gold nanoparticles of 2–5 nm were prepared using the didodecyldimethyl ammonium bromide (DDAB) cationic surfactant [115] and the variation in colloid size was achieved by altering the ratio of the concentration of surfactant to gold salt and using different reducing agents. Gold nanoparticles ranging from 5.9 to 26 nm has been obtained [129] using a CTAB/n-butanol/n-octane/water system. A decrease in the size of the particles was observed with decreasing $[AuCl_4]^-$ and increasing $[BH_4]^-$ concentration, probably due to the faster rate of reaction. In addition, the strong adhesion of the cationic surfactant on the surface of the gold nanoparticles leading to a stable dispersion of the colloids in toluene was demonstrated in contrast to the same experiment performed using the nonionic surfactant [115] which resulted in precipitation of the colloids upon dilution as the surfactant was extracted into the new phase.

The change in shape of nanoparticles prepared in a CTAB/octane + 1-butanol/H_2O w/o microemulsion system was studied by Lin et al. [127]. While in the absence of any additive, single crystals with many different shapes (e.g., cylinder, triangle, spherical, hexagon, trapezoid, pentagon) and a broad size distribution (7–40 nm) were obtained, addition of dodecanethiol resulted in formation of monodispersed gold nanoparticles. Dodecanethiol thus prevented the aggregation and growth of bare gold particles, by effective capping of the gold surface. A small modification in the quaternary microemulsions of the CTAB/n-pentanol/hexane/water system, i.e., using n-pentanol instead of n-butanol and hexane (comparatively higher vapor pressure solvent) instead of octane [127], resulted in 7 nm gold nanoparticles [134]. The stronger van der Waals interaction of n-pentanol with CTAB (as compared to with n-butanol), allows the formation of more compact and stable interfacial film on the Au nanoparticles and prevents their aggregation (without the addition of added capping agent). Hence, n-pentanol has been proposed as a better co-surfactant with CTAB than n-butanol.

In a series of publications Esumi and co-workers [131, 132, 135] have shown that the photochemical reduction of the gold salts in the presence of the cationic surfactant micelles formed high aspect ratio gold particles because of the templating effect of rod-like micelles. Similar results were obtained when the experiments [133] were conducted in the presence of electrolyte, NaCl, in cationic surfactant solutions (CTAC and DTAC) above the CMC leading to formation of linear thread-like aggregates. While in the presence of NaCl, CTAC forms worm-like micelles [136, 137], DTAC, because of its shorter alkyl chain does not form rod-like micelles [138] even at high concentrations of salt. Using TEM and fluorescence spectroscopy it is shown that, the rod-like or thread-like particle morphologies do not originate from a templating mechanism and that the formation of thread-like gold particles occurs primarily through a combination of crystal aggregation and specific crystal face stabilization.

Gold nanoparticles have also been prepared in a w/o microemulsions system using gemini surfactant 2-hydroxy-1,3-bis (octadecyldimethylammonium) and propane dibromide as the stabilizer [139]. Others have prepared gold nanoparticles at different W_0 using mixed w/o microemulsions composed of AOT/isooctane and AOT/phosphatidyulcholine/isooctane [140]. In these systems, it was observed that an increase in W_0 results in higher optical absorbance from the gold solutions resulting in darker suspensions. Synthesis of octadecylamine capped 3–6 nm gold nanoparticles in reverse micelles using microwave irradiation has also been reported [141].

6.3
Gold Nanoparticle Synthesis in Other Surfactant Aggregate-Based Systems

In a novel approach, metal nanoparticles and nanostructures have been synthesized via decomposition of an insoluble metal–organic precursor compound in a dynamic mixed amphiphile LB film at the gas/liquid interface [18]. The advantage of this approach is the ability to manipulate the monolayer during the nanoparticle growth processes in addition, to the freedom to carry out the synthesis of nanoparticles in reaction media at different physical states. The method also allows for the deposition of nanoparticulate monolayers onto the solid substrates. Khomutov et al. [142] prepared gold nanoparticles from water-insoluble precursor molecules $Au(P(C_6H_5)_3)Cl$ in a mixed LB film with arachidic acid or octadecyl amine, as surfactants at the gas/aqueous interface of a sodium borohydride solution. Gold nanoparticles have also been photo-chemically generated in multilayer LB-films of positively charged amphiphiles deposited from an aqueous $HAuCl_4$ subphase [143]. With the aim to prevent nanoparticle aggregation upon formation, Johnson and others [144] have used surfactants containing polar aromatic groups, alkylthiol molecules etc., instead of straight alkyl chain surfactants. The presence of polar groups on the surface is desirable for applications in nonlinear optics and also for the incorporation of the nanoparticles into an organic matrix. It has been further demonstrated that the presence of such surfactants in the nanoparticles does not hinder the self-assembling process.

Employing a two-phase method, gold nanoparticles with average diameters from 9 to 15 nm have been synthesized in toluene using cationic surfactants of variable chain lengths bearing the general formula CnTAC ($n = 10$, 12, 14 and 18) [145]. Here $AuCl_4^-$ is transferred from an aqueous solution to toluene using a phase transfer reagent and then reduced in the presence of a series of cationic surfactants. The average particle size is found to increase from 9 to 15 nm with decreasing chain length. This is in contrast to earlier observations [146] and has been explained in terms of a dynamic formation mechanism [147, 148]. Accordingly, an increase in chain length of the surfactant hinders the diffusion of the gold ions/atoms leading to a decrease in the growth stage of the particles followed by subsequent capping, leading to smaller sizes. Surfactants have also been used as phase transfer agents in a biosynthetic method for the preparation of gold nanoparticles. Sastry and co-workers [149] have described the synthesis of gold nanoparticles (15 to 25 nm) using *Emblica officinalis* (amla, Indian Gooseberry) fruit extract as the reducing agent. The synthesized Au nanoparticles were phase transferred into an organic solution using a cationic surfactant octadecylamine. Metal nanoparticles with a narrow size distribution have also been prepared by sonolysis of aqueous solutions of metal using surfactants such as SDS, polyethylene glycol monostearate, etc. to stabilize formed particles [150].

Shape anisotropy introduces new optical properties in gold (and silver) nanoparticles such as longitudinal plasmon resonance bands in the visible and the near-IR part of the electromagnetic spectrum. Preparation of anisotropic gold particles (e.g., gold nanorods) has been achieved using cationic surfactants. Nanorod formation is explained on the basis of either the formation of rod-like micelles (i.e., soft template) in the solution, which are used as nanoreactors for the reduction of gold or by inducing 1D growth of a material by binding preferentially to a specific crystal face and inhibiting growth along that surface [151](Fig. 6). For instance, gold nanorods have been prepared via electrochemical oxidation/reduction [152, 153] in a simple two-electrode type cell using surfactants [154, 155]. Gold (anode) and platinum (cathode) electrodes are immersed in an electrolytic solution consisting of a cationic surfactant $C_{16}TAB$ and a rod-inducing co-surfactant tetraoctylammonium bromide (TC_8AB). While $C_{16}TAB$ serves not only as the supporting electrolyte but also as the stabilizer for nanoparticles to prevent their further growth TC_8AB (and several analogous surfactants, such as $TC_{10}AB$), a much more hydrophobic surfactant, has rod-inducing capability. During the synthesis (under ultrasonication and controlled temperature), the bulk gold is converted from the anode to form gold nanoparticles most probably at the interfacial region of the cathodic surface.

Another methodology that employs the surfactant as a stabilizer is the seed-mediated method [156]. Here, appropriate amounts of precursor ions are reduced over the preformed "seed" or "germ" (i.e., small particles formed by suitable, usually strong, reducing agents). The reducing agent used in the second stage of "seed"-mediated growth is generally a weaker one, viz., NH_2OH, ascorbate ion, etc. [157, 158] which should reduce only the ions that are adsorbed onto the "seed" surface. The final size of the particles would depend on the size of the "seed" and the amount of the precursor ions to be reduced on them [159]. Thus, the smaller is the starting seed; the lower will be the desired size limit of the particles. Sau et al. [160] prepared particles over a broad size range. Various reports are now available for the prepar-

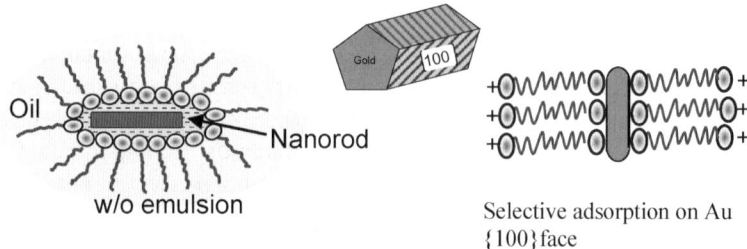

Fig. 6 Surfactant-directed growth for synthesis of nanorods

ation of gold nanoparticles by the seeding method [161, 162]. In a three-step seed-mediated growth method using different surfactants, alkyltrimethylammonium bromides (CnTAB, n = 10, 12, 14, 16, and 18) and cetylpyridinium chloride ($C_{16}PC$) it was found that as the length of the surfactant chain increased, the resulting gold nanoparticles' aspect ratio increased: the aspect ratio being 1 (for $C_{10}TAB$), 5 ± 2 ($C_{12}TAB$), 17 ± 3 ($C_{14}TAB$), and 23 ± 4 ($C_{16}TAB$) [163]. More recently, a modification of the seed-mediated method was used by Loo et al. [164] to achieve gold coating on the silica nanoparticles and the particles were employed for imaging applications and therapy.

In an example of nanoparticle formation from a top-down approach, namely laser ablation, the role of surfactant in controlling particle size has been highlighted [165, 166]. In the laser ablation process particles are generated from bulk metal and thus there is no need of a reducing agent. Typically the size distribution of the nanoparticles tends to be broadened because the coagulation processes of atoms cannot be controlled. A modification known as "laser-induced size reduction" has been developed to modulate the size and change the geometrical structure of gold nanoparticles formed via laser ablation. This method takes advantage of the surface plasmon peak centered at 520 nm that is characteristic of gold nanoparticulates [165, 167, 168]. Mafune et al. [166] have demonstrated laser-induced size reduction at various laser fluences and surfactant concentrations to obtain a desired size with a slightly broader size distribution than the gold nanoparticles prepared by the conventional wet-chemistry technique. In another report by the same group [165] the size distribution of the nanoparticles produced was found to shift to a smaller size with an increase in surfactant concentration showing the dependence of the nanoparticle abundance on surfactant concentration. In an extension [169] of the same scheme ~ 20 nm gold nanoparticles were prepared in water via laser ablation against a gold metal plate. It was also observed that gold nano-networks and much smaller gold particle could be formed by modulating the laser fluence and by the addition of SDS to the solution.

7
Quantum Dots

Quantum dots (QD) are luminescent semiconductor nanocrystals that have several advantageous properties in comparison to organic dyes. These include enhanced stability to photobleaching and high quantum yield. The QDs can be classified as generally belonging to the elements combined from the II and VI groups (e.g., CdS, CdSe, CdTe), or the elements combined from the III and V groups (e.g., GaAs, InP, InAs). Both the excitation and the emission of QDs depend on the nanocrystallite size. This is briefly explained as follows. In extended semiconductors, the overlap of atomic orbitals leads to the formation of valence and conduction bands separated by an energy gap. When an excited electron is promoted from a filled valence band to a largely empty conduction band, it leads to the formation of an electron–hole pair. The spatial separation (Bohr radius) of such an electron–hole pair ("exciton") is typically of the order of 1–10 nm for most semiconductors (i.e., of the same order as the particle dimension). This results in a quantum size effect. In QDs, the excitons are confined in a way similar to a particle-in-the-box problem leading to a finite band gap and discretization of energy levels. QDs are also classified as being "Type I" or "Type II". Type I QDs have both the electrons and holes located inside of the QD core, whereas Type II QDs have the electrons and holes spatially separated between a core and shell layer. Fluorescence of semiconductor nanocrystals is due to the radiative recombination of an excited electron–hole pair. The band gap and thus their optical properties are size-dependent and governed by quantum effects [170, 171].

QDs are finding increasing application in bioimaging because of the following unique advantages over conventional contrast agents:

1. The emission spectra of QDs can be tuned across a wide range by changing the size and composition of the QD core, e.g., from UV (ZnS [172]) to near infrared (CdS/HgS/CdS [173], InP, InAs [174]) through the visible range (CdE, with E = S, Se, Te [175]).
2. Typically QDs have broad excitation and narrow emission spectra. This permits a single excitation wavelength to excite QDs of different colors and reduces spectral overlap.
3. QDs have very large molar extinction coefficients and high quantum yields, up to 85% [176], making them bright fluorescent probes [6].
4. QDs have long fluorescence lifetimes on the order of 20–50 ns, which allows them to be distinguished from background and other fluorophores for increased sensitivity of detection [7].
5. QDs are highly photostable [177]. The photostable nature results from the shell protecting the core, which is generally a large band gap semiconductor material [178, 179]. For instance, InAs cores have been covered by different shells made of ZnS, ZnSe, GaAs [178], etc.
6. QDs have been shown to be quite stable to metabolic degradation (with the exception of problems associated with heavy metal leachants) [177, 180].
7. QDs can be conjugated to the linker [181] (e.g., avidin, protein A or protein G, or a secondary antibody) by covalent binding, passive adsorption, multivalent chelation or by electrostatic interactions [182, 183].
8. Coating of the QDs with suitable materials such as amphiphilic polymer and silica improves their water dispersability and makes them suitable for biological applications. From our research group, water-soluble silica-overcoated CdS:Mn/ZnS highly luminescent and photostable semiconductor QDs have been prepared using the w/o microemulsion approach [47, 184, 185].

7.1
Quantum Dot Synthesis
Using Nonionic Surfactant-Based Microemulsions

W/o microemulsion method has been successful for preparing QDs and other semiconductor nanoparticulates. The preparation and properties of these particles have been extensively reviewed [114, 186].

The most commonly synthesized QD is cadmium sulfide (CdS), for which quantum confinement effects on band gap absorption have been discussed extensively [170, 187, 188]. Wang [189] et al. have reported preparation of 5 nm CdS nanoparticles by a novel and simple one-step, solid-state reaction in the presence of a nonionic surfactant, $C_{18}H_{37}O(CH_2CH_2O)_{10}H$ ($C_{18}EO_{10}$). It is suggested that the surfactant $C_{18}EO_{10}$ forms a "shell" surrounding the CdS particles thus forming a physical and spacious obstacle between the particles and preventing particle aggregation. Gan et al. prepared 5 nm particles of Mn-doped ZnS in a microemulsion system consisting of nonionic surfactants NP-5 and NP-9 [190].

Different shapes of CdS nanomaterials such as quasi-nanospheres, nanoshuttles and nanotubes have been prepared in the w/o microemulsion system of nonionic surfactant (Tween-80, polyetheneoxy(20)octadecyl ether, polyoxyethylele(9)dodecyl ether ($C_{12}E_9$) or Tx-100)/n-pentanol/aqueous solution/cyclohexane [191]. Another kind of QDs, PbS, have also been prepared in nonionic $C_{12}E_9$/water/cyclohexane microemulsions [192].

7.2
Quantum Dot Synthesis Using Ionic Surfactant-Based Microemulsions

AOT is again the most commonly used surfactant to form w/o microemulsions. CdS [193], CdS/ZnS [194], Cd(OH)$_2$, CdO [195], CdSe [196], CdMnSe [197], CdMnS [198], ZnSe [199], ZnS [200], ZnS:Mn [201], ZnS:Mn/ZnS [202], ZnS/CdS/ZnS [203] and other nanoparticles have been prepared in AOT w/o microemulsions. CdS nanoparticles have been prepared by using either the micellar exchange process or by bubbling hydrogen sulfide amongst the micellar droplets in a solution containing cadmium ions [3, 204, 205]. In a typical procedure two heptane solutions of AOT are prepared. An aqueous standard solution of precursor reagents, e.g., Cd(ClO$_4$)$_2$ is added to one heptane solution, while an aqueous solution of Na$_2$S is added to the other. In most cases the W_0 ratio is kept the same, usually less than 10 (e.g., 4 [206] or 6 [207]), and the nanoparticles are obtained by mixing the micellar solutions. CdS nanoparticles (up to 5 nm [206]) are formed in transparent, slightly yellow colored solution. Nakanishi et al. [206] have further shown that the size of CdS nanoparticles can be controlled easily by changing the W_0 ratio. By increasing the amount of water in w/o microemulsions, an increase in the average size of the particles and size distribution have been observed. CdS nanoparticles with an absorption maximum at 310 nm (corresponding to a particle size of about 1 nm) were obtained at $W_0 = 1$, while those showing a shoulder in the absorption spectrum at 450 nm (corresponding to a particle size of about 5 nm) were obtained with $W_0 = 10$. The size of CdS nanoparticles agreed with the expected diameter of the "water pool" in the w/o microemulsions [208, 209]. CdS nanoparticles of up to 3.4 nm were prepared similarly in a w/o microemulsion system consisting of AOT/isooctane/water and using $Cd(NO_3)_2$ and Na_2S as precursor salt materials by varying the ratio of feed reactant concentrations, $y = [S^{2-}]/[Cd^{2+}]$ (parameter y) and embedding them into mesoporous silica [210]. A higher degree of monodispersity and smaller size has been obtained by using either very high concentrations of surfactant ([AOT] = 0.5 M) [204, 205, 210] or a very small water content [211, 212].

Semiconductor nanoclusters trapped in AOT w/o microemulsions are reported to exhibit longer excited state lifetimes (about 10–100 ns) than those in aqueous solution or in monophasic organic solvents [213]. Clearly the surfactant–nanoparticle interaction is very important not only in restricting growth but also in extending the lifetimes of the excited states. Tata et al. [214] have shown that the removal of water from the micelles leads to a strong increase in fluorescence intensity, and the addition of specific quencher, 4-hydroxythiophenol, leads to variations in quenching efficiencies.

A seemingly trivial variation in the procedure can affect the particle crystallinity and optical properties for the CdS nanoparticles prepared in a AOT/heptane/water [215] system. It is reported that on mixing different volumes (5 ml and 15 ml) of w/o microemulsion solutions, while keeping all other parameters (viz. AOT concentration, aqueous salts concentration, W_0 (= 4), time of mixing and temperature) constant nanocrystallites with similar size (3.8 and 4 nm) but different crystalline structures (namely, cubic and hexagonal) of CdS are obtained.

Pileni [216]et. al have prepared CdS nanoparticles using normal ($Cd(NO_3)_2$/ ± HMP/AOT/isooctane/water) and functionalized micelles ($Cd(AOT)_2$/AOT/ isooctane/water), in the presence and absence of surface-capping agent sodium hexametaphosphate (HMP). Using normal w/o microemulsions it is found that the size of the particles increases with the water content and the presence of HMP as a protecting agent allows a reduction in the size of the particle. With functionalized micelles, the particles formed are more monodisperse than those obtained by using cadmium nitrate in the absence and in the presence of a protecting polymer. However, in the latter case, it is possible to prepare smaller particles. Moreover, it is observed that the size of CdS is always smaller when one of the two reactants are in excess showing that the crystallization process is faster when one of the reacting species is in excess [217]. It is also seen that the rate of formation of CdS particles depends strongly on the chemical nature of the continuous phase and the temperature of reaction. The kinetic studies [212] of quantum-sized CdS particles prepared in water-in-oil microemulsions of AOT show that the rate changes with the nature of the oil according to the order cyclohexane < n-heptane < n-decane. The kinetics of droplet communication in different oil solvents have also been determined by indicator reactions and rationalized in terms of emulsion stability and natural droplet curvature [32].

The ability to modify the surface of the CdS nanoparticle within the cavity of the w/o microemulsions has been extended for their immobilization. CdS nanoparticles (e.g., prepared in AOT/isooctane w/o microemulsions) were immobilized onto thiol-modified aluminosilicate particles [218] and thiol-modified alumina [219] by a simple addition of thiol-modified aluminosilicates and alumina, respectively, in the micellar solution. The resulting CdS nanoparticles–aluminosilicate composites were used as photocatalysts for H_2 generation from 2-propanol aqueous solution.

By using a combination of surfactant with AOT such as zwitterionic phospholipid L-α-phosphatidylcholine (lecithin) [220] spherical and nonspherical CdS nanoparticles have been prepared. Here advantage is taken of the curvature formed by different surfactants in aqueous and oil phases. While AOT tends to have a spontaneous curvature that is concave toward water [68] leading to the formation of spherical w/o microemulsions above the critical micelle concentration [221], lecithin has a significantly larger head group (thus a smaller packing parameter of 0.6) [222] and in aqueous systems, it develops interfaces with minimal curvatures. CdS particles of different morphology have been precipitated in the modified droplet shape of water-in-oil microemulsions formed using AOT and lecithin as surfactants.

Lead sulfide (PbS) is well known for having the smallest band gap of 0.41 eV and a large excitation Bohr radius of 18 nm in the bulk form (galena) [223]. PbS nanoparticles (used in the IR detector [224], solar absorber [225], Pb^{2+} ion-selective sensors [226], and photography [227]) were similarly prepared in inverse micelles [228] following the Steigerwald et al. method [208] for cadmium sulfide particles. Contrary to the general method of controlling the particle size by varying W_0 (which was unsuccessful possibly because of the faster growth rate of the PbS particles in solution) different-sized PbS nanoparticles were obtained by sampling portions at different lengths of time, e.g., starting from colorless (immediately after preparation, < 1.2 nm), light yellow (3 days, 1.3 nm), yellow (7 days, 1.9 nm), orange (15 days, 2.5 nm), to brown (60 days, 4.2 nm). The sizes of these different colored PbS particles in solution are in good agreement with those prepared in Nafion by Miyoshi et al. [229].

Diluted magnetic semiconductors (DMS) [230] are alloys in which magnetic ions (e.g., Mn^{2+}, Fe^{2+}) are diluted in nonmagnetic $A^{II}B^{VI}$ semiconductors (such as CdS, CdTe). DMS bulk materials exhibit a number of interesting magneto optical effects such as large Zeeman splitting of carriers [231] and bound magnetic polarons [232]. Optical and fluorescence properties due to isolated Mn^{2+} ions in tetrahedrical coordination have been studied that were attributed to a quantum size effect [233, 234]. $Cd_{1-y}Mn_yS$ nanoparticles have been prepared by coprecipitation on mixing of two micellar solutions (AOT as surfactant, $W_0 = 40$) one containing S^{2-} ions and the other, a mixed micellar solution made of $Cd(AOT)_2$, $Mn(AOT)_2$, and $Na(AOT)$. Syntheses were performed at various ratios of $Mn(AOT)_2$ and $Cd(AOT)_2$ and in the presence of an excess of sulfur ions [235]. Surface capping with dodecanethiol was achieved by adding it to the micellar solution containing the crystallite. A selective surface reaction occurs between the thio derivative and the cadmium and manganese ions at the interface protecting its surface. $Cd_{1-y}Mn_yS$ nanocrystals of average size, 1.8 to 4 nm, and different compositions (y: 0.03 to 0.3) have been prepared similarly using AOT as the surfactant [236, 237].

ZnSe-capped CdSe core-shell QDs [238] have been synthesized in an AOT/water/heptane microemulsion system using cadmium and zinc perchlorate as precursors and bis(trimethylsily1) selenide (bisTMS-Se) as a source for Se. The core is first formed by the addition of bisTMS-Se to a micellar solution containing the cation and grown to a size determined by W_0. The capping layer was formed by addition of the cation (from which the second layer was to be grown) prepared in a micellar solution of larger W_0 (W_0'). The final water-to-surfactant ratio, (and hence the particle size-determining water pool size), was the average of W_0, and W_0' weighted by the volume of each micellar solution used. Additional bisTMS-Se added to the resulting micellar solution to initiate deposition of the second material around the core particles. CdSe-capped ZnSe nanoparticles were also prepared similarly. Kortan et al. have attached an organic ligand to a composite semiconductor of CdSe grown on a ZnS seed (and vice versa) [211].

The w/o microemulsion-based synthesis of high-quality CdS-doped Mn/ZnS (CdS:Mn/ZnS) core-shell QDs [184, 185, 239] have been reported. Yang et al. [239, 240] reported a synthesis of manganese-doped cadmium sulfide core and zinc sulfide shell (CdS:Mn/ZnS) QDs using a AOT/water/heptane w/o microemulsion system. From our group Santra et al. [184, 185, 239] recently prepared highly luminescent and photostable CdS:Mn/ZnS core/shell QDs (Fig. 7). To create a water-soluble quantum dot of a desired surface functionality, CdS:Mn/ZnS quantum dots were synthesized in a w/o microemulsion system, and then they were consecutively coated with a very thin silica layer (approx. 2.5 nm thick) within the same microemulsion. The water droplet serves as a nanosized reactor for the controlled hydrolysis and condensation of a silica precursor, TEOS, using ammonium hydroxide (NH_4OH) [196, 198, 203, 240, 241]. However, to allow bioconjugation of QDs, 3-(aminopropyl) triethoxysilane (APTS) was added along with TEOS in the w/o microemulsion system to introduce free amine groups on the particle surface. It was observed that on creating the amino functionality, the zeta potential drops from -40 mV to -0.5 mV. This adversely affected the dispersibility of QDs in aqueous medium. Santra et al. improved the QDs dispersibility by adding 3-(trihydroxysilyl) propyl methylphosphonate (TH-PMP) along with TEOS and APTS in a microemulsion system to obtain highly dispersed QDs. These aminated QDs were conjugated with TAT (a cell penetrating peptide) for an efficient bioimaging of brain tumor tissues [184, 185].

Khomane et al. prepared dodecanethiol-capped CdS QDs of 4 nm size by using a Winsor II microemulsion system [242], which are soluble in solvents such as n-heptane, toluene, n-hexane, thus demonstrating the dual role of the anionic surfactant, viz., forming the microemulsion and facilitating the extraction of oppositely charged ions from the aqueous to the organic phase.

In contrast to most of the examples using the w/o microemulsion systems Fan et al. [243] reported a facile synthesis of water dispersible QDs using an o/w microemulsion system. A concentrated suspension of QDs in chloroform was added to an aqueous solution containing a mixture of surfactants or phospholipids. It is reported that after a vigorous stirring an oil-in-water microemulsion was formed. Thereafter, chloroform was evaporated from the microemulsion, which led to QDs dispersed in the aqueous phase by an interfacial process driven by the hydrophobic van der Waals interactions between the primary alkane of the stabilizing ligand and the secondary alkane of the surfactant, resulting in thermodynamically defined interdigitated bilayer structures. Further, it was reported that such water-dispersed QDs retain the original hydrophobic QDs optical properties.

In a novel approach the Fendler group reported the preparation of mercaptopropionic acid-capped CdS nanoparticles [244] of sizes ranging from 1.8 to 5.4 nm using a self-reproducing w/o microemulsion approach. The w/o microemulsions were prepared by mixing aqueous solutions of sodium octanoate, NaOA, with isooctane and octanol (in a fixed ratio) (v/v) and then adding water to obtain a desired W_0 value. The precursor solutions were prepared by adding appropriate amounts of aqueous solution of LiOH or Cd(ClO$_4$)$_2$ or Pb(NO$_3$)$_2$ or Zn(NO$_3$)$_2$. Addition of calculated amounts of octyloctanoate (O–OL) and H$_2$S caused hydrolysis of O–OL in the w/o microemulsions. The products of the O–OL hydrolysis are oc-

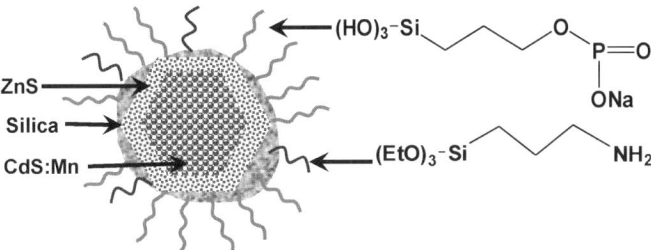

Fig. 7 Schematic representation of CdS : Mn/ZnS core shell quantum dot surface functionalized with silane compounds

tanoate ion, OA, and octanol, OL, that act as surfactant and co-surfactant respectively, forming a w/o microemulsion in situ that is defined as a self-reproducing microemulsion. The formation of these microemulsions induces a change in the hydrodynamic radius. In the absence of or with a lesser amount of water, additional amounts of OA can only be aggregated into smaller-sized w/o microemulsions. Additionally, Hirai et al. reported the preparation of a semiconductor nanoparticle–polymer composite by direct w/o microemulsion polymerization using a polymerizable surfactant cetyl-p-vinylbenzyldimethylammonium chloride (CVDAC), where nanoparticles were incorporated successfully into the polymerized CVDAC and were shown to retain their quantum size effect [201].

Cationic surfactant systems such as the CTAB/co-surfactant/alkane/water system have also been used to prepare QDs. It is possible here to change the water droplets dimension and surface character by either varying the alcohol or the water content. However, contradictory results are present in the literature regarding the ability to control the particle size by varying the water pool radius in CTAB microemulsions. Zhang et al. [245] prepared nanoparticles from about 4 to 8 nm, with increasing W_0 from 5.01 to 24.45, indicating that the w/o microemulsion droplet can be utilized as a micro reactor to confine the growth of nanoparticles. On the other hand, Agostiano [246] and others [247] have shown that the size of semiconductor particles seems to be poorly correlated with the aqueous droplet dimension in the quaternary microemulsion.

The role of the co-surfactant on the stability of quaternary microemulsion and the size of the semiconductor nanoparticles has also been investigated [246, 247]. The co-surfactant can affect the interfacial properties of the microemulsion in two primary ways. First, it can absorb to the interfacial film and modify the surfactant packing parameter thereby modulating the radius of curvature of the microemulsion droplet. Second, it can provide stability to the system by its exchange dynamics. When alcohols are used as co-surfactants, the interface stability is inversely proportional to the alcohol alkyl chain. While in a ternary microemulsion, formed using AOT/alkane, the droplet interface has a high rigidity in the CTAB system, the alkyl chain of pentanol increases the interfacial curvature by increasing the surfactant packing parameter, thus favoring the formation of small droplets with low water content. At higher water and surfactant concentrations, the addition of water to the microemulsion mainly leads to the formation of more and smaller droplets [248] until a phase separation occurs. This difference in the interfacial rigidity and microemulsion stability results in the availability of a wider range of water pool radii and thus a better control over size of the precipitated particle in the AOT/alkane system as compared to the CTAB/co-surfactant/alkane system. For instance, by varying the W_0 from 10 to 30 in AOT microemulsions, an increase of 35 Å in the droplets' radii is observed, whereas a corresponding increase of only 10 Å is observed in the

CTAB microemulsion [247]. The increase in the particle size is due to the ability of alcohols to reduce the rigidity of the w/o interface thereby increasing the exchange rates of micelles, allowing exchange of solubilized matter and thus facilitating Ostwald ripening leading to larger-sized particles. It is also shown that the tendency of the particles to sediment increases, when the amount of alcohol in the micellar interface decreases and at higher co-surfactant concentration, a stable microemulsion is obtained even at high values of W_0. It is also demonstrated that pentanol (and also thiol) acts as a capping agent at higher concentration for the alcohol. The addition of thiols [246] is found to increase re-dispersibility in organic solvents such as ethanol and pyridine.

In order to improve the luminescence behaviors and obtain better quantum yields, Zhang et al. [245] have suggested a reflux treatment by diluting w/o microemulsions of CdS nanoparticles with the same w/o microemulsions but substituting the reactant solution with H_2O. The water in the w/o microemulsion droplets was removed by the co-surfactant (n-hexanol), the trap sites on the nanoparticle surface decreased improving the crystallinity and thus the fluorescence efficiency.

In an attempt to organize the nanoparticles on surfaces and in three-dimensional geometries Simmons et al. [220] have shown that nanoparticles (CdS) synthesized in AOT water-in-oil microemulsions can be spatially compartmentalized into the strands of a rigid organogel. The driving force has been proposed to be hydrogen bonding between the phenolic species and the surfactant, which leads to a spontaneous phase transition from the liquid microemulsion state to the organogel state. Fairly monodisperse nanoparticles of CdS have been prepared using a polymer surfactant gel matrix [249] comprised of a chloride salt of N,N-dimethyl-N-methyl derivative of hydroxymethyl cellulose and anionic surfactant SDS to prevent particle aggregation.

Some miscellaneous examples of QD synthesis using surfactant systems are described below. A layer-by-layer [206] structure of dithiol self-assembled monolayers (SAM) and CdS mono- and multilayer nanoparticles were fabricated on a gold substrate covered with alkanedithiol. SAMs were formed by an alternate immersion of the substrate into ethanolic solutions of dithiol, and dispersion of CdS nanoparticles (ca. 3 nm in diameter), the latter of which was prepared in AOT/H_2O/heptane w/o microemulsions.

Electroporation [250] in synthetic unilamellar dioleoylphosphatidylcholine vesicles of 178 nm diameter has been utilized for the preparation of PbS QDs. 3–10 Å sized QDs are prepared by initiation of the opening of the pores between the Pb^{2+} ions originally entrapped in the vesicles and S^{2-} ions placed in the bulk [251]. 2.7 to 5 nm CdS nanoparticles have been prepared in a mixed vesicle dispersion formed using ionic surfactants, i.e., decyltrimethylammonium bromide and sodium 10-undecenoate [252]. The particle formation was achieved by exposing the vesicles containing the Cd (as a Cd complex of EDTA) to H_2S.

8
Conclusion

Microemulsion-mediated synthesis of nanoparticles for bioimaging offers a widely applicable approach to produce a variety of engineered monodispersed optical nanoprobes (e.g., bioconjugated dye-doped silica nanoparticles, functionalized gold nanoparticles and QDs). Recent years have seen a rapid development in controlling nanoparticle size, size distribution, surface functionalization and tailored multimodal properties. Unfortunately, the clinical use of these particles has been limited. Concerns with regards to nanoparticle toxicity and the need for cross-disciplinary researchers to address these concerns have had limited progress. In the future, we anticipate that concerted national and international efforts in nanotoxicology will provide guidelines for particle design to avoid toxicity issues—thereby bringing the use of luminescent nanoparticles to the forefront of clinical medicine. The design of bioimaging nanoparticles that carry gated energy release of therapeutic agents in addition to diagnostic applications is envisioned to provide new tools for simultaneous imaging and therapy.

Acknowledgements The authors acknowledge the financial support of the Particle Engineering Research Center (PERC) at the University of Florida, the National Science Foundation (NSF Grant EEC-94-02989, NSF-NIRT Grant EEC-0506560), National Institutes of Health (Grant 1-P20-RR020654-01) and the Industrial Partners of the PERC for support of this research. Any opinions, findings and conclusions or recommendations expressed in this material are those of the author(s) and do not necessarily reflect those of the National Science Foundation.

References

1. Mahmood U, Weissleder R (2002) Some tools for molecular imaging. Acad Radiol 9:629–631
2. Murray CB, Kagan CR, Bawendi MG (2000) Synthesis and characterization of monodisperse nanocrystals and close-packed nanocrystal assemblies. Ann Rev Mater Sci 30:545–610
3. Feldheim DL, Foss CA (2002) Metal nanoparticles: synthesis, characterization, and applications. Marcel Dekker, New York
4. Cushing BL, Kolesnichenko VL, O'Connor CJ (2004) Recent advances in the liquid-phase syntheses of inorganic nanoparticles. Chem Rev 104:3893–3946
5. Yin Y, Alivisatos AP (2005) Colloidal nanocrystal synthesis and the organic–inorganic interface. Nature 437:664–670
6. Chan WC, Maxwell DJ, Gao X, Bailey RE, Han M, Nie S (2002) Luminescent quantum dots for multiplexed biological detection and imaging. Curr Opin Biotechnol 13:40–46
7. Bruchez M Jr, Moronne M, Gin P, Weiss S, Alivisatos AP (1998) Semiconductor nanocrystals as fluorescent biological labels. Science 281:2013–2016
8. Licha K, Riefke B, Ntziachristos V, Becker A, Chance B, Semmler W (2000) Hydrophilic cyanine dyes as contrast agents for near-infrared tumor imaging: synthe-

sis, photophysical properties and spectroscopic in vivo characterization. Photochem Photobiol 72:392–398
9. Pham W, Lai WF, Weissleder R, Tung C H (2003) High efficiency synthesis of a bioconjugatable near-infrared fluorochrome. Bioconjug Chem 14:1048–1051
10. Huber MM, Staubli AB, Kustedjo K, Gray MHB, Shih J, Fraser SE, Jacobs RE, Meade TJ (1998) Fluorescently detectable magnetic resonance imaging agents. Bioconjug Chem 9:242–249
11. Santra S, Bagwe RP, Dutta D, Stanley JT, Walter GA, Tan W, Moudgil BM, Mericle RA (2005) Synthesis and characterization of fluorescent, radio-opaque, and paramagnetic silica nanoparticles for multimodal bioimaging applications. Adv Mater 17:2165–2169
12. Gao XH, Cui YY, Levenson RM, Chung LWK, Nie SM (2004) In vivo cancer targeting and imaging with semiconductor quantum dots. Nat Biotechnol 22:969–976
13. Pellegrino T, Manna L, Kudera S, Liedl T, Koktysh D, Rogach AL, Keller S, Radler J, Natile G, Parak WJ (2004) Hydrophobic nanocrystals coated with an amphiphilic polymer shell: A general route to water soluble nanocrystals. Nano Lett 4:703–707
14. Abe M, Scamehorn JF (2005) Mixed surfactant systems. Marcel Dekker, New York
15. Wu SX, Zeng HX, Schelly ZA (2005) Growth of uncapped, subnanometer size gold clusters prepared via electroporation of vesicles. J Phys Chem B 109:18715–18718
16. Fendler JH (1982) Membrane mimetic chemistry: characterizations and applications of micelles, microemulsions, monolayers, bilayers, vesicles, host–guest systems, and polyions. Wiley, New York
17. Fendler JH (1996) Nanoparticles at air/water interfaces. Curr Opin Colloid Interface Sci 1:202–207
18. Khomutov GB, Obydenov AY, Yakovenko SA, Soldatov ES, Trifonov AS, Khanin VV, Gubin SP (1999) Synthesis of nanoparticles in Langmuir monolayer. Mater Sci Eng C 8-9:309–318
19. Ottova A, Tvarozek V, Racek J, Sabo J, Ziegler W, Hianik T, Tien HT (1997) Self-assembled BLMs: biomembrane models and biosensor applications. Supramol Sci 4:101–112
20. Pileni MP (1993) Reverse Micelles as Microreactors. J Phys Chem 97:6961–6973
21. Pillai V, Kumar P, Hou M J, Ayyub P, Shah DO (1995) Preparation of Nanoparticles of Silver-Halides, Superconductors and Magnetic-Materials Using Water-in-Oil Microemulsions as Nano-Reactors. Adv Colloid Interface Sci 55:241–269
22. Shah DO (1998) Micelles, microemulsions, and monolayers: science and technology. Marcel Dekker, New York
23. Holmberg K (2003) Surfactants and polymers in aqueous solution. John Wiley & Sons, Chichester; Hoboken, NJ
24. Pileni MP (1989) Structure and reactivity in reverse micelles. Elsevier, Amsterdam, New York
25. Mittal KL, Kumar P (1999) Handbook of microemulsion science and technology. Marcel Dekker, New York
26. Capek I (2004) Preparation of metal nanoparticles in water-in-oil (w/o) microemulsions. Adv Colloid Interface Sci 110:49–74
27. Lopez-Quintela MA (2003) Synthesis of nanomaterials in microemulsions: formation mechanisms and growth control. Curr Opin Colloid Interface Sci 8:137–144
28. Lopez-Quintela MA, Tojo C, Blanco MC, Rio LG, Leis JR (2004) Microemulsion dynamics and reactions in microemulsions. Curr Opin Colloid Interface Sci 9:264–278
29. Maitra A (1984) Determination of Size Parameters of Water Aerosol Ot Oil Reverse Micelles from Their Nuclear Magnetic-Resonance Data. J Phys Chem 88:5122–5125

30. Evans D F, Wennerstrom H K (1999) The colloidal domain: where physics, chemistry, biology, and technology meet. Wiley-VCH, New York
31. Eicke HF, Shepherd JCW, Steinemann A (1976) Exchange of Solubilized Water and Aqueous-Electrolyte Solutions between Micelles in Apolar Media. J Colloid Interface Sci 56:168–176
32. Fletcher PDI, Howe AM, Robinson BH (1987) The Kinetics of Solubilisate Exchange between Water Droplets of a Water-in-Oil Microemulsion. J Chem Soc Faraday Trans I 83:985–1006
33. Robinson BH, Toprakcioglu C, Dore JC, Chieux P (1984) Small-Angle Neutron-Scattering Study of Microemulsions Stabilized by Aerosol-Ot.1. Solvent and Concentration Variation. J Chem Soc Faraday Trans I 80:13–27
34. Huang JS (1985) Surfactant Interactions in Oil Continuous Microemulsions. J Chem Phys 82:480–484
35. Huang J S, Safran S A, Kim M W, Grest G S, Kotlarchyk M, Quirke N (1984) Attractive Interactions in Micelles and Microemulsions. Phys Rev Lett 53:592–595
36. Jain TK, Cassin G, Badiali JP, Pileni MP (1996) Relation between exchange process and structure of AOT reverse micellar system. Langmuir 12:2408–2411
37. Rosi NL, Mirkin CA (2005) Nanostructures in biodiagnostics. Chem Rev 105:1547–1562
38. Smith PW (2002) Fluorescence emission-based detection and diagnosis of malignancy. J Cell Biochem Suppl 39:54–59
39. Qhobosheane M, Santra S, Zhang P, Tan WH (2001) Biochemically functionalized silica nanoparticles. Analyst 126:1274–1278
40. Santra S, Zhang P, Wang KM, Tapec R, Tan WH (2001) Conjugation of biomolecules with luminophore-doped silica nanoparticles for photostable biomarkers. Anal Chem 73:4988–4993
41. Santra S, Wang KM, Tapec R, Tan WH (2001) Development of novel dye-doped silica nanoparticles for biomarker application. J Biomed Optics 6:160–166
42. Santra S, Dutta D, Moudgil BM (2005) Functional dye-doped silica nanoparticles for bioimaging, diagnostics and therapeutics. Food Bioprod Process 83:136–140
43. He XX, Duan JH, Wang KM, Tan WH, Lin X, He CM (2004) A novel fluorescent label based on organic dye-doped silica nanoparticles for HepG liver cancer cell recognition. J Nanosci Nanotechnol 4:585–589
44. Lian W, Litherland SA, Badrane H, Tan WH, Wu DH, Baker HV, Gulig PA, Lim DV, Jin SG (2004) Ultrasensitive detection of biomolecules with fluorescent dye-doped nanoparticles. Anal Biochem 334:135–144
45. Santra S, Liesenfeld B, Bertolino C, Dutta D, Cao Z, Tan WH, Moudgil BM, Mericle RA (2006) Fluorescence lifetime measurements to determine the core-shell nanostructure of FITC-doped silica nanoparticles: An optical approach to evaluate nanoparticle photostability. J Luminesc 117:75–82
46. Zhao XJ, Bagwe RP, Tan WH (2004) Development of organic-dye-doped silica nanoparticles in a reverse microemulsion. Adv Mater 16:173
47. Santra S, Yang H, Dutta D, Stanley JT, Holloway PH, Tan WH, Moudgil BM, Mericle RA (2004) TAT conjugated, FITC doped silica nanoparticles for bioimaging applications. Chem Commun, pp 2810–2811
48. Lettinga MP, van Zandvoort MAMJ, van Kats CM, Philipse AP (2000) Phosphorescent colloidal silica spheres as tracers for rotational diffusion studies. Langmuir 16:6156–6165
49. Tapec R, Zhao XJJ, Tan WH (2002) Development of organic dye-doped silica nanoparticles for bioanalysis and biosensors. J Nanosci Nanotechnol 2:405–409

50. Senarath-Yapa MD, Saavedra SS, Aspinwall CA, Roberts DL (2004) Poly(lipid)-coated, bye doped silica nanoparticles for biological sensing applications. Abstr Pap Am Chem Soc 227:U849–U849
51. Qian KJ, Zhang L, Yang ML, He PG, Fang YZ (2004) Preparation of luminol-doped nanoparticle and its application in DNA hybridization analysis. Chinese J Chem 22:702–707
52. Schmidt H (1988) Chemistry of Material Preparation by the Sol-Gel Process. J Non-Crystalline Solids 100:51–64
53. Ulrich DR (1988) Prospects of Sol–Gel Processes. J Non-Crystalline Solids 100:174–193
54. Chang CL, Fogler HS (1996) Kinetics of silica particle formation in nonionic w/o microemulsions from TEOS. AIChE J 42:3153–3163
55. Osseoasare K, Arriagada FJ (1990) Preparation of SiO_2 Nanoparticles in a Nonionic Reverse Micellar System. Colloids Surf 50:321–339
56. Arriagada FJ, Osseoasare K (1992) Phase and Dispersion Stability Effects in the Synthesis of Silica Nanoparticles in a Nonionic Reverse Microemulsion. Colloids Surf 69:105–115
57. Chang CL, Fogler HS (1997) Controlled formation of silica particles from tetraethyl orthosilicate in nonionic water-in-oil microemulsions. Langmuir 13:3295–3307
58. Bagwe RP, Yang CY, Hilliard LR, Tan WH (2004) Optimization of dye-doped silica nanoparticles prepared using a reverse microemulsion method. Langmuir 20:8336–8342
59. Arriagada FJ, Osseoasare K (1994) Silica Nanoparticles Produced in Aerosol Ot Reverse Microemulsions – Effect of Benzyl Alcohol on Particle-Size and Polydispersity. J Dispers Sci Technol 15:59–71
60. Arriagada FJ, Osseo-Asare K (1999) Synthesis of nanosize silica in a nonionic water-in-oil microemulsion: Effects of the water/surfactant molar ratio and ammonia concentration. J Colloid Interface Sci 211:210–220
61. Yang HH, Qu HY, Lin P, Li HS, Ding MT, Xu JG (2003) Nanometer fluorescent hybrid silica particle as ultrasensitive and photostable biological labels. Analyst 128:462–466
62. Jain TK, Roy I, De TK, Maitra A (1998) Nanometer silica particles encapsulating active compounds: A novel ceramic drug carrier. J Am Chem Soc 120:11092–11095
63. Yang W, Zhang CG, Qu HY, Yang HH, Xu JZ (2004) Novel fluorescent silica nanoparticle probe for ultrasensitive immunoassays. Anal Chim Acta 503:163–169
64. Wang L, Yang C, Tan WH (2005) Dual-luminophore-doped silica nanoparticles for multiplexed signaling. Nano Lett 5:37–43
65. Ye ZQ, Tan MQ, Wang GL, Yuan JL (2004) Novel fluorescent europium chelate-doped silica nanoparticles: preparation, characterization and time-resolved fluorometric application. J Mater Chem 14:851–856
66. Fontell K, Khan A, Lindstrom B, Maciejewska D, Puangngern S (1991) Phase-Equilibria and Structures in Ternary-Systems of a Cationic Surfactant (C16tabr or (C16ta)2so4), Alcohol, and Water. Colloid Polym Sci 269:727–742
67. Israelachvili JN, Mitchell DJ, Ninham BW (1977) Theory of Self-Assembly of Lipid Bilayers and Vesicles. Biochim Biophys Acta 470:185–201
68. De TK, Maitra A (1995) Solution Behavior of Aerosol Ot in Nonpolar-Solvents. Adv Colloid Interface Sci 59:95–193
69. Nazario LMM, Hatton TA, Crespo JPSG (1996) Nonionic cosurfactants in AOT reversed micelles: Effect on percolation, size, and solubilization site. Langmuir 12:6326–6335

70. Lang J, Jada A, Malliaris A (1988) Structure and Dynamics of Water-in-Oil Droplets Stabilized by Sodium Bis(2-Ethylhexyl) Sulfosuccinate. J Phys Chem 92:1946–1953
71. Arriagada FJ, Osseoasare K (1995) Synthesis of Nanosize Silica in Aerosol Ot Reverse Microemulsions. J Colloid Interface Sci 170:8–17
72. Bergna HE (1990) American Chemical Society. Division of Colloid and Surface Chemistry. and American Chemical Society. Meeting The Colloid chemistry of silica: developed from a symposium sponsored by the Division of Colloid and Surface Chemistry at the 200th National Meeting of the American Chemical Society, Washington, DC, August 26–31, 1990, The Society, Washington, DC, 1994
73. Ma ZN, Friberg SE, Neogi P (1988) Observation of Temporary Liquid-Crystals in Water-in-Oil Microemulsion Systems. Colloids Surf 33:249–258
74. Friberg SE, Yang CC, Sjoblom J (1992) Amphiphilic Association Structures and the Microemulsion Gel Method for Ceramics – Influence on Original Phase Regions by Hydrolysis and Condensation of Silicon Tetraethoxide. Langmuir 8:372–376
75. Friberg SE, Ahmed AU, Yang CC, Ahuja S, Bodesha SS (1992) Gelation of a Microemulsion by Silica Formed Insitu. J Mater Chem 2:257–258
76. Marchand KE, Tarret M, Lechaire JP, Normand L, Kasztelan S, Cseri T (2003) Investigation of AOT-based microemulsions for the controlled synthesis of MoSx nanoparticles: an electron microscopy study. Colloids Surf A 214:239–248
77. Bagwe RP, Khilar KC (2000) Effects of intermicellar exchange rate on the formation of silver nanoparticles in reverse microemulsions of AOT. Langmuir 16:905–910
78. Ekwall P, Mandell L, Fontell K (1969) Cetyltrimethylammonium Bromide-Hexanol-Water System. J Colloid Interface Sci 29:639
79. Israelachvili JN, Mitchell DJ, Ninham BW (1976) Theory of Self-Assembly of Hydrocarbon Amphiphiles into Micelles and Bilayers. J Chem Soc Faraday Trans II 72:1525–1568
80. Liu JC, Ikushima Y, Shervani Z (2004) Investigation on the solubilization of organic dyes and micro-polarity in AOT water-in-CO2 microemulsions with fluorinated cosurfactant by using UV-Vis spectroscopy. J Supercrit Fluids 32:97–103
81. Shen D, Zhang R, Han BX, Dong Y, Wu WZ, Zhang JL, Li JC, Jiang T, Liu ZM (2004) Enhancement of the solubilization capacity of water in Triton X-100/cyclohexane/water system by compressed gases. Chem Eur J 10:5123–5128
82. Song LY, Ge XW, Zhang ZC (2005) Interfacial fabrication of silica hollow particles in a reverse emulsion system. Chem Lett 34:1314–1315
83. Lin YS, Hung Y, Su JK, Lee R, Chang C, Lin ML, Mou YC (2004) Gadolinium(III)-incorporated nanosized mesoporous silica as potential magnetic resonance imaging contrast agents. J Phys Chem B 108:15608–15611
84. Yin W, Zhang MS (2003) Characterization of nanosized Tb-MCM-41 synthesized by the sol–gel-assisted self-assembly method. J Alloys Compounds 360:231–235
85. Matos JR, Mercuri LP, Jaroniec M, Kruk M, Sakamoto Y, Terasaki O (2001) Synthesis and characterization of europium-doped ordered mesoporous silicas. J Mater Chem 11:2580–2586
86. Suzuki K, Ikari K, Imai H (2004) Synthesis of silica nanoparticles having a well-ordered mesostructure using a double surfactant system. J Am Chem Soc 126:462–463
87. Kresge CT, Leonowicz ME, Roth WJ, Vartuli JC, Beck JS (1992) Ordered Mesoporous Molecular-Sieves Synthesized by a Liquid–Crystal Template Mechanism. Nature 359:710–712
88. Nooney RI, Thirunavukkarasu D, Chen YM, Josephs R, Ostafin AE (2002) Synthesis of nanoscale mesoporous silica spheres with controlled particle size. Chem Mater 14:4721–4728 .

89. Haskouri JE, Cabrera S, Caldes M, Guillem C, Latorre J, Beltran A, Beltran D, Marcos MD, Amoros P (2002) Surfactant-assisted synthesis of the SBA-8 mesoporous silica by using nonrigid commercial alkyltrimethyl ammonium surfactants. Chem Mater 14:2637–2643
90. Czuryszkiewicz T, Rosenholm J, Kleitz F, Linden M (2002) Synthesis and characterization of mesoscopically ordered surfactant/cosurfactant templated metal oxides. Impact of Zeolites and Other Porous Materials on the New Technologies at the Beginning of the New Millennium, Book Series: Studies in Surface Science and Catalysis, Pts A and B 142:1117–1124
91. Bohren CF, Huffman DR (1983) Absorption and scattering of light by small particles. Wiley, New York
92. Hayat MA (1989) Colloidal gold: principles, methods, and applications. Academic Press, San Diego
93. Khlebtsov NG, Trachuk LA, Mel'nikov AG (2005) The effect of the size, shape, and structure of metal nanoparticles on the dependence of their optical properties on the refractive index of a disperse medium. Optics Spectrosc 98:77–83
94. Murphy CJ, Jana NR (2002) Controlling the aspect ratio of inorganic nanorods and nanowires. Adv Mater 14:80–82
95. Ghosh SK, Nath S, Kundu S, Esumi K, Pal T (2004) Solvent and ligand effects on the localized surface plasmon resonance (LSPR) of gold colloids. J Phys Chem B 108:13963–13971
96. Oldenburg SJ, Averitt RD, Westcott SL, Halas NJ (1998) Nanoengineering of optical resonances. Chem Phys Lett 288:243–247
97. Novak JP, Nickerson C, Franzen S, Feldheim DL (2001) Purification of molecularly bridged metal nanoparticle arrays by centrifugation and size exclusion chromatography. Anal Chem 73:5758–5761
98. Marinakos SM, Novak JP, Brousseau LC, House AB, Edeki EM, Feldhaus JC, Feldheim DL (1999) Gold particles as templates for the synthesis of hollow polymer capsules. Control of capsule dimensions and guest encapsulation. J Am Chem Soc 121:8518–8522
99. Underwood S, Mulvaney P (1994) Effect of the Solution Refractive-Index on the Color of Gold Colloids. Langmuir 10:3427–3430
100. Mrksich M (2000) A surface chemistry approach to studying cell adhesion. Chem Soc Rev 29:267–273
101. Bright RM, Walter DG, Musick MD, Jackson MA, Allison KJ, Natan MJ (1996) Chemical and electrochemical Ag deposition onto preformed Au colloid monolayers: Approaches to uniformly-sized surface features with Ag-like optical properties. Langmuir 12:810–817
102. Hermanson GT (1996) Bioconjugate techniques. Academic Press, San Diego
103. O'Neal DP, Hirsch LR, Halas NJ, Payne JD, West J (2004) Photo-thermal tumor ablation in mice using near infrared-absorbing nanoparticles. Cancer Lett 209:171–176
104. Loo C, Lowery A, Halas N, West J, Drezek R (2005) Immunotargeted nanoshells for integrated cancer imaging and therapy. Nano Lett 5:709–711
105. Hirsch LR, Jackson JB, Lee A, Halas NJ, West J (2003) A whole blood immunoassay using gold nanoshells. Anal Chem 75:2377–2381
106. Elghanian R, Storhoff JJ, Mucic RC, Letsinger RL, Mirkin CA (1997) Selective colorimetric detection of polynucleotides based on the distance-dependent optical properties of gold nanoparticles. Science 277:1078–1081

107. Storhoff JJ, Lazarides AA, Mucic RC, Mirkin CA, Letsinger RL, Schatz GC (2000) What controls the optical properties of DNA-linked gold nanoparticle assemblies? J Am Chem Soc 122:4640–4650
108. Taton TA, Mirkin CA, Letsinger RL (2000) Scanometric DNA array detection with nanoparticle probes. Science 289:1757–1760
109. Beesley JE (1989) Colloidal gold: a new perspective for cytochemical marking. Oxford University Press, Oxford; Royal Microscopical Society, New York Oxford
110. Schmid G, Chi LF (1998) Metal clusters and colloids. Adv Mater 10:515–526
111. Schmid G, Corain B (2003) Nanoparticulated gold: Syntheses, structures, electronics, and reactivities. Eur J Inorg Chem 3081-3098
112. Daniel MC, Astruc D (2004) Gold nanoparticles: Assembly, supramolecular chemistry, quantum-size-related properties, and applications toward biology, catalysis, and nanotechnology. Chem Rev 104:293–346
113. Boutonnet M, Kizling J, Stenius P (1982) The Preparation of Monodisperse Colloidal Metal Particles from Micro-Emulsions. Colloids Surf 5:209–225
114. Fendler JH (1987) Atomic and Molecular Clusters in Membrane Mimetic Chemistry. Chem Rev 87:877–899
115. Wilcoxon JP, Williamson RL, Baughman R (1993) Optical-Properties of Gold Colloids Formed in Inverse Micelles. J Chem Phys 98:9933–9950
116. Spirin MG, Brichkin SB, Razumov VF (2005) Synthesis and stabilization of gold nanoparticles in reverse micelles of aerosol OT and triton X-100. Colloid J 67:485–490
117. Herrera AP, Resto O, Briano JG, Rinaldi C (2005) Synthesis and agglomeration of gold nanoparticles in reverse micelles. Nanotechnology 16:S618–S625
118. Sato H, Ohtsu T, Komasawa I (2000) Atomic force microscopy study of ultrafine particles prepared in reverse micelles. J Colloid Interface Sci 230:200–204
119. Chiang CL (2000) Controlled growth of gold nanoparticles in aerosol-OT/sorbitan monooleate/isooctane mixed reverse micelles. J Colloid Interface Sci 230:60–66
120. Pomogailo AD, Rozenberg AS, Uflyand IE (2000) Nanochastitsy metallov v polimerakh (Metal Nanoparticles in Polymers). Khimiya, Moscow
121. Brichkin SB, Razumov VF, Spirin MG (2000) Kolloidn Zh 62:12
122. Pal A (2004) Photochemical synthesis of gold nanoparticles via controlled nucleation using a bioactive molecule. Mater Lett 58:529–534
123. Chiang CL (2001) Controlled growth of gold nanoparticles in AOT/C12E4/isooctane mixed reverse micelles. J Colloid Interface Sci 239:334–341
124. Neuman RD, Ibrahim TH (1999) Novel structural model of reversed micelles: The open water-channel model. Langmuir 15:10–12
125. Kumar R, Maitra AN, Patanjali PK, Sharma P (2005) Hollow gold nanoparticles encapsulating horseradish peroxidase. Biomaterials 26:6743–6753
126. Manna A, Imae T, Yogo T, Aoi K, Okazaki M (2002) Synthesis of gold nanoparticles in a Winsor II type microemulsion and their characterization. J Colloid Interface Sci 256:297–303
127. Lin J, Zhou WL, O'Connor CJ (2001) Formation of ordered arrays of gold nanoparticles from CTAB reverse micelles. Mater Lett 49:282–286
128. Barnickel P, Wokaun A (1990) Synthesis of Metal Colloids in Inverse Microemulsions. Mol Phys 69:1–9
129. Porta F, Prati L, Rossi M, Scari G (2002) Synthesis of Au(0) nanoparticles from W/O microemulsions. Colloids Surf A 211:43–48
130. Giustini M, Palazzo G, Colafemmina G, DellaMonica M, Giomini M, Ceglie A (1996) Microstructure and dynamics of the water-in-oil CTAB/n-pentanol/n-hexane/water

microemulsion: A spectroscopic and conductivity study. J Phys Chem 100:3190–3198
131. Torigoe K, Esumi K (1992) Preparation of Colloidal Gold by Photoreduction of Aucl4–Cationic Surfactant Complexes. Langmuir 8:59–63
132. Kameo A, Suzuki A, Torigoe K, Esumi K (2001) Fiber-like Gold Particles Prepared in Cationic Micelles by UV Irradiation: Effect of Alkyl Chain Length of Cationic Surfactant on Particle Size. J Colloid Interface Sci 241:89–292
133. Leontidis E, Kleitou K, Kyprianidou-Leodidou T, Bekiari V, Lianos P (2002) Gold colloids from cationic surfactant solutions. 1. Mechanisms that control particle morphology. Langmuir 18:3659–3668
134. Chen FX, Xu GQ, Hor TSA (2003) Preparation and assembly of colloidal gold nanoparticles in CTAB-stabilized reverse microemulsion. Mater Lett 57:3282–3286
135. Esumi K, Matsuhisa K, Torigoe K (1995) Preparation of Rodlike Gold Particles by UV Irradiation Using Cationic Micelles as a Template. Langmuir 11:3285–3287
136. Imae T, Ikeda S (1987) Characteristics of Rodlike Micelles of Cetyltrimethylammonium Chloride in Aqueous Nacl Solutions – Their Flexibility and the Scaling Laws in Dilute and Semidilute Regimes. Colloid Polym Sci 265:1090–1098
137. Lee YS, Surjadi D, Rathman JF (1996) Effects of aluminate and silicate on the structure of quaternary ammonium surfactant aggregates. Langmuir 12:6202–6210
138. Ozeki S, Ikeda S (1981) The Stability of Spherical Micelles of Dodecyltrimethylammonium Chloride in Aqueous Nacl Solutions. Bull Chem Soc Japan 54:552–555
139. Xu J, Han X, Liu H L, Hu Y (2005) Synthesis of monodisperse gold nanoparticles stabilized by gemini surfactant in reverse micelles. J Dispers Sci Technol 26:473–476
140. Azene H, Sigers S, Johnson V (2003) Formation and characterization of gold nanoparticles in dioctyl sulfosuccinate/isooctane and dioctyl sulfosuccinate/phosphatidylcholine/isooctane mixed reverse micelles. Abstr Pap Am Chem Soc 225:U23–U23
141. Shen M, Du YK, Rong HL, Li JR, Jiang L (2005) Preparation of hydrophobic gold nanoparticles with safe organic solvents by microwave irradiation method. Colloids Surf A 257–258:439–443
142. Khomutov GB, Gubin SP (2002) Interfacial synthesis of noble metal nanoparticles. Mater Sci Eng C 22:141–146
143. Ravaine S, Fanucci GE, Seip CT, Adair JH, Talham DR (1998) Photochemical generation of gold nanoparticles in Langmuir–Blodgett films. Langmuir 14:708–713
144. Johnson SR, Evans SD, Mahon SW, Ulman A (1997) Synthesis and characterisation of surfactant-stabilised gold nanoparticles. Supramol Sci 4:329–333
145. Praharaj S, Ghosh SK, Nath S, Kundu S, Panigrahi S, Basu S, Pal T (2005) Size-selective synthesis and stabilization of gold organosol in CnTAC: Enhanced molecular fluorescence from gold-bound fluorophores. J Phys Chem B 109:13166–13174
146. Prasad BLV, Stoeva SI, Sorensen CM, Klabunde KJ (2002) Digestive ripening of thiolated gold nanoparticles: The effect of alkyl chain length. Langmuir 18:7515–7520
147. Mafune F, Kohno J, Takeda Y, Kondow T, Sawabe H (2000) Structure and stability of silver nanoparticles in aqueous solution produced by laser ablation. J Phys Chem B 104:8333–8337
148. Mafune F, Kohno J, Takeda Y, Kondow T, Sawabe H (2000) Formation and size control of sliver nanoparticles by laser ablation in aqueous solution. J Phys Chem B 104:9111–9117
149. Ankamwar B, Damle C, Ahmad A, Sastry M (2005) Biosynthesis of gold and silver nanoparticles using Emblica officinalis fruit extract, their phase transfer and transmetallation in an organic solution. J Nanosci Nanotechnol 5:1665–1671

150. Maeda Y, Okitsu K, Inoue H, Nishimura R, Mizukoshi Y, Nakui H (2004) Preparation of nanoparticles by reducing intermediate radicals formed in sonolytical pyrolysis of surfactants. Res Chem Intermediat 30:775–783
151. Murphy CJ, Sau TK, Gole A, Orendorff CJ (2005) Surfactant-directed synthesis and optical properties of one-dimensional plasmonic metallic nanostructures. MRS Bull 30:349–355
152. Reetz MT, Helbig W (1994) Size-Selective Synthesis of Nanostructured Transition-Metal Clusters. J Am Chem Soc 116:7401–7402
153. Reetz MT, Helbig W, Quaiser SA, Stimming U, Breuer N, Vogel R (1995) Visualization of Surfactants on Nanostructured Palladium Clusters by a Combination of STM and High-Resolution TEM. Science 267:367–369
154. Yu YY, Chang SS, Lee CL, Wang CRC (1997) Gold nanorods: Electrochemical synthesis and optical properties. J Phys Chem B 101:6661–6664
155. Chang SS, Shih CW, Chen CD, Lai WC, Wang CRC (1999) The shape transition of gold nanorods. Langmuir 15:701–709
156. Zsigmondy R, Thiessen PA (1925) Das kolloide Gold. Akademische Verlagsgesellschaft mbh, Leipzig
157. Brown KR, Walter DG, Natan MJ (2000) Seeding of colloidal Au nanoparticle solutions, 2. Improved control of particle size and shape. Chem Mater 12:306–313
158. Goia DV, Matijevic E (1998) Preparation of monodispersed metal particles. New J Chem 22:1203–1215
159. Schmid G (1992) Large Clusters and Colloids – Metals in the Embryonic State. Chem Rev 92:1709–1727
160. Sau TK, Pal A, Jana NR, Wang ZL, Pal T (2001) Size controlled synthesis of gold nanoparticles using photochemically prepared seed particles. J Nanoparticle Res 3:257–261
161. Lu LH, Wang HS, Zhou YH, Xi SQ, Zhang HJ, Jiawen HBM, Zhao B (2002) Seed-mediated growth of large, monodisperse core-shell gold-silver nanoparticles with Ag-like optical properties. Chem Commun, pp 144–145
162. Jana NR, Gearheart L, Murphy CJ (2001) Evidence for seed-mediated nucleation in the chemical reduction of gold salts to gold nanoparticles. Chem Mater 13:2313–2322
163. Gao JX, Bender CM, Murphy CJ (2003) Dependence of the gold nanorod aspect ratio on the nature of the directing surfactant in aqueous solution. Langmuir 19:9065–9070
164. Loo C, Lin A, Hirsch L, Lee MH, Barton J, Halas N, West J, Drezek R (2004) Nanoshell-enabled photonics-based imaging and therapy of cancer. Technol Cancer Res Treat 3:33–40
165. Mafune F, Kohno J, Takeda Y, Kondow T, Sawabe H (2001) Formation of gold nanoparticles by laser ablation in aqueous solution of surfactant. J Phys Chem B 105:5114–5120
166. Mafune F, KohnoJ Y, Takeda Y, Kondow T (2002) Full physical preparation of size-selected gold nanoparticles in solution: Laser ablation and laser-induced size control. J Phys Chem B 106:7575–7577
167. Niidome Y, Hori A, Sato T, Yamada S (2000) Enormous size growth of thiol-passivated gold nanoparticles induced by near-IR laser light. Chem Lett 4:310–311
168. Kamat PV, Flumiani M, Hartland GV (1998) Picosecond dynamics of silver nanoclusters. Photoejection of electrons and fragmentation. J Phys Chem B 102:3123–3128
169. Mafune F, Kohno J, Takeda Y, Kondow T (2003) Formation of gold nanonetworks and small gold nanoparticles by irradiation of intense pulsed laser onto gold nanoparticles. J Phys Chem B 107:12589–12596

170. Alivisatos AP (1996) Perspectives on the physical chemistry of semiconductor nanocrystals. J Phys Chem 100:13226–13239
171. Gaponenko SV (1998) Optical properties of semiconductor nanocrystals. Cambridge Unviersity Press, Cambridge, New York
172. Weller H, Koch U, Gutierrez M, Henglein A (1984) Photochemistry of Colloidal Metal Sulfides, 7. Absorption and Fluorescence of Extremely Small Zns Particles - the World of the Neglected Dimensions. Ber Bunsenges 88:649–656
173. Eychmuller A, Mews A, Weller H (1993) A Quantum-Dot Quantum-Well - Cds/Hgs/Cds. Chem Phys Lett 208:59–62
174. Guzelian AA, Katari JEB, Kadavanich AV, Banin U, Hamad K, Juban E, Alivisatos AP, Wolters RH, Arnold CC, Heath JR (1996) Synthesis of size-selected, surface-passivated InP nanocrystals. J Phys Chem 100:7212–7219
175. Mattoussi H, Mauro JM, Goldman ER, Anderson GP, Sundar VC, Mikulec FV, Bawendi MG (2000) Self-assembly of CdSe-ZnS quantum dot bioconjugates using an engineered recombinant protein. J Am Chem Soc 122:12142–12150
176. Peng ZA, Peng XG (2001) Formation of high-quality CdTe, CdSe, and CdS nanocrystals using CdO as precursor. J Am Chem Soc 123:183–184
177. Jaiswal JK, Mattoussi H, Mauro JM, Simon SM (2003) Long-term multiple color imaging of live cells using quantum dot bioconjugates. Nat Biotechnol 21:47–51
178. Cao YW, Banin U (2000) Growth and properties of semiconductor core/shell nanocrystals with InAs cores. J Am Chem Soc 122:9692–9702
179. Micic OI, Smith BB, Nozik AJ (2000) Core-shell quantum dots of lattice-matched ZnCdSe2 shells on InP cores: Experiment and theory. J Phys Chem B 104:12149–12156
180. Ballou B, Lagerholm BC, Ernst LA, Bruchez MP, Waggoner AS (2004) Noninvasive imaging of quantum dots in mice. Bioconjug Chem 15:79–86
181. Goldman ER, Anderson GP, Tran PT, Mattoussi H, Charles PT, Mauro JM (2002) Conjugation of luminescent quantum dots with antibodies using an engineered adaptor protein to provide new reagents for fluoroimmunoassays. Anal Chem 74:841–847
182. Gerion D, Pinaud F, Williams SC, Parak WJ, Zanchet D, Weiss S, Alivisatos AP (2001) Synthesis and properties of biocompatible water-soluble silica-coated CdSe/ZnS semiconductor quantum dots. J Phys Chem B 105:8861–8871
183. Dubertret B, Skourides P, Norris DJ, Noireaux V, Brivanlou AH, Libchaber A (2002) In vivo imaging of quantum dots encapsulated in phospholipid micelles. Science 298:1759–1762
184. Santra S, Yang H, Holloway PH, Stanley JT, Mericle RA (2005) Synthesis of water-dispersible fluorescent, radio-opaque, and paramagnetic CdS:Mn/ZnS quantum dots: a multifunctional probe for bioimaging. J Am Chem Soc 127:1656–1657
185. Santra S, Yang H, Stanley JT, Holloway PH, Moudgil BM, Walter G, Mericle RA (2005) Rapid and effective labeling of brain tissue using TAT-conjugated CdS:Mn/ZnS quantum dots. Chem Commun, pp 3144–3146
186. Esteves ACC, Trindade T (2002) Synthetic studies on II/VI semiconductor quantum dots. Curr Opin Solid State Mater Sci 6:347–353
187. Brus LE (1984) Electron–Electron and Electron–Hole Interactions in Small Semiconductor Crystallites - the Size Dependence of the Lowest Excited Electronic State. J Chem Phys 80:4403–4409
188. Henglein A (1989) Small-Particle Research - Physicochemical Properties of Extremely Small Colloidal Metal and Semiconductor Particles. Chem Rev 89:1861–1873

189. Wang WZ, Liu ZH, Zheng CL, Xu CK, Liu YK, Wang GH (2003) Synthesis of US nanoparticles by a novel and simple one-step, solid-state reaction in the presence of a nonionic surfactant. Mater Lett 57:2755–2760
190. Gan LM, Liu B, Chew CH, Xu SJ, Chua SJ, Loy GL, Xu GQ (1997) Enhanced photoluminescence and characterization of Mn-doped ZnS nanocrystallites synthesized in microemulsion. Langmuir 13:6427–6431
191. Yang XH, Wu QS, Li L, Ding YP, Zhang GX (2005) Controlled synthesis of the semiconductor CdS quasi-nanospheres, nanoshuttles, nanowires and nanotubes by the reverse micelle systems with different surfactants. Colloids Surf A 264:172–178
192. Zhang B, Li GH, Zhang J, Zhang Y, Zhang LD (2003) Synthesis and characterization of PbS nanocrystals in water/C12E9/cyclohexane microemulsions. Nanotechnology 14:443–446
193. Robinson BH, Towey TF, Zourab S, Visser AJWG, Vanhoek A (1991) Characterization of Cadmium-Sulfide Colloids in Reverse Micelles. Colloids Surf 61:175–188
194. Qi LM, Ma JM, Cheng HM, Zhao ZG (1996) Synthesis and characterization of mixed CdS-ZnS nanoparticles in reverse micelles. Colloids Surf A 111:195–202
195. Zou BS, Little RB, Wang JP, El-Sayed MA (1999) Effect of different capping environments on the optical properties of CdS nanoparticles in reverse micelles. Int J Quantum Chem 72:439–450
196. Seddon AB, Ou DL (1998) CdSe quantum dot doped amine-functionalized Ormosils. J Sol–Gel Sci Technol 13:623–628
197. Guo B, Pang Q, Yang C, Ge W, Yang S, Wang J (2005) Reverse micelles synthesis and optical characterization of manganese doped CdSe quantum dots. Department of Physics, Hong Kong University of Science and Technology, Hong Kong, Peop. Rep. China. AIP Conference Proceedings, 772 (Physics of Semiconductors, Part A), pp 605–606
198. Kim D, Miyamoto M, Nakayama M (2005) Photoluminescence properties of CdS and CdMnS quantum dots prepared by a reverse-micelle method. J Electron Microsc 54:I31–I34
199. Quinlan FT, Kuther J, Tremel W, Knoll W, Risbud S, Stroeve P (2000) Reverse micelle synthesis and characterization of ZnSe nanoparticles. Langmuir 16:4049–4051
200. Calandra P, Goffredi M, Liveri VT (1999) Study of the growth of ZnS nanoparticles in water/AOT/n-heptane microemulsions by UV-absorption spectroscopy. Colloids Surf A 160:9–13
201. Hirai T, Watanabe T, Komasawa I (2000) Preparation of semiconductor nanoparticle-polymer composites by direct reverse micelle polymerization using polymerizable surfactants. J Phys Chem B 104:8962–8966
202. Cao LX, Zhang JH, Ren SL, Huang SH (2002) Luminescence enhancement of core-shell ZnS:Mn/ZnS nanoparticles. Appl Phys Lett 80:4300–4302
203. Cao LX, Huang SH, Shulin E (2004) ZnS/CdS/ZnS quantum dot quantum well produced in inverted micelles. J Colloid Interface Sci 273:478–482
204. Lianos P, Thomas JK (1986) Cadmium-Sulfide of Small Dimensions Produced in Inverted Micelles. Chem Phys Lett 125:299–302
205. Lianos P, Thomas JK (1987) Small Cds Particles in Inverted Micelles. J Colloid Interface Sci 117:505–512
206. Nakanishi T, Ohtani B, Uosaki K (1998) Fabrication and characterization of CdS-nanoparticle mono- and multilayers on a self-assembled monolayer of alkanedithiols on gold. J Phys Chem B 102:1571–1577

207. Tsuruoka T, Akamatsu K, Nawafune H (2004) Synthesis, surface modification, and multilayer construction of mixed-monolayer-protected CdS nanoparticles. Langmuir 20:11169–11174
208. Steigerwald ML, Alivisatos AP, Gibson JM, Harris TD, Kortan R, Muller AJ, Thayer AM, Duncan TM, Douglass DC, Brus LE (1988) Surface Derivatization and Isolation of Semiconductor Cluster Molecules. J Am Chem Soc 110:3046–3050
209. Zulauf M, Eicke HF (1979) Inverted Micelles and Microemulsions in the Ternary-System H2o-Aerosol-Ot-Isooctane as Studied by Photon Correlation Spectroscopy. J Phys Chem 83:480–486
210. Hirai T, Okubo H, Komasawa I (2001) Incorporation of CdS nanoparticles formed in reverse micelles into mesoporous silica. J Colloid Interface Sci 235:358–364
211. Kortan AR, Hull R, Opila RL, Bawendi MG, Steigerwald ML, Carroll PJ, Brus LE (1990) Nucleation and Growth of Cdse on Zns Quantum Crystallite Seeds, and Vice Versa, in Inverse Micelle Media. J Am Chem Soc 112:1327–1332
212. Towey TF, Khanlodhi A, Robinson BH (1990) Kinetics and Mechanism of Formation of Quantum-Sized Cadmium-Sulfide Particles in Water Aerosol-Ot Oil Microemulsions. J Chem Soc Faraday Trans 86:3757–3762
213. Meyer M, Wallberg C, Kurihara K, Fendler JH (1984) Photosensitized Charge Separation and Hydrogen-Production in Reversed Micelle Entrapped Platinized Colloidal Cadmium-Sulfide. J Chem Soc Chem Commun, pp 90–91
214. Tata M, Banerjee S, John VT, Waguespack Y, McPherson GL (1997) Fluorescence quenching of CdS nanocrystallites in AOT water-in-oil microemulsions. Colloids Surf A 127:39–46
215. Bunker CE, Harruff BA, Pathak P, Payzant A, Allard LF, Sun YP (2004) Formation of cadmium sulfide nanoparticles in reverse micelles: Extreme sensitivity to preparation procedure. Langmuir 20:5642–5644
216. Pileni MP, Motte L, Petit C (1992) Synthesis of Cadmium-Sulfide Insitu in Reverse Micelles – Influence of the Preparation Modes on Size, Polydispersity, and Photochemical Reactions. Chem Mater 4:338–345
217. Fischer CH, Weller H, Fojtik A, Lumepereira C, Janata E, Henglein A (1986) Photochemistry of Colloidal Semiconductors, 10. Exclusion Chromatography and Stop Flow Experiments on the Formation of Extremely Small Cds Particles. Ber Bunsenges 90:46–49
218. Hirai T, Bando Y (2005) Immobilization of CdS nanoparticles formed in reverse micelles onto aluminosilicate supports and their photocatalytic properties. J Colloid Interface Sci 288:513–516
219. Hirai T, Bando Y, Komasawa I (2002) Immobilization of CdS nanoparticles formed in reverse micelles onto alumina particles and their photocatalytic properties. J Phys Chem B 106:8967–8970
220. Simmons B, Li CS, John VT, McPherson GL, Taylor C, Schwartz DK, Maskos K (2002) Spatial compartmentalization of nanoparticles into strands of a self-assembled organogel. Nano Lett 2:1037–1042
221. Kotlarchyk M, Stephens RB, Huang JS (1988) Study of Schultz Distribution to Model Polydispersity of Microemulsion Droplets. J Phys Chem 92:1533–1538
222. Israelachvili JN (1991) Intermolecular and surface forces. Academic Press, London, San Diego
223. Zhou Y, Itoh H, Uemura T, Naka K, Chujo Y (2002) Preparation, optical spectroscopy, and electrochemical studies of novel pi-conjugated polymer-protected stable PbS colloidal nanoparticles in a nonaqueous solution. Langmuir 18:5287–5292

224. Gadenne P, Yagil Y, Deutscher G (1989) Transmittance and Reflectance Insitu Measurements of Semicontinuous Gold-Films During Deposition. J Appl Phys 66:3019–3025
225. Chaudhuri TK (1992) A Solar Thermophotovoltaic Converter Using Pbs Photovoltaic Cells. Int J Energ Res 16:481–487
226. Wu XS, Pan LJ, Zou G, Liu JP, He PS (2004) Preparation of PbS/poly(acrylic acid) nanocrystal micropatterns by soft lithography. Chinese J Chem Phys 17:641–644
227. Nair PK, Nair MTS (1992) Chemically Deposited Zns Thin-Films – Application as Substrate for Chemically Deposited Bi2s3, Cuxs and Pbs Thin-Films. Semiconductor Sci Technol 7:239–244
228. Eastoe J, Cox AR (1995) Formation of Pbs Nanoclusters Using Reversed Micelles of Lead and Sodium Aerosol-Ot. Colloids Surf A 101:63–76
229. Miyoshi H, Yamachika M, Yoneyama H, Mori H (1990) Photochemical Properties of PbS Microcrystallites Prepared in Nafion. J Chem Soc Faraday Trans 86:815–818
230. Aggarwal RL, Furdyna JK, Von S (1987) Molnar and Materials Research Society. Diluted magnetic (semimagnetic) semiconductors. Materials Research Society, Pittsburgh
231. Bastard G, Rigaux C Mycielski A (1977) Giant Spin Splitting Induced by Exchange Interactions in Hg1-Kmnkte Mixed-Crystals. Phys Status Solidi B 79:585–593
232. Dietl T, Spalek J (1982) Effect of Fluctuations of Magnetization on the Bound Magnetic Polaron – Comparison with Experiment. Phys Rev Lett 48:355–358
233. Bhargava RN, Gallagher D, Hong X, Nurmikko A (1994) Optical-Properties of Manganese-Doped Nanocrystals of Zns. Phys Rev Lett 72:416–419
234. Wang Y, Herron N, Moller K, Bein T (1991) 3-Dimensionally Confined Diluted Magnetic Semiconductor Clusters – Zn1-Xmnxs. Solid State Commun 77:33–38
235. Feltin N, Levy L, Ingert D, Pileni MP (1999) Magnetic properties of 4-nm Cd1-yMnyS nanoparticles differing by their compositions, y. J Phys Chem B 103:4–10
236. Levy L, Feltin N, Ingert D, Pileni MP (1997) Three dimensionally diluted magnetic semiconductor clusters Cd1-yMnyS with a range of sizes and compositions: Dependence of spectroscopic properties on the synthesis mode. J Phys Chem B 101:9153–9160
237. Levy L, Hochepied JF, Pileni MP (1996) Control of the size and composition of three dimensionally diluted magnetic semiconductor clusters. J Phys Chem 100:18322–18326
238. Hoener CF, Allan KA, Bard AJ, Campion A, Fox MA, Mallouk TE, Webber SE, White JM (1992) Demonstration of a Shell Core Structure in Layered CdSe-ZnSe Small Particles by X-Ray Photoelectron and Auger Spectroscopies. J Phys Chem 96:3812–3817
239. Yang H S, Santra S Holloway PH (2005) Syntheses and applications of Mn-doped II-VI semiconductor nanocrystals. J Nanosci Nanotechnol 5:1364–1375
240. Yang HS, Holloway PH, Santra S (2004) Water-soluble silica-overcoated CdS : Mn/ZnS semiconductor quantum dots. J Chem Phys 121:7421–7426
241. Pang Q, Guo BC, Wang JN, Yang SH, Wang YQ, Ge WK, Gong ML (2004) Synthesis of Cd1-xMnxS quantum dots via reverse micelles and its photoluminescence performance. Chem J Chinese Universities 25:1593–1596
242. Khomane RB, Manna A, Mandale AB, Kulkarni BD (2002) Synthesis and characterization of dodecanethiol-capped cadmium sulfide nanoparticles in a Winsor II microemulsion of diethyl ether/AOT/water. Langmuir 18:8237–8240
243. Fan HY, Leve HY, Scullin C, Gabaldon J, Tallant D, Bunge S, Boyle T, Wilson MC, Brinker CJ (2005) Surfactant-assisted synthesis of water-soluble and biocompatible semiconductor quantum dot micelles. Nano Lett 5:645–648

244. Dutta P, Fendler JH (2002) Preparation of cadmium sulfide nanoparticles in self-reproducing reversed micelles. J Colloid Interface Sci 247:47–53
245. Zhang J, Sun LD, Liao CS, Yan CH (2002) Size control and photoluminescence enhancement of CdS nanoparticles prepared via reverse micelle method. Solid State Commun 124:45–48
246. Agostiano A, Catalano M, Curri ML, Della Monica M, Manna L, Vasanelli L (2000) Synthesis and structural characterisation of CdS nanoparticles prepared in a four-components "water-in-oil" microemulsion. Micron 31:253–258
247. Curri ML, Agostiano A, Manna L, Della Monica M, Catalano M, Chiavarone L, Spagnolo V, Lugara M (2000) Synthesis and characterization of CdS nanoclusters in a quarternary microemulsion: The role of the cosurfactant. J Phys Chem B 104:8391–8397
248. Maidment LJ, Chen V, Warr GG (1997) Effect of added cosurfactant on ternary microemulsion structure and dynamics. Colloids Surf A 130:311–319
249. Chakraborty I, Moulik SP (2004) Preparation and characterization of nanoscale semiconductor particles of ZnS, CdS, and PbCrO4 in polymer-surfactant gel matrix. J Dispers Sci Technol 25:849–859
250. Neumann E, Sowers AE, Jordan CA (1989) Electroporation and electrofusion in cell biology. Plenum Press, New York
251. Wu SX, Zeng HX, Schelly ZA (2005) Preparation of ultrasmall, uncapped PbS quantum dots via electroporation of vesicles. Langmuir 21:686–691
252. Yu WL, Pei J, Huang W, Zhao GX (1997) Formation of CdS nanoparticles in mixed cationic–anionic surfactant vesicle system. Mater Chem Phys 49:87–92

Directed Synthesis of Micro-Sized Nanoplatelets of Gold from a Chemically Active Mixed Surfactant Mesophase

Jayashri Sarkar[1] · Ganapathiraman Ramanath[2] · Vijay John[3] · Arijit Bose[1] (✉)

[1] Department of Chemical Engineering, University of Rhode Island, RI USA
bose@egr.uri.edu

[2] Department of Material Science and Engineering, Rensselaer Polytechnic Institute, NY USA

[3] Department of Chemical Engineering, Tulane University, LA USA

1	Introduction	235
2	Materials	238
2.1	Experiments	238
2.1.1	Facetted Particles (Experiment A)	238
2.1.2	Spherical Particles (Experiment B)	239
2.1.3	Free Precipitation (Experiment C)	240
3	Results and Discussion	241
4	Conclusions	246
	References	246

Abstract We report the synthesis of gold nanoparticles of controllable size and morphology from ordered mesophase templates comprised of iso-octane, sodium bis(2-ethylhexyl) sulfosuccinate (AOT) and lecithin along with an aqueous phase containing auric acid ($HAuCl_4$), the gold precursor. Highly facetted nanoparticles are formed by the reduction of $HAuCl_4$ directly by the dioctyl sulfosuccinate termini of AOT. In sharp contrast, rapid reduction of the gold precursor by the addition of sodium borohydride $NaBH_4$ in the aqueous phase results in spherical nanoparticles. The size of the nanoparticles can be adjusted by varying the auric acid concentration as well as the volume fraction of the aqueous phase. The value of this technique is the ease with which nanoscale particles of different shape and size can be formed, with concomitant impact on their physical and chemical properties.

Keywords Facetted nanoparticles · Reduction · Surfactants

1
Introduction

Metal nanoparticles often exhibit completely different properties than their bulk counterparts [1]. These properties can be tuned by manipulating the particle structure, shape, size and surface chemistry [2–4]. For example, bulk gold is a chemically inert electrical conductor, whereas gold nanoparticles are efficient nanocatalysts, and exhibit particle size- and shape-dependent plasmon resonance signatures that can span the entire visible spectrum [5, 6]. These features are attractive for realizing refractive index-tunable nanocomposite media, and sensors based on surface plasmon resonance (SPR) [7]. Nano triangles have shown greater SPR responses than nano-hemispheres due to larger sensing volumes and are more sensitive to multilayer adsorbates than the nano-hemispheres [7]. Nanoprisms possess specific quadruple resonance [8, 9]. The size-dependent optical properties of the nanoparticles alter their ability to polarize light and thus influence the dielectric properties of the medium. Branched nanoparticles can have potential applications as interconnects for nanocircuits and nanodevices fabricated using "bottom-up" self-assembly routes [10]. Hence, there is widespread interest in precisely controlling nanoparticle size, morphology, structure and chemistry, to obtain novel physical properties and chemical responses, and to exploit their applications in areas like catalysis [2, 3, 11], sensing [12], nano-lithography [13], biolabeling [14], photovoltaics, electrochemistry [11, 15–22] and biology [23–28].

Gold nanoparticles are synthesized and stabilized against aggregation by reducing the gold precursor in the presence of ionic or non-ionic surfactants. The surfactants cap the nanoparticles and prevent agglomeration by steric stabilization [29, 30]. Surface capping has been achieved using different surfactants [21, 29, 31], polymers [32], and block copolymers [33–36].

Several methods have been attempted for attaining control over the morphology of the forming nanoparticles. Different templates varying between hard ones such as porous alumina, [37] carbon nanotubes [38] and polycarbonate membranes [39], or soft ones such as organized surfactant [40, 41], have been used for the synthesis of nanorods. Murphy et al. have developed a two-step technique for the synthesis of gold and silver nanorods using a seeding method [42–45]. In this technique, gold and silver seed crystals are allowed to grow in the presence of a surfactant by reduction of $HAuCl_4$ with a mild reducing agent. Various electrochemical and photochemical approaches have been developed for nanorod synthesis [46–48]. Synthesis of branched nanoparticles with bipod, tripod, tetrapod and pentapod morphologies was accomplished using different methods including the seeded growth approach [10, 44, 49, 50].

Synthesis of anisotropic gold nanoparticles in mixed surfactant systems has also been reported [29, 51]. Chiang synthesized gold nanoparticles in reverse micelles of isoocane/AOT/ tetraethylene glycol monododecyl ether ($C_{12}E_4$) with gold chloride solution and hydrazine as the aqueous phase [29, 51]. Anisotropic nanoparticles were produced when the molar ratio of hydrazine to gold chloride was less than 0.5. Triangular silver nanoplates were synthesized by the reduction of silver salt using ascorbic acid (a mild reducing agent) in the presence of CTAB [29, 51].

Mirkin et al. used photo-induction for synthesizing silver (Ag) nanoprisms by conversion of spheres to large prisms [52]. A colloidal solution of Ag nanoparticles irradiated with a narrow band light source resulted in the formation of nanoprisms with a bimodal size distribution. The bimodal size distribution of the nanoparticles was attributed to the "edge-selective particle fusion" mechanism, where the size of the nanoparticle of larger dimension was equivalent to the dimensions of four smaller nanoparticles. Sastry et al. synthesized triangular gold nanoparticles using lemon grass extract as a reducing agent [27]. The particle size increased with time and a mixture of both spherical and plate-like particles developed in the same system. The

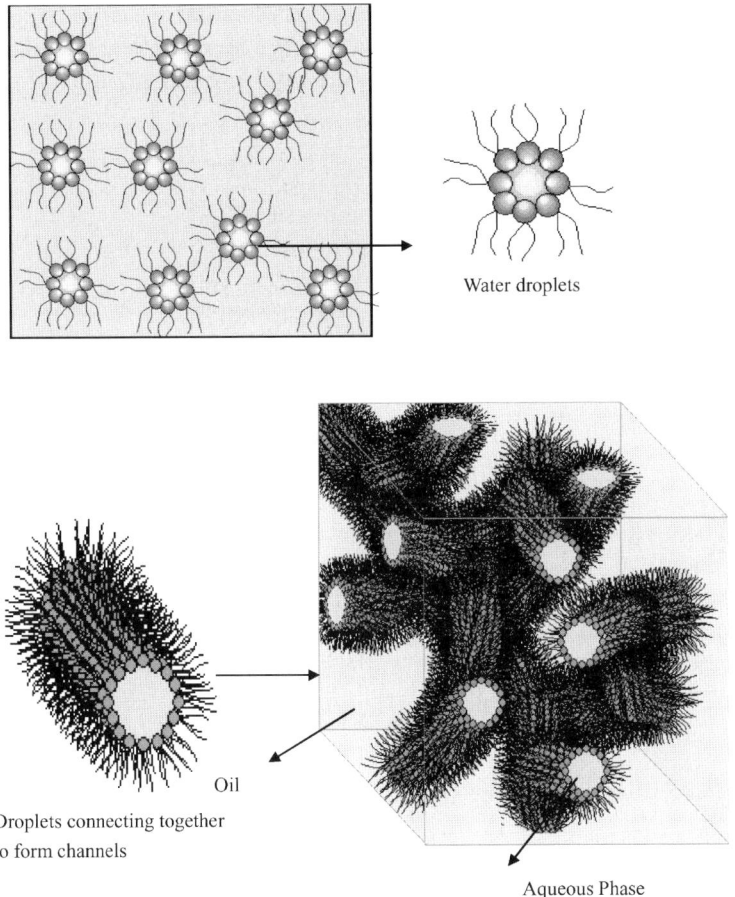

Fig. 1 Formation of reverse micelles in a self-assembled mixed surfactant system. The addition of water tends to link these droplets to form a highly viscous bi-continuous microemulsion with aqueous and isooctane nanochannels separated by the surfactants

triangular plates were estimated to be of 45% yield of the total material. Dong et al. showed nanoplate synthesis using reduction by aspartates [53].

In this paper, we report the synthesis of highly facetted and spherical gold particles in a surfactant mesophase. The formation of the mesophase is illustrated in Fig. 1. The addition of an aqueous solution to a solution of AOT and lecithin in isooctane first results in the formation of a water-in-oil microemulsion. Further addition of the aqueous component results in a highly viscous bicontinuous phase, with the aqueous and organic nanochannels separated by the surfactants. These organized structures form the templates for particle growth. When the surfactant head group donates electrons to reduce the metal ions, the formation of gold is slow. The surfactants preferentially absorb on to certain planes and hinder their growth, and anisotropic nanoparticles are formed. If a strong reducing agent is added, the gold formation is much more rapid, and the particles formed are close to spherical. Modulation of the metal ion reduction kinetics, and preferential adsorption of surfactants has a strong impact on the resulting morphology of the gold nanoparticles.

2
Materials

Isooctane, AOT and $HAuCl_4 \cdot 3H_2O$ were acquired from Sigma–Aldrich, lecithin (L-α-phosphatidylcholine, 95% Plant Soy) from Avanti Polar Lipids.

2.1
Experiments

The surfactants AOT (0.8 M) and Lecithin (0.4 M) were dissolved in isooctane. The aqueous phase, consisting of either a 10^{-3} M $HAuCl_4$ solution, or alternate aliquots of 10^{-3} M $HAuCl_4$ and $NaBH_4$ (10^{-2} M), was then added to the surfactant-containing organic phase until the desired W_0 (moles water/moles AOT) was reached. The reactions proceeded at room temperature. We carried out three sets of experiments to synthesize nanoparticles of different morphologies.

2.1.1
Facetted Particles (Experiment A)

$HAuCl_4$ (0.01 M) solution was added in installments to the Isooctane/AOT (0.8 M)/Lecithin (0.4 M) system. Each installment increased the W_0 by 10. The system turned into a highly viscous gel phase at $W_0 = 70$ [54], and the reaction was allowed to proceed at room temperature. Over about 15 minutes, the color of the solution transitioned from yellow to green and ultimately

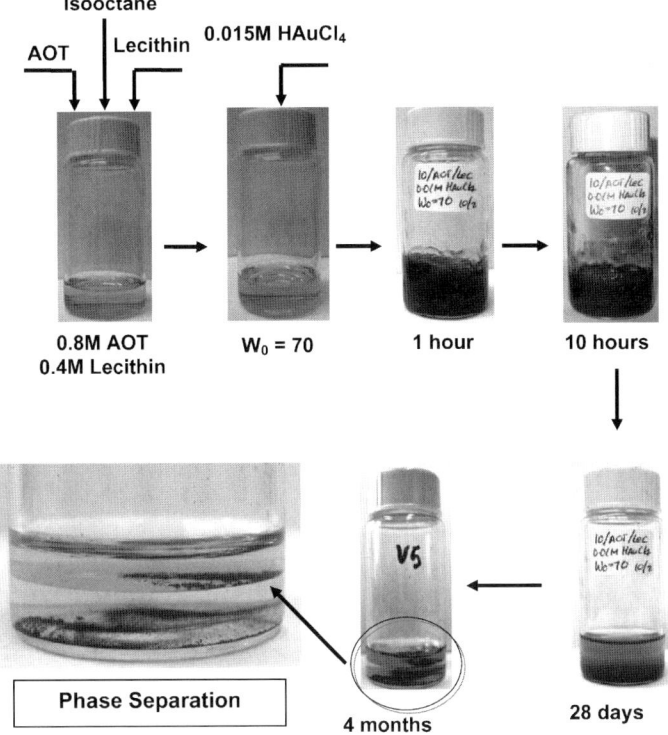

Fig. 2 Synthesis of facetted nanoparticles showing change in the color and nature of the sample with time and showing the phase separation observed in the isooctane/AOT (0.8 M)/Lecithin (0.4 M)/HAuCl$_4$ (0.01 M) sample after 4 months

to reddish brown, upon incremental HAuCl$_4$ addition. After 45 days, phase separation was clearly observed in the system. After 4 months the particles had settled at the bottom of the vial and two distinct phases were observed (Fig. 2). This was an indication that the gold chloride reduction was occurring due to interaction with the surfactant. As the surfactant is consumed, the isooctane–aqueous system becomes unstable, and phase-separates into aqueous and organic regions.

2.1.2
Spherical Particles (Experiment B)

NaBH$_4$ (0.01 M) and HAuCl$_4$ (0.015 M) solution were added in alternate installments to a solution of isooctane/AOT (0.8 M)/Lecithin (0.4 M). Each installment increased the W_0 by 10. The system was taken to various W_0 values (from 40 to 100) by altering the aqueous phase content (to maintain the stoichiometric ratio of the reactants) and the reaction was allowed to proceed at

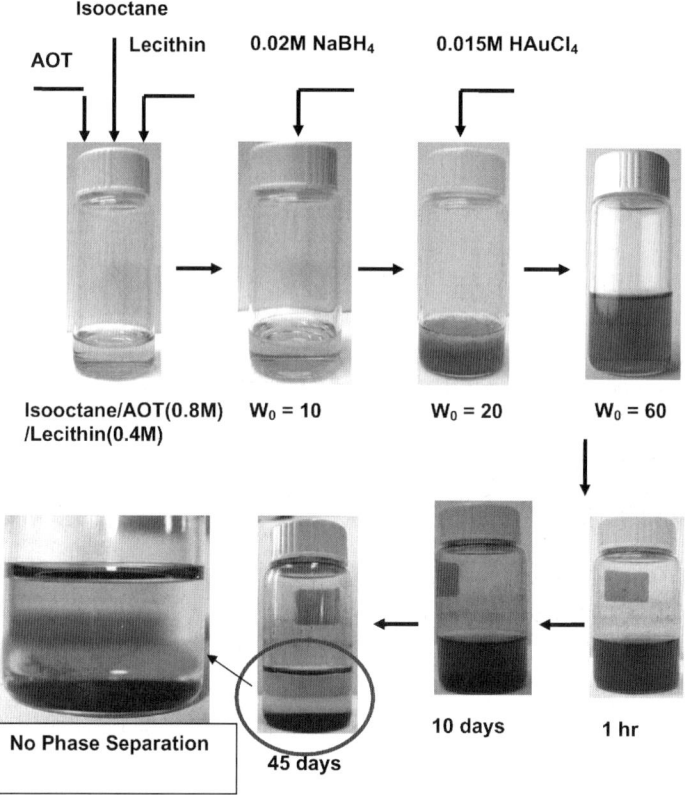

Fig. 3 Synthesis of spherical nanoparticles showing changes in the sample with time. There was no phase separation observed in the isooctane/AOT (0.8 M)/Lecithin (0.4 M)/NaBH$_4$ (0.01 M)/HAuCl$_4$ (0.01 M) sample after 45 days

room temperature (Fig. 3). The color of the solution turned red indicating the formation of gold nanoparticles.

2.1.3
Free Precipitation (Experiment C)

In order to observe the size and shape of the particles formed by free precipitation, a control sample was prepared with 0.015 M of HAuCl$_4$ solution to which gradually the reducing agent NaBH$_4$ (0.01 M) was added drop wise and mixed well. The yellow solution gradually transformed into a purple solution, indicating gold nanoparticle formation.

These nanoparticles were washed with isooctane to remove the excess surfactant. Ethanol was then added into the isooctane and nanoparticle mixture. The nanoparticles assembled at the interface. They were extracted from the interface and re-suspended in ethanol before being observed and analyzed.

3
Results and Discussion

The TEM results for experiment A reveals nanoplates of gold (see Fig. 4) with triangular and hexagonal facets and some that are nearly spherical in shape. Since the contrast from the particles at a wide range of tilt angles is monotonic, the triangular and hexagonal particles are thin (~ 20 nm), and allow electron transparency only in one dimension. The electron diffraction for a single triangular nanoplate (see Fig. 4c) suggests that these plate-like, highly

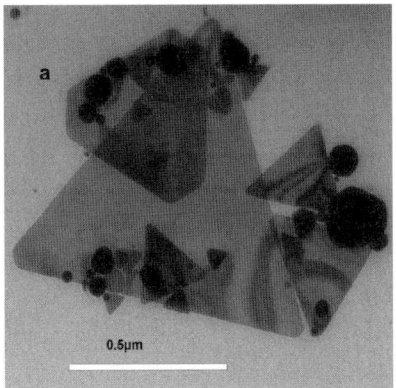
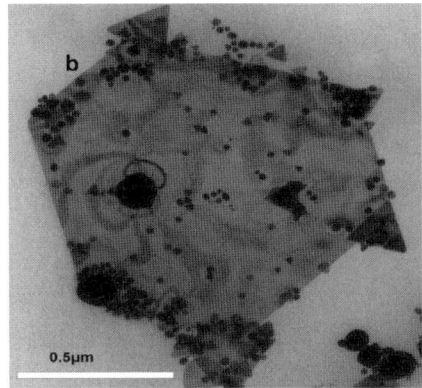

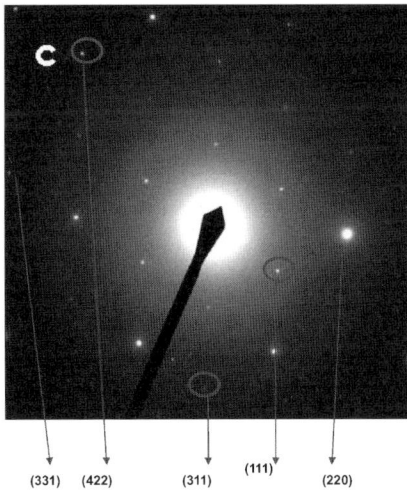

Fig. 4 Transmission electron micrographs of **a** highly facetted mostly triangular gold particles, **b** a hexagonal particle, **c** electron diffraction pattern of the triangular particle showing that it is a single crystal. Diffraction from the (111), (220), ($\bar{3}$11), (331), (422) planes are identified

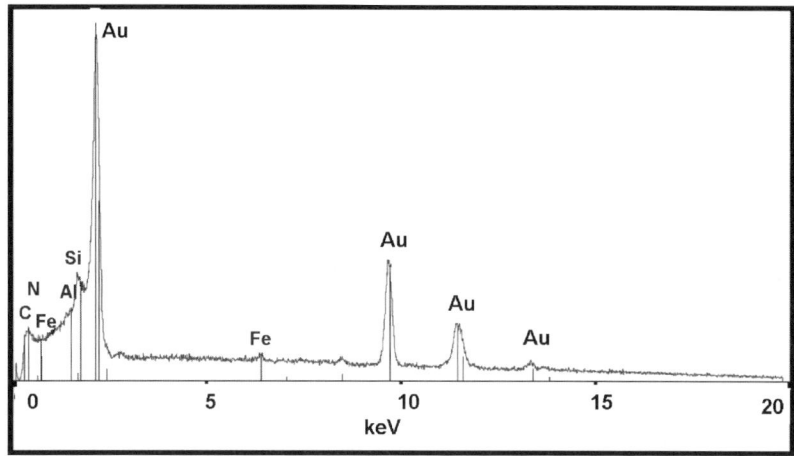

Fig. 5 Energy dispersive X-ray analysis for the facetted gold particles showing that the highly facetted triangular nanoparticles are metallic gold

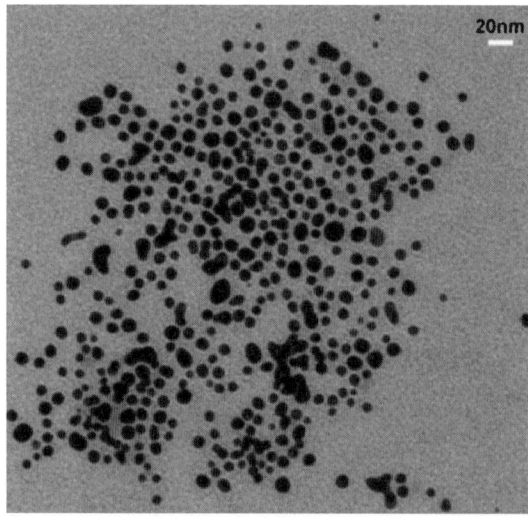

Fig. 6 Spherical gold nanoparticles formed by the reaction of the reducing agent, $NaBH_4$ with $HAuCl_4$ in the Isooctane/AOT (0.8 M)/Lecithin (0.4 M) system W_0 60

facetted structures are single crystals. The lateral dimensions of the particles range from 0.1–1 µm, with a wide dispersity. Composition analysis for these facetted particles using EDS showed that these particles were metallic gold (Fig. 5).

In contrast, carrying out the synthesis by adding (0.01 M) $NaBH_4$ to the mesophase mixture (experiment B) suppressed faceting and resulted in predominantly spherical particles of diameter in the range of 3.5–9 nm for W_0 60 (Fig. 6). In these experiments, there was no phase separation observed in the surfactant mesophase, indicating that surfactants were not being consumed. The reduction of the gold ions by $NaBH_4$ is the main mechanism for metal nanoparticle formation. The lack of any observable agglomeration is caused partly by the high viscosity of the underlying template that lowers the mobility of the particles, as well as a stabilizing absorbed layer of surfactant on the metal particles.

Adding $NaBH_4$ in a drop-by-drop fashion directly to the $HAuCl_4$ solution (experiment C) resulted in irregularly shaped nanoparticles with a size range

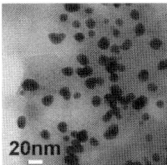

Fig. 7 Irregularly shaped particles form during the "free precipitation" reaction of $HAuCl_4$ with $NaBH_4$

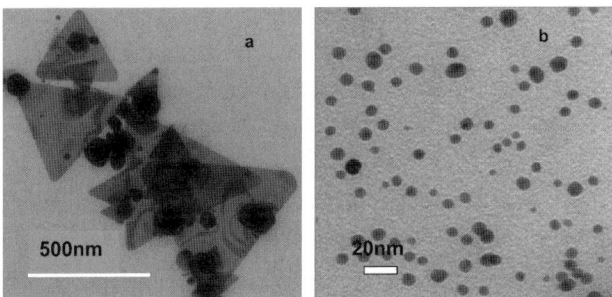

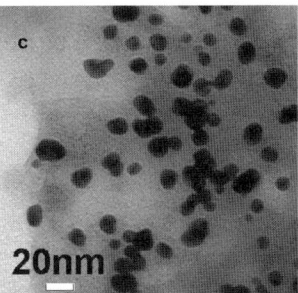

Fig. 8 a Facetted particles formed without reducing agent, **b** spherical particle produced by the use of a reducing agent, **c** irregular shaped particles obtained by free precipitation

of 10–20 nm (Fig. 7). Figure 8 compares the different nanoparticles obtained using alternate strategies for synthesis.

We proposed an empirical model to explain the formation of the facetted nanoparticles (see Fig. 9). In experiment A, the dioctyl sulfosuccinate termini of AOT have enough electron donating power to reduce the $HAuCl_4$. Following this electron donation, the surfactants adsorb preferentially onto specific crystallographic planes, and restrict growth of those planes. The faster growing planes disappear because they "grow out". The diffraction patterns obtained for a single triangular plate were indexed for the (111), (220), ($\bar{3}$11), (331), (422) crystal planes. Tilting in bright field caused the sample to become dark, again indicative of the single-crystal nature of the particles. Since the

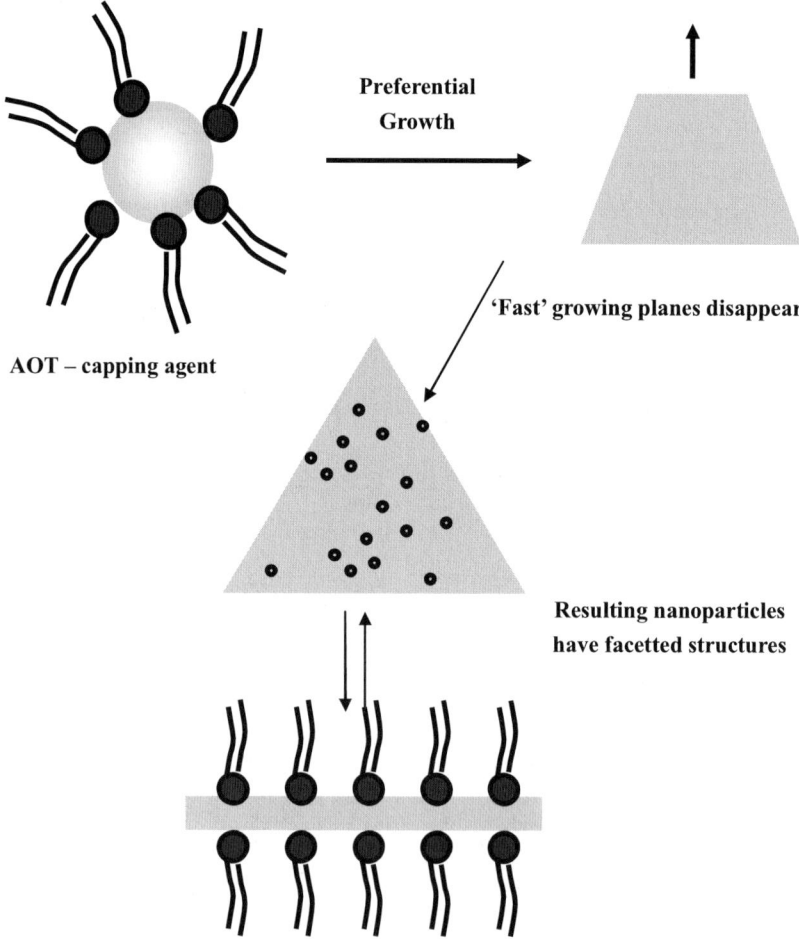

Fig. 9 Schematic of the proposed theory behind the formation of the facetted nanoparticles

crystal faces of the molecularly passivated planes are pinned, growth mainly occurs by step flow through attachment of atoms to step edges. As AOT is consumed for metal ion reduction and capping of the platelets, the mesophase surfactant mixture becomes unstable and phase separates. The surfactant-capped gold particles remain at the oil–water interface in the phase-separated system as the hydrophobic tail group of the surfactant favors remaining in the oil medium. Once the particles grow enough, their weight increases, capillary forces are no longer capable of retaining the particles at the interface and they sediment.

The TEM results obtained for the nanoparticles of the experiments in set (B), revealed that the particle size could be controlled by controlling the W_0 of the sample (Fig. 10). For W_0 60, the particle size was approximately 5 to 9 nm. For W_0 100 the nanoparticles were 15–20 nm. In order to determine their three-dimensional structure, stage tilting experiments were performed. Figure 11 shows the structure of the nanoparticles at $0°$ and $20°$ stage tilt, respectively. There was no difference in the shape of the particles indicating that the particles are essentially spherical and not disc-type structures.

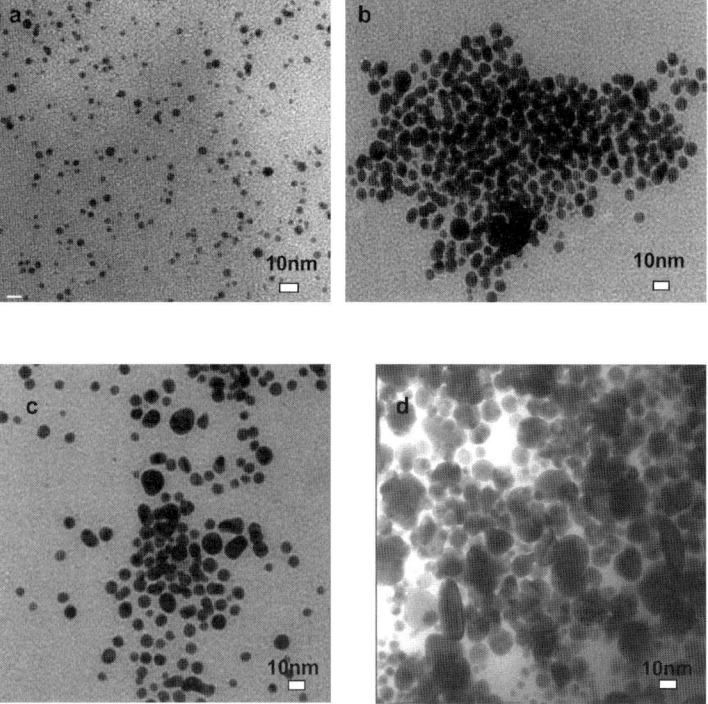

Fig. 10 Transmission electron micrographs showing the control of the size of the gold nanoparticles synthesized using $NaBH_4$ (0.02 M) as the reducing agent by altering the W_0.
a $W_0 = 40$, **b** $W_0 = 50$, **c** $W_0 = 60$, **d** $W_0 = 100$

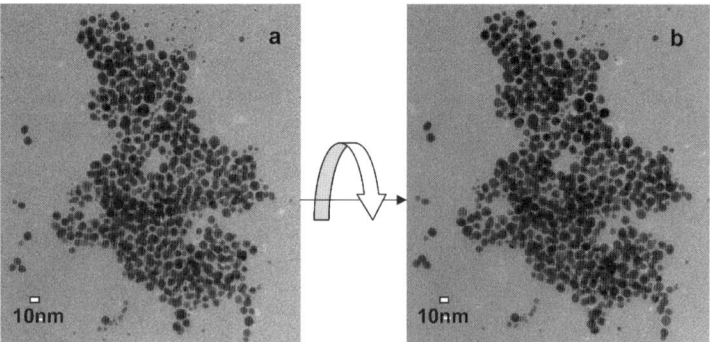

Fig. 11 Stage tilting of the nanoparticles showing that they are spherical in nature, **a** 0°, **b** +20°

4
Conclusions

Gold nanoparticles of different shapes were synthesized in the same system with and without the use of reducing agent. The surfactants act as shape regulators due to their selective adsorption to the nanoparticle surfaces resulting in different morphologies of the nanoparticles. Size control over the capped nanoparticles was obtained by altering the aqueous phase content. More relevant information can be found in [55–62].

Acknowledgements We express our sincere gratitude to University of Rhode Island Transportation Center and Honda Motors for their financial support for this work. We would like to thank Paul Johnson, Evan Wujick, for their support throughout the work.

References

1. Favier F, Walter EC, Zach MP, Benter T, Penner RM (2001) Hydrogen Sensors and Switches from Electrodeposited Palladium Mesowire Arrays. Science 293:2227–2231
2. Link S, El-Sayed M (2000) Shape and Size Dependence of Radiative, Non-Radiative and Photochemical Properties of Gold Nanocrystals. Int Rev Phys Chem 19:409–453
3. El-Sayed MA (2001) Some Interesting Properties of Metals Confined in Time and Nanometer Space of Different Shapes. Acc Chem Res 34(4):257–264
4. Sun Y, Mayers B, Xia Y (2003) Transformation of Silver Nanospheres into Nanobelts and triangular Nanopaltes therough a Thermal Precess. Nano Lett 3:675–679
5. Maillard M, Giorgio S, Pileni M-P (2003) Tuning the Size of Silver Nanodisks with Similar Aspect Ratios: Synthesis and Optical Properties. J Phys Chem 107:2466–2470
6. Schatz GC, Hupp JT, Kelley KL, Hao E (2002) Synthesis of Silver Nanodisks using Polystyrene Mesospheres as Templates. J Am Chem Soc 124:15182–15183
7. Van Duyne RP, Haes AJ, Zou S, Schatz GC (2004) A Nanoscale Optical Biosensor: The Long Range Distance Dependence of the Localized Surface Plasmon Resonance of Noble Metal Nanoparticles. J Phys Chem B 108:109–116

8. Mirkin CA, Millstone JE, Park S, Shuford KL, Qin L, Schatz GC (2005) Observation of a Quadrupole Plasmon Mode for a Colloidal Solution of Gold Nanoprisms. J Am Chem Soc 127:5312–5313
9. Khlebtsov NG, Trachuk LA, Mel'nikov AG (2005) The Effect of the Size, Shape, and Structure of Metal Nanoparticles on the Dependence of Their Optical Properties on the Refractive Index of a Disperse Medium. Opt Spectrosc 98:77–83
10. Chen S, Wang ZL, Ballato J, Fougler SH, Carroll DL (2003) Monopod, Tripod and Tetrapod Gold nanoparticles. J Am Chem Soc 125:16186–16187
11. Pradhan N, Pal A, Pal T (2001) Catalytic Reduction of Aromatic Nitro Compounds by Coinage Metal Nanoparticles. Langmuir 17:1800–1802
12. Han M, Gao X, Su JZ, Nie S (2001) Quantum-dot-tagged microbeads for multiplexed optical coding of biomolecules. Nat Biotechnol 16:631–635
13. Yun WS et al. (2005) Size-Controlled Synthesis of Machinable Single Crystalline Gold Nanoplates. Chem Mater 17(22):5558–5561
14. Nicewarner-Pena SR et al. (2001) Submicrometer Metallic Barcodes. Science 294:137–141
15. Jaramillo TF, Baeck S-H, Cuenya BR, McFarland EW (2003) Catalytic Activity of Supported Au Nanoparticles Deposited from Block Copolymer Micelles. J Am Chem Soc 125:7148–7149
16. Xin X, Luo G, Zhao R (2005) Advances in preparation and application of supported gold nano-particles catalyst with high catalytic activity. Shiyou Huagong 34:898–902
17. Gates BC, Fierro-Gonzalez JC (2005) Catalysis by supported gold: Roles of cationic and zerovalent gold. Abstracts of Papers, 229th ACS National Meeting, San Diego, CA, USA, March 13–17, 2005, PETR-021
18. Maye MM, Luo J, Han L, Kariuki N, Rab Z, Khan N, Naslund HR, Zhong C-J (2004) Gold and alloy nanoparticle catalysts in fuel cell reactions. Div Fuel Chem 49:938–939
19. Raj CR, Okajima T, Ohsaka T (2003) Gold nanoparticle arrays for the voltammetric sensing of dopamine. J Electroanal Chem 543:127–133
20. Wang E, Cheng W, Dong S (2002) Gold nanoparticles As Fine Tuners of Electrochemical Properties of Electrode/Solution Interface. Langmuir 18(25):9947–9952
21. Evans SD, Johnson SR, Mahon SW, Ulman A (1997) Synthesis and Characterization of Surfactant-Stabilized Gold Nanoparticles. Supramol Sci 4:329–333
22. Lev O, Neiman B, Eli G (2001) Use of Gold Nanoparticles to Enhance Capillary Electrophoresis. Anal Chem 73:5220–5227
23. Jin R, Cao Y, Mirkin CA, Kelly KL, Schatz GC, Zheng JG (2001) Photoinduced Conversion of Silver Nanospheres to Nanoprisms. Science 294:1901–1903
24. Huynh WU, Dittmer JJ, Alivisatos AP (2004) Hybrid Nanorod-Polymer Solar Cells. Science 295(5564):2425–2427
25. McConnell WP et al. (2000) Electronic and Optical Properties of Chemically Modified Metal Nanoparticles and Molecularly Bridged Nanoparticle Arrays. J Phys Chem B 104:8925–8930
26. Hayat M (1989) Gold: Principles, Methods, and Applications, Vol 1. Academic, London
27. Sastry M, Shankar SS, Rai A, Ankamwar B, Singh A, Ahmad A (2004) Biological Synthesis of Triangular Gold Nanoprisms. Nat Mater 3:482–488
28. Regan MR, Banerjee IA (2005) Bioinspired Preparation of Germania Supported Nanocatalysts. Abstracts, 33rd Northeast Regional Meeting of the American Chemical Society, Fairfield, CT, USA, July 14–17, ONSUB-041
29. Chiang CL (2001) Controlled Growth of Gold Nanoparticles in AOT/C12E4/Isooctane Mixed Reverse Micelles. J Colloid Int Sci 239:334–341

30. Sau TK, Pal A, Jana NR, Wang ZL, Pal T (2001) Size Controlled Synthesis of Gold Nanoparticles Using Photochemically Prepared Seed Particles. J Nanoparticle Res 3:257–261
31. Chiang C-L, Hsu M-B, Lai L-B (2004) Control of Nucleation and Growth of Gold Nanoparticles in AOT/Span80/Isooctane Mixed Reverse Micelles. J Solid State Chem 177:3891–3895
32. Sakai T, Alexandridis P (2005) Spontaneous Formation of Gold Nanoparticles in Poly(ethylene oxide)-Poly(propylene oxide) Solutions: Solvent Quality and Polymer Structure Effects. Langmuir 21:8019–8025
33. Alexandridis P (2004) Nanoparticle Synthesis and Colloidal Stabilization using Amphiphilic Block Copolymer Solutions. Abstracts, 32nd Northeast Regional Meeting of the American Chemical Society, Rochester, NY, USA, October 31–November 3, GEN-119
34. Alexandridis P, Sakai T (2004) Amphiphilic block copolymer solutions as media for the facile synthesis and colloidal stabilization of metal nanoparticles. Abstracts of Papers, 228th ACS National Meeting, Philadelphia, PA, USA, August 22–26, 2004, PMSE-510
35. Alexandridis P, Sakai T (2005) Amphiphilic block copolymer-templated nanoparticle synthesis and stabilization. Abstracts of Papers, 229th ACS National Meeting, San Diego, CA, USA, March 13–17, 2005, COLL-474
36. Sakai T, Alexandridis P (2004) Single-step synthesis and stabilization of metal nanoparticles in aqueous Pluronic block copolymer solutions at ambient temperature. Langmuir 20:8426–8430
37. Van der Zande BMI, Böhmer MR, Fokkink LGJ, Schönenberger C (2000) Colloidal Dispersions of Gold Rods: Synthesis and Optical Properties. Langmuir 16:451–458
38. Govindaraj A, Satishkumar BC, Nath M, Rao CNR (2000) Metal Nanowires and Intercalated Metal Layers in Single-Walled Carbon Nanotube Bundles. Chem Mater 12:202–205
39. Cepak VM, Martin CR (1998) Preparation and Stability of Template-Synthesized Metal Nanorod Sols in Organic Solvents. J Phys Chem B 102:9985–9990
40. Esumi K, Matsuhisa K, Torigoe K (1995) Preparation of Rod like Gold Particles by UV Irradiation Using Cationic Micelles as a Template. Langmuir 11:3285–3287
41. Murphy CJ, Jana NR, Gearheart L (2001) Wet Chemical Synthesis of High Aspect Ratio Cylindrical Gold Nanorods. J Phys Chem B 105:4065–4067
42. Murphy CJ, Jana NR (2002) Controlling the Aspect Ratio of Inorganic Nanorods and Nanowires. Adv Mater 14:80–82
43. Cao L, Liu Z, Zhu T (2006) Formation Mechanisms of Non-Spherical Gold Nanoparticles During Seeding Growth: Roles of Anion Adsorption and Reduction Rate. J Colloid Interface Sci 293:67–69
44. Huang MH, Kuo C-H (2005) Synthesis of Branched Gold Nanocrystals by a Seeding Growth Approach. Langmuir 21:2012–2016
45. Murphy CJ, Sau TK, Gole AM, Orendorff CJ, Gao J, Gou L, Hunyadi SE, Li T (2005) Anisotropic Metal Nanoparticles: Synthesis, Assembly and Optical Applications. J Phys Chem B 109:13857–13870
46. Hsu H-Y, El-Sayed M, Eustis S (2004) Photochemical Synthesis of Gold nanoparticles with Interesting Shapes. NIN REU Res Accomp, pp 68–69
47. Yang P, Song JH, Kim F (2002) Photochemical Synthesis of Gold Nanorods. J Am Chem Soc 124:14316–14317
48. Wang CRC, Yu Y-Y, Chang S-S, Lee C-L (1997) Gold Nanorods: Electrochemical Synthesis and Optical Properties. J Phys Chem B 101:6661–6664

49. Hupp JT, Schatz GC, Hao E, Bailey RC, Li S (2003) Synthesis and Optical Properties of "Branched" Gold Nanocrystals. Nano Lett 4:327–330
50. Willner I, Shlyahovsky B, Pavlov V, Popov I, Xiao Y (2005) Shape and Color of Au Nanoparticles Follow Biocatalytic Processes. Langmuir 21:5659–5662
51. Chen S, Carroll DL (2002) Synthesis and Characterization of Truncated Triangular Silver Nanoplates. Nano Lett 2:1003–1007
52. Mirkin CA, Jin R, Cao YC, Hao E, Metraux GS, Schatz GC (2003) Controlling Anisotropic Nanoparticle Growth Through Plasmon Excitation. Nature 425:487–490
53. Dong S, Jin Y, Shao Y (2004) Synthesis of Gold Nanoplates by Asparate Reduction of Gold Chloride. Chem Commun
54. Simmons BA, Irvin GC, Agarwal V, Bose A, John VT, McPherson GL, Balsara NP (2002) Small Angle Neutron Scattering Study of Microstructural Transitions in a Surfactant-Based Gel Mesophase. Langmuir 18:624–632
55. Halder A, Ravishankar N (2007) Ultrafine single-crystalline gold nanowire arrays by oriented attachment. Adv Mater 19(14):1854–1858
56. Halder A, Ravishankar N (2006) Gold Nanostructures from Cube-Shaped Crystalline Intermediates. J Phys Chem B 110(13):6595–6600
57. Wang L-Y, Chen X, Zhan J, Sui Z-M, Zhao JK, Sun ZW (2004) Controllable morphology formation of gold nano- and micro-plates in amphiphilic block copolymer-based liquid crystalline phase. Chem Lett 33(6):720–721
58. Uwada T, Asahi T, Masuhara H, Ibano D, Fujishiro M, Tominaga T (2007) Multiple resonance modes in localized surface plasmon of single hexagonal/triangular gold nanoplates. Chem Lett 36(2):318–319
59. Huang W-L, Chen C-H, Huang MH (2007) Investigation of the Growth Process of Gold Nanoplates Formed by Thermal Aqueous Solution Approach and the Synthesis of Ultra-Small Gold Nanoplates. J Phys Chem C 111(6):2533–2538
60. Wei H, Wang EK (2007) Submicrometre scale single-crystalline gold plates of nanometre thickness: synthesis through a nucleobase process and growth mechanism. Nanotechnology 18(29):295603/1–295603/5
61. Kawasaki H, Yonezawa T, Nishimura K, Arakawa R (2007) Fabrication of submillimeter-sized gold plates from thermal decomposition of $HAuCl_4$ in two-component ionic liquids. Chem Lett 36(8):1038–1039
62. Luo YL (2007) Preparation of single-crystalline gold microplates on a large scale by heating a $HAuCl_4$-tartaric acid aqueous solution. Mater Lett 61(1):134–136

Colloid Chemistry: The Fascinating World of Microscopic Order

Amelie Zapf · Heinz Hoffmann (✉)

University of Bayreuth, Germany
heinz.hoffmann@nmbgmbh.de

Abstract In this work, we explain how small surfactant molecules with dimensions of one to two nanometers, when dissolved in water with a concentration of a few weight percent, organize themselves spontaneously into various structures with a long-range order over macroscopic dimensions of several centimeters, even though the molecules are all in the liquid state. It follows that two molecules that are more than a million times their main length apart still point, on average, in the same direction in three dimensional space.

The organized structures give to the aqueous phases new macroscopic properties like iridescent colors, viscoelasticity, gel character, a yield stress, and, between crossed polarizers, beautifully colored patterns that make the order in the samples visible. The self-organization of the surfactant molecules is simply a result of the hydrophobic and electrostatic interaction between the individual molecules and the micellar structures. The size of the micellar structures, as in the case of small unilamellar vesicles, can be extremely monodisperse, even though one vesicle consists of hundreds of surfactant molecules.

Nanotechnology became a buzzword during the past decade and, today, remains a focal point of developing technologies. The ability to create things that are smaller and smaller has always been a source of fascination for engineers and scientists alike. However, most people continue to associate nanotechnology with state-of-the-art computer processors and atomic force microscopes, overlooking the fact that the microscopic structures discovered through colloid chemistry predate these recent scientific advances by decades.

The results of colloid science play a crucial role in our everyday lives: From hair spray to shower gel, from the lacquer we are using to repaint our front doors to the LCD displays on our new computers, colloidal and liquid crystalline systems are omnipresent in our immediate surroundings.

Colloid chemistry investigates substance mixtures. These substance mixtures can be heterogenous, such as emulsions (in which tiny droplets of one liquid are dispersed in another), suspensions (consisting of a fine dispersion of solid particles in a liquid volume phase), and aerosols (in which liquid droplets are dispersed in the gas phase). However, there are also homogenous mixtures in which the solute is present in larger, supermolecular aggregates. These homogenous mixtures include micellar solutions and liquid crystalline phases that are formed when the molecule of the solute is amphiphilic. Amphiphilic molecules generally have the shape of elongated chains, one end of which mixes with water, and is thus called hydrophilic (water-loving). The other end of the chain is oil-mixable and therefore called hydrophobic (water-fearing). When such a substance is dissolved in water, globular, rodlike, or disklike aggregates, called micelles, are formed in which the hydrophilic ends point to the outside, creating a shell around a core made of the hydrophobic tails.

Inside this core, substances insoluble in pure water can be enclosed and dissolved, which makes for the detergent properties of amphiphilics. Disklike micelles can also fuse with one another to form liquid crystals of extended planar lamellae consisting of two layers of amphiphilic molecules with their respective hydrophilic headgroups pointing toward the outside of the bilayers, which are oriented parallel to one another or interconnected to form large networks. Surfactant molecules within these structures are in dynamic equilibrium with those in the solvent – there is a continuous exchange of molecules entering and exiting the structures.

Most substances colloid chemists concern themselves with, however, are inconspicuous to the naked eye: they are solutions ranging from water-clear to extremely turbid. Some of these solutions, however, have very interesting properties, like the iridescent phases that are obtained when lamellar liquid crystalline phases are diluted with water until the distance between neighboring lamellae is in the size range of the wavelength of visible light [1]. Incident light is then refracted at the lattice of the liquid crystal and the solution appears colored, much like a CD does when light is shone upon it, although it is, in fact, entirely colorless. Furthermore, the colors depend on the angle from which the sample is viewed. Thus, a solution can appear green on one side, and red on the other. A collection of such phases can be seen in Fig. 1.

Another example of highly-ordered structure in a lamellar phase is obtained when the lamellae are oriented parallel to the walls of the test tube. Thus, all membranes form concentric cylinders [2]. This is all the more remarkable because the membranes themselves are only 2–3 nm thick, whereas the distance between two lamellae is about 20–50 nm. Although the surfactant molecules themselves are in a fluid state, their arrangement in the sample is of extremely high order, as Fig. 2 (taken between crossed polarizing foils) shows. The "stripes" in the photograph represent the cylindrical lamellar stacks. Polarizers are widely used in the investigation of liquid crystalline phases, as structures that are oriented in space appear bright between crossed polarizers. The waves of the light that can pass through such foils are all swinging in one direction. Thus, if the foils are "crossed" at an angle of 90°, no light can pass these foils if its wave is not rotated by passing through ordered structures between the polarizers.

When colloidal systems are placed under a microscope, a whole new world of sometimes astounding beauty opens up. Optical microscopes allow for

Colloid Chemistry: The Fascinating World of Microscopic Order

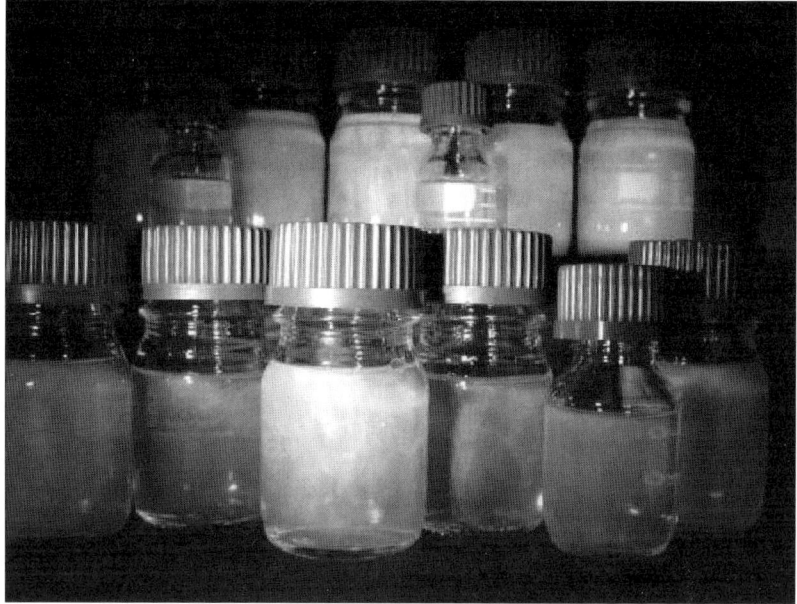

Fig. 1 A selection of iridescent surfactant solutions

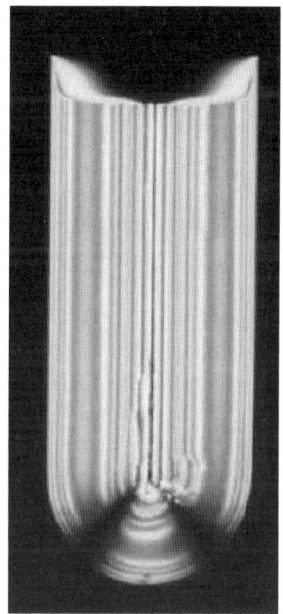

Fig. 2 A highly-ordered lamellar phase between crossed polarizers

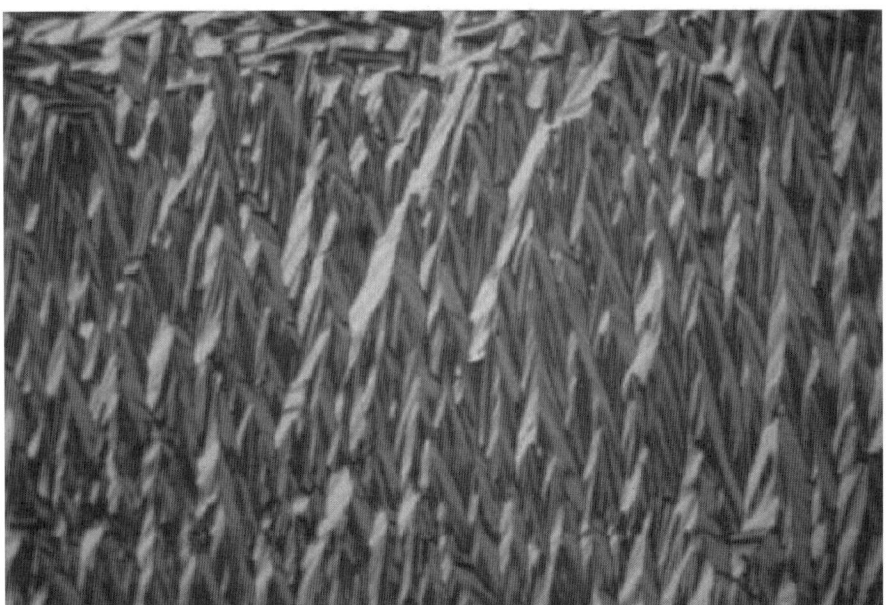

Fig. 3 Lamellar focal conics under the polarization microscope using color contrast by means of a λ mask

a magnification of up to about 1000 times. In this size range, larger colloidal particles start to become visible. The polarization microscope combines the resolution of the optical microscope with the benefits of crossed polarizers.

Lamellar focal conics show a fascinating highly-ordered structure when observed under the polarization microscope. This texture consists of surfactant bilayers that are shaped like ice cream cones and stuck inside of one another. These stacks of cones are quite densely packed in the solution and, under the polarization microscope, create extended regions of amazing regularity. Figure 3 shows a photograph of such a system, taken with a polarization microscope with a λ mask to achieve color contrast.

What is most amazing of all in this picture is the degree of microscopic order present in a solution that appears quite unexceptional to the unaided eye. Usually, we associate "beauty" and aesthetic appeal with symmetry and regular shapes, just as in the examples of the ordered lamellar phase and lamellar focal conics. However, sometimes also asymmetric shapes have that special quality about them that conveys what we call beauty. Figure 4 shows a water-rich foam composed of dish soap with coconut oil. It consists of tightly-packed bubbles of very different sizes that create an asymmetric pattern of astounding beauty [3].

When lamellar phases are sheared, e.g. by flowing through a narrow tube, the membranes are disrupted and the resulting fragments close to form spherical shells, termed vesicles [4]. These vesicles can consist of a single shell

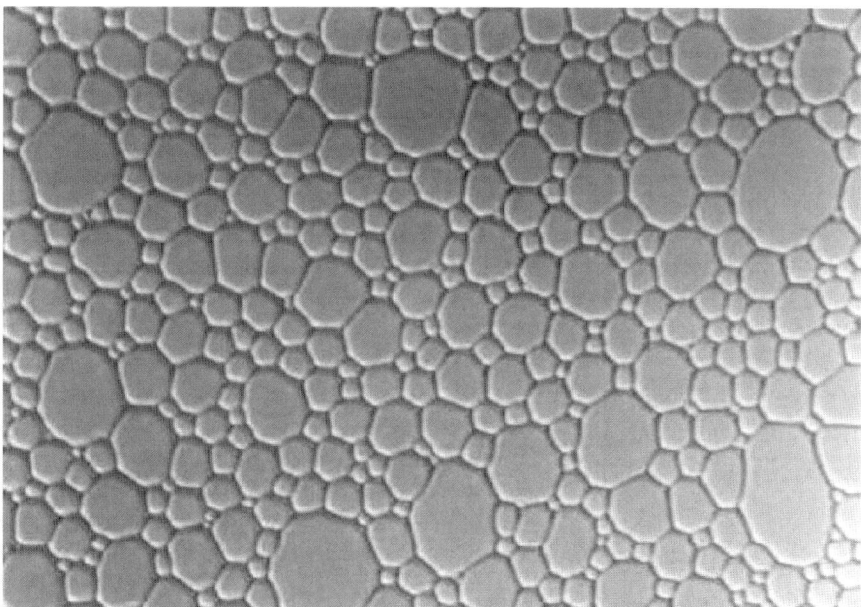

Fig. 4 A water-filled microscopic foam composed of dish soap, coconut oil, and water

or several, layered inside of one another, like the shells of an onion. They were first described in 1970 in dispersions of double-chain phospholipids, but have subsequently been found in a variety of mixtures of single-chain surfactants with cosurfactants or other surfactants. Frequently, vesicles are metastable and, over time, change their size or revert to the planar lamellar phase once again.

The invention of the electron microscope in the 1930s by Knoll and Ruska cleared the way for scientists to take an even closer look at vesicles and other colloidal structures [5]. Improving the resolution of the optical microscope roughly by the factor that the optical microscope improved that of the unaided eye, the finer structures of colloidal systems became visible. With the electron microscope, single bilayers can be made visible and the distance between lamellae can be determined. Thus, the structure of a given system can be determined to up to 1/10 000 000 of a millimeter, which is about the distance of six atoms in a molecule. The most impressive results are obtained with the freeze fracture and cryo-TEM methods [6].

Figure 5 shows small multilamellar vesicles under the electron microscope, visible as concentric spheres. Note the fascinating texture of the spheres – the actually fluid bilayer appears structured, frozen in time at the moment of preparation.

If the electrostatic forces between charged surfactant head groups are sufficiently high, vesicles can also be a thermodynamically stable phase and be

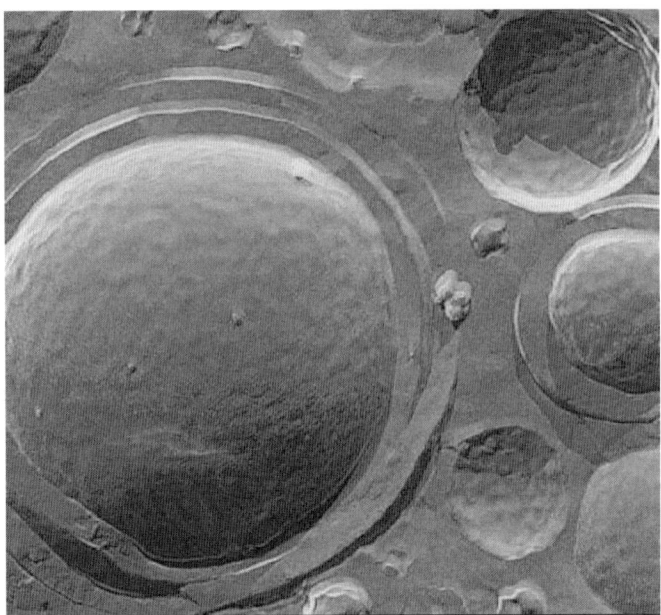

Fig. 5 Electron micrograph of small multilamellar vesicles

formed without shear. This can be proven by altering the head group charge of a micellar solution at rest, e.g. by addition of acid, whereupon the formation of vesicles occurs. Vesicles prepared in this manner often have a very uniform size throughout the sample so that they can form a cubic crystal lattice with a high degree of long-range order (cf. Figure 6) [7].

Vesicles are, first of all, aesthetically pleasing, but are also of considerable technical value, as they enclose a volume of solvent and separate it from the bulk solvent. Thus, they can be used as mini-reactors for different applications. The volume enclosed in vesicles can be custom-tailored to the application over more than nine orders of magnitude, as the vesicle radius can be varied from about 40 nm to 10 µm.

Another application of vesicles is as an agent of drug delivery in medicine. The pharmaceutical is enclosed inside the vesicle (also termed liposome) that fuses with the cell membrane and delivers its contents directly into the cell.

A third technical application is to control the thickness of solutions by means of vesicles: A solution of tightly-packed vesicles behaves like a viscoelastic fluid. Frequently, such solutions have a yield stress value, which means that a certain amount of force is necessary to make these solutions flow. Air bubbles in such fluids do not rise to the top (cf. Figure 7). Such a behavior is obtainable with an addition of only 1–2% of surfactant.

Colloid chemists do not only rely on optical methods and rheology to obtain information about the structure of their samples. Spectroscopy, in which

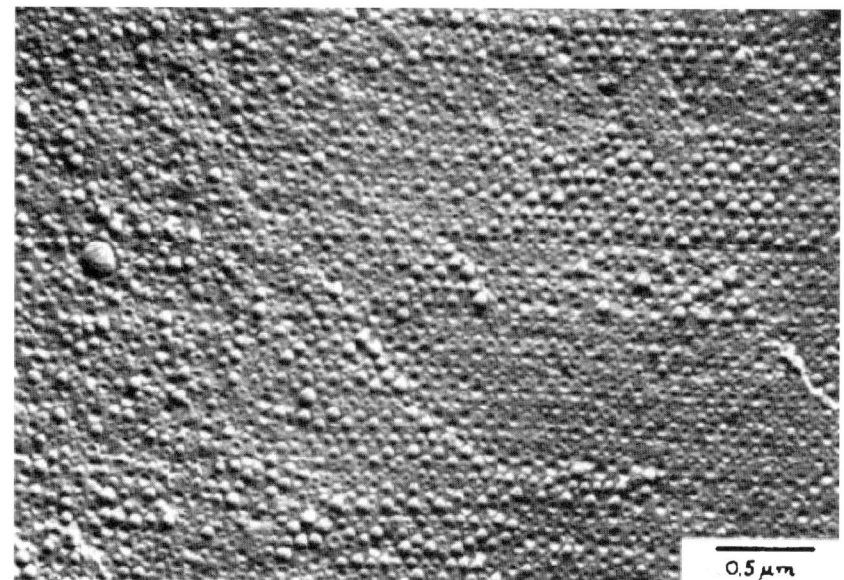

Fig. 6 A cubic phase composed of small unilamellar vesicles

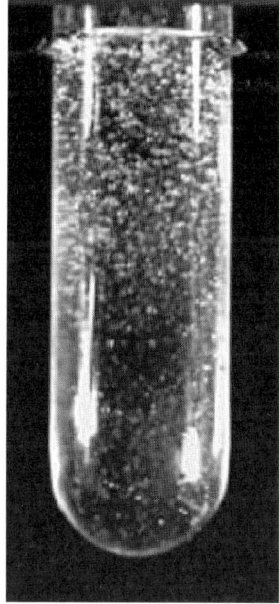

Fig. 7 Shower gel with a yield stress value. The air bubbles suspended in the solution do not rise to the top [8]

the interaction of radiation of varying wavelength with the sample is studied, is of equal importance to deliver structural data about the substance investigated. A particularly interesting field is the scattering of neutrons or X-rays on the sample, as their wavelength is so short that it allows for a very high resolution [9]. However, the scattering spectra obtained with these techniques are not actual images of the sample, but images representing the structure of the sample.

From these spectra, one can calculate information about the microscopic structure using mathematical transformations. Nonetheless, the scattering spectra preserve the symmetry of the system and are little works of art in themselves. Figure 8 shows an example, the neutron scattering pattern of a typical lamellar phase. Note the asymmetry of the spectrum: it indicates that there is a preferred ordering direction in the system. From the distance of the bright circles to the center, the distance between two lamellae can be calculated.

However, the methods presented in this article do not give the "big picture" of techniques available to colloid scientists. A host of methods had to be left out for brevity's sake, including fields as important as rheology [10], which concerns itself with the flow characteristics of liquids, measurements of the electric conductivity, and surface tension, to name but a few.

Combined, these techniques provide a powerful means to determine the relationship between microscopic structure and macroscopic properties of

Fig. 8 Small-angle neutron scattering pattern of a lamellar phase

colloidal systems; thus leading scientists to new insights into the interaction of supermolecular aggregates with one another and enabling the industry to fine-tune the physico-chemical properties of their products to the customer's needs. The improvements made in the last years on products like detergents, paints, cosmetic articles, crop protection agents and even pharmaceuticals, most notably with new ways of drug delivery using liposomes, would be unthinkable without the continued progress of colloid science. In fact, the standard of living we are accustomed to would not be possible had it not been for the progress of colloid chemistry. When one adds to this unquestionable significance the rich scientific and aesthetic rewards that colloid chemists can gain from their work, one must come to the conclusion that colloid science, with its long history and wide scope of industrial application, remains a fascinating field for professional chemists and the larger public alike.

References

1. Thunig C, Hoffmann H, Platz G (1989) Iridescent colors in surfactant solutions. Progr Colloid Polym Sci 79:297–307
2. Wolf C (2007) Thesis, Bayreuth
3. Hoffmann H, Ebert G (1988) Surfactants, micelles and fascinating phenomena. Angew Chem Int Ed 27:902–912
4. Escalante J, Gradzielski M, Mortensen K, Hoffmann H (2000) The shear induced transition of an originally undisturbed lamellar phase to a vesicle phase. Langmuir 16(23):8653
5. Knoll M, Ruska E (1932) Das Elektronenmikroskop. Z Phys 78:318–339
6. Bellare JR, Davis HT, Scriven LE, Talmon Y (1988) Controlled environment vitrification system: an improved sample preparation technique. J Electron Microsc Techn 10:87–111
7. Gradzielski M, Bergmeier M, Müller M, Hoffmann H (1997) Novel gel phase: A cubic phase of densely packed monodisperse, unilamellar vesicles. J Phys Chem B 101(10):1719–1722
8. Hoffmann H, Rauscher A (1993) Aggregating systems with a yield stress value. Colloid Polym Sci 271:390–395
9. Liu YC, Baglioni P, Teixeira J, Chen SH (1994) Structure and Interaction of lithium dodecyl sulfate micelles in the presence of Li-specific macrocyclic cage: a study by SANS. J Phys Chem 98(40):10208–10215
10. Hoffmann H (1994) Viscoelastic surfactant solutions. In: Herb CA, Prud'homme RK (eds) ACS-Symposium Ser 578, pp 1–31

Subject Index

Acetals, cyclic 75
Adsorption isotherm 164
Adsorption theories 47
Aerosol-T (AOT) 17
–/graphite 173
Aggregation number (hemimicellar number) 179
Alcohol ethoxylate tetra(ethylene glycol)monooctyl ether 67
Alcohol ethoxylates 60
Alkali dodecylsulfates 26, 44
Alkanolamines 69
Alkene-1-sulfonate 155
Alkyl benzene sulfonate 155
Alkyl chloroformate 73
Alkyl olefinic sulfonate 155
Alkyl sulfates 58
Alkylbenzene sulfonates 59
Alkylphenol ethoxylates 59
Amide bonds 74
Amide surfactants 75
Aquatic toxicity 60

Betaine esters 70
Biodegradation 61
Bis-2-hydroxyethylammonium chloride, dicetylester 69
Butler 27

CAC 152
Cadmium sulfide (CdS) 213
Candida antarctica lipase B (CALB) 66, 73
Capillary pressure, meniscus profile 136
Carbonate bonds 73
Carbonate surfactants 73
CdS nanoparticles, mercaptopropionic acid-capped 217
CdSe core-shell QDs, ZnSe-capped 216
CMC 152

Cohesive energy ratio (CER) 85
Collective behavior 91
Contact line displacement 134
Contact line position 132
Critical aggregation concentration (CAC) 152
Critical micelle concentration (CMC) 152
Critical solloids (CSC) 152
Cyclic acetals 75
Cyclic ketals 77

De-oiling experiment 122
Di(hydrogenated tallow)dimethylammonium chloride 69
Di(tallow fatty acid)ester 68
Dihydroxyethyldimethylammonium chloride, di(tallow fatty acid)ester 68
Diluted magnetic semiconductors (DMS) 215
2,2-Dimethylhexanoic acid 67
Dinaphthyl propane (DNP) 182
1,3-Dioxane 75
1,3-Dioxolane 75
Disjoining pressure isotherms 131
Dodecyl betainate, hydroxyethylcellulose (HM-HEC) 72
Dodecylsulfate surfactants, alkali counterions 45
Dodecyltrimethylammonium bromide (DTAB) 157
n-Dodecyl-β-D-maltoside 177
Dye-doped silica nanoparticles 195
–, synthesis, ionic surfactant-based microemulsions 200

Egg-yolk phosphatidylcholine/water 17
Electrolytes, inorganic 34
Emulsifier 145

Environmental regulations 60
Eotvos number 129
Equilibrium adsorption 25
Ester bonds 64
Ester quats 59, 68
Ethoxylated nonionics 87
Ethylene oxide number (EON) 84
Excimer fluorescence 149

Facetted particles 238
Fatty acid ethoxylates 64
Fatty alcohol ethoxylates 59
Fluorescence 148
Fluorescence probing 143
–, surfactant assemblies 151
Fluorescence spectroscopy 147
Fluorescent probes, characteristics 150
–, extrinsic 151
Fluorescent silica, bioimaging 190
Free precipitation 240
Frumkin adsorption isotherm 36, 38

Gas–liquid interface 25
Gold nanoparticles 203
–, bioimaging 190
–, synthesis, ionic surfactant-based microemulsions 205
–, –, nonionic surfactant-based microemulsions 204
–, –, surfactant aggregate-based systems 209
Gouy–Chapman theory 34

Hemimicelles 143, 147
Hexadecane/squalane 128
Hydrodynamic radius 194
Hydrolysable surfactants 63
Hydrophilic–lipophilic balance (HLB) 85
Hydrophilic–lipophilic deviation (HLD) 85, 87
Hydroxyethylcellulose, hydrophobically modified (HM-HEC) 72

Imaging agents 190
Inherently biodegradable 62
Inorganic electrolytes 34
Interfacial thermodynamics 25
Ionic surfactants 25, 43
Iridescent surfactants 253

Ketal-based surfactants 77
Ketals, cyclic 77

Lamellar focal conics 254
Langmuir/Frumkin adsorption isotherms 26
Lead sulfide 215
Lecithin 215, 238
Linear alkylbenzene sulfonates (LAS/LABS) 59
Linear mixing rule 84
Lipases 66
Liposomes 146
Lucassen–Reynders 27

Marangoni effect 119, 128
Methyldiethanolamine 69
Micelle fluidity (nanoviscosity) 143, 181
Micelle volume fraction, meniscus profile 133
Micelles 143
–, reversed 145
Microemulsions 145, 196
Micropolarity, probe 176
Mixing rules 92
–, deviations 101
Mixtures, anionic/nonionic 101
–, behavior, quantitative measurement 86
–, partitioning 95
Mucor miehei lipase (MML) 66, 73
Myelinic figures 3, 16

Nanoparticle formation, microemulsions 193
Nanoparticles, bio-imaging 191
–, facetted 235
Nanoviscosity 181
NEODOL 25-7 10
Noncollective behavior 85
Nonionic surfactants 13, 25
–, adsorption 27
–, mixtures 9
Nonionic/anionic surfactants 13
Nonylphenol ethoxylates 59

Octaethyleneglycol-n-alkyl ethers 26, 38
Octyl phenol ethoxylates 59
Oil displacement, micellar nanofluid 128
Optical contrast agents 190

Optimum formulation 86
Organized assemblies, solution/interfaces 143
Ornstein–Zernike equation 131
Ortho esters 78

Packing parameter 59
Particle size, meniscus profile 135
pH sensitive systems 102
Phase inversion temperature (PIT) 85, 87, 128
Phosphatidic acid 147
Phosphatidylcholine (lecithin) 147, 215, 238
Phosphatidylethanolamine 147
Phosphatidylinositol 147
Phosphoglycerides 146
Phospholipids 3
Phosphorescence 149
Photostability 190
Polarity parameter 143
Poly(acrylic acid) (PAA) 157
Poly(ethylene glycol), methyl-capped 78
Polyoxyethylated alkyl ethers 154
Polyoxyethylated sodium lauryl sulfonate 155
Poly(propylene glycol) (PPG) 78
Pyrene 143
–, emission spectrum 153
Pyrene labeled poly(acrylic acid) (PPAA) 175

Quantum dot synthesis, ionic surfactant-based microemulsions 213
–, nonionic surfactant-based microemulsions 213
Quantum dots 211

Ready biodegradation 62

Silwet L-77 122
Sodium alkyl sulfates 26
Sodium bis(2-ethylhexyl) sulfosuccinate/water 18
Sodium dodecyl [poly(oxyethylene)](2,4)sulfonate 155
Sodium dodecyl sulfate (SDS) 58

Sodium hexadecyl n-sulfate 44
Sodium oleate 14
Solid–liquid interfaces, aggregation 164
Solloids 147
Solubilization site, probe 176
Spherical particles 239
Spreading, trisiloxane 119
Spreading coefficients 118, 120
Stable quats 59
Stern adsorption isotherm 36
Superspreaders 117
Surface potential 25
Surface tension 25
Surfactant affinity difference (SAD) 85, 87
Surfactant aggregation, interface 33
Surfactant assemblies 144
Surfactant dissolution 3
Surfactant orientation, interface 32
Surfactant penetration experiments 3
Surfactant system-based nanoparticle synthesis 192
Surfactant–oil–water (SOW) systems 85
Surfactants, pure nonionic 6
Szyszkowski–Langmuir adsorption theory 26, 38

Tergitol 15-S-7 10
Tetradecyltrimethylammonium chloride (TTAC) 171
Tetraethylene glycol 67
Tetraethylene glycol mono-n-octyl ether 46
Tetraethylorthosilicate (TEOS) 194
TPeAB (tetrapentyl ammonium bromide) 42, 47
Triethanolamin 69
Trisiloxane solution, hydrophobic surface 119
Trisiloxane-N-alkyl pyrrolidinone 120

Vertical film climbing 126
Vesicles 146

Wetting 118
Winsor diagrams 86

Zeolite 4A 12

Printing: Krips bv, Meppel, The Netherlands
Binding: Stürtz, Würzburg, Germany